FIBRE REINFORCED COMPOSITES 1986

Conference Planning Panel

D Holt, PhD (Chairman)
Westland Helicopters plc
Yeovil
Somerset

R A Downey, PhD, BSc, MPRI
Pilkington Reinforcements Limited
Merseyside

D Kewley, PhD
British Leyland Technology Limited
Warwick

R C Haines, BSc, FPRI
Dunlop Slazenger International Limited
West Yorkshire

F L Matthew, BSc(Eng), ACGI, CEng, MRAeS, FPRI
Imperial College
London

J Johnson, PhD
Manufacturing Technology Composites R and D
Rolls-Royce Plastics and Composites Laboratory
Derby

J Humphries, CEng, MIMechE
Pilkington Brothers
Lancashire

W S Faville, BSc, MSc
Faville Consultants Limited
London

P A Ramsden, BSc(Eng), ACGI, MRAeS
British Aerospace plc
Lancashire

G W Stockdale
The Plastics and Rubber Institute
London

J A Quinn, CEng, MIProdE
Bridon Composites Limited
Cheshire

A G Gibson, BSc, MSc, PhD, FPRI
University of Liverpool

J R H Whittenbury
Akzo Chemie
London

THE SECOND INTERNATIONAL CONFERENCE ON
FIBRE REINFORCED COMPOSITES 1986

Sponsored by
The Engineering Manufacturing Industries Division
of The Institution of Mechanical Engineers

Co-sponsored by
The Plastics and Rubber Institute
The Institution of Production Engineers
The Royal Aeronautical Society

8–10 April 1986
The University of Liverpool
Mossley Hill
Liverpool

Published for
The Institution of Mechanical Engineers
by Mechanical Engineering Publications Limited
LONDON

Proceedings of the Institution of Mechanical Engineers

© The Institution of Mechanical Engineers 1986

ISBN 0 85298 589 4 ✓

The Publishers are not responsible for any statement made in this publication. Data, discussion, and conclusions developed by authors are for information only and are not intended for use without independent substantiating investigation on the part of potential users.

Printed by Waveney Print Services Ltd, Beccles, Suffolk.

CONTENTS

The Institution of Mechanical Engineers

The primary purpose of the 76,000-member Institution of Mechanical Engineers, formed in 1847, has always been and remains the promotion of standards of excellence in British mechanical engineering and a high level of professional development, competence and conduct among aspiring and practising members. Membership of IMechE is highly regarded by employers, both within the UK and overseas, who recognise that its carefully monitored academic training and responsibility standards are second to none. Indeed they offer incontrovertible evidence of a sound formation and continuing development in career progression.

In pursuit of its aim of attracting suitably qualified youngsters into the profession — in adequate numbers to meet the country's future needs — and of assisting established Chartered Mechanical Engineers to update their knowledge of technological developments — in areas such as CADCAM, robotics and FMS, for example — the IMechE offers a comprehensive range of services and activities. Among these, to name but a few, are symposia, courses, conferences, lectures, competitions, surveys, publications, awards and prizes. A Library containing 150,000 books and periodicals and an Information Service which uses a computer terminal linked to databases in Europe and the USA are among the facilities provided by the Institution.

If you wish to know more about the membership requirements or about the Institution's activities listed above — or have a friend or relative who might be interested — telephone or write to IMechE in the first instance and ask for a copy of our colour 'at a glance' leaflet. This provides fuller details and the contact points — both at the London HQ and IMechE's Bury St Edmunds office — for various aspects of the organisation's operation. Specifically it contains a tear-off slip through which more information on any of the membership grades (Student, Graduate, Associate Member, Member and Fellow) may be obtained.

Corporate members of the Institution are able to use the coveted letters 'CEng, MIMechE' or 'CEng, FIMechE' after their name, designations instantly recognised by, and highly acceptable to, employers in the field of engineering. There is no way other than by membership through which they can be obtained!

Thermoplastic composites in woven fabric form

U MEASURIA, BSc and F N COGSWELL
Imperial Chemical Industries (ICI) plc, Wilton, Middlesborough, Cleveland

SYNOPSIS : For thermoplastic composites we can identify at least three woven product forms based on: woven impregnated single tows, directly impregnated woven fabrics, and interpenetrated fibre systems. These complement and extend the usefulness of the archetype form of preimpregnated tape based on continuous unidirectional fibres. Each of those woven product forms is itself capable of an indefinite diversity. In this paper we seek to define the basic property profile of such woven composites in a simple biaxial structure and to relate those properties to the microstructure of the composite and its method of manufacture.

1   INTRODUCTION

The development of Aromatic Polymer Composite (APC) based on carbon fibre reinforced polyether etherketone (PEEK) has concentrated on the archetype product - a high loading of continuous collimated fibres wetted by resin to form a thin preimpregnated tape. In this development three criteria have been key elements of our materials philosophy : first, the product must be made in a continuous process; second, the fibres should, as far as possible, be completely wetted by the resin phase; and finally, the product chemistry, morphology and structure should be immutable under normal processing conditions.

With the potential advantages of composites based on thermoplastics matrices now firmly established (1,2,3) and considerable experience with processing of such materials now available (4), we seek to make a full range of material forms available to the designer. In particular, with the uniaxial product, APC-2, currently made in widths up to 220 mm we appreciate the desirability of a "broadgoods" product based on woven fabric. In seeking a route to a woven "broadgoods" product we wish to retain all the virtues of APC-2 and full compatibility with that product in respect of processing technology and product performance.

Such materials, sometimes described as "Textile Structural Composites" (5) are well known in the field of thermoset composites representing at present about one half of the total market, and their mechanical performance has recently been subjected to fundamental scientific analysis (5,6). The field of "Textile Composites" ultimately includes three dimensionally woven, knitted and braided structures. In this paper we shall treat only of simple two dimensional fabric forms, in comparison with cross-plied laminates. All product forms evaluated here are based on similar grades of high strength carbon fibre (fibre modulus 230 GPa fibre strength 3600 MPa) at approximately 60% by volume of fibre.

2   QUALITATIVE COMPARISON BETWEEN CROSS PLIED LAMINATES AND WOVEN FABRIC FORMS

In making a qualitative comparison between structures built up from laminated uniaxial preimpregnated tapes and from woven fabric the latter form can be identified as having several disadvantages (7):

(i)   They are usually based on more expensive 3000 filament tows to ensure a thin uniform sheet.

(ii)   They involve an additional manufacturing stage - weaving.

(iii)   Because the fibres are necessarily kinked

the product cannot be expected to yield its full potential strength (6).

and   Although it is possible to produce
(iv)   0/45° weaves as well as 0/90, the options for designed anisotropy are reduced.

These are powerful economic and design factors which require equally powerful technological and pragmatic arguments to be countered.

A preference for a woven cloth reinforcement usually cites one or more of the following advantages.

(i)   Enhanced product toughness in respect of delamination processes - a product based on woven cloth has fewer layers (8) and the kinking of the fibres tends to deflect interlaminar cracking in such structures to give a higher fracture toughness.

(ii)   The ability to make thin, single ply,

laminates with balanced properties - a single ply of woven fabric may be as thick as two plies of collimated fibre but you need four plies of uniaxial product to balance a 0/90 lay up.

(iii) The close spacing of the fibre tows. The nature of the weave contrains lateral movement and 'fibre wash' so that enhanced thickness uniformity is possible.

(iv) The convenience of handling a "broad-goods" product is a factor which should not be under-estimated. Since each ply is inspected during lay up, inspection time is effectively halved if you lay two plies at once.

(v) An associated aspect of handling is that any tendency to splitting is suppressed - with woven product there is no weak transverse direction.

(vi) The absence of a weak transverse direction ply may reduce crack initiation or rapidly arrest such cracks at the next fibre cross over with potential for enhanced fatigue life and damage tolerance.

(vii) Because the fibres are slightly crimped by the weave in a controlled manner, mis-handling during fabrication does not result in uncontrolled fibre wrinkling.

(viii) The drapeability of certain weaves is a special factor enhancing certain processing operations.

It should be emphasised that most advocates of woven fabric would wish to use selective reinforcement with uniaxial product in combination with the weave so that woven products are seen as accessory to, rather than replacement of, uniaxial product forms.

3   THERMOPLASTIC WOVEN FABRIC PRODUCT FORMS

We can identify four distinct forms for a thermoplastic product based on a woven fabric. In principle all such forms should be capable of giving comparable mechanical performance as finished structures, they differ in their method of manufacture and the processes by which they can be formed into shapes.

3.1 Film Stacked Composite Materials

For many years film stacking technology (9,10, 11) has been widely used in the preparation of thermoplastic composites, and this technique has been most conveniently used with the reinforcement in woven fabric form. The preferred fabrication technology involves presizing the fabric with a thermoplastic resin from solution and then sandwiching such layers between layers of film and applying pressure until the laminate is fully consolidated. However, as MacMahon (3) points out the pressure not only forces the resin into the fibre bed but also forces the fibres together. This effect was also observed by Post and Van Dreumel (12). The microsection of such mouldings shows the reinforcement and

resin in distinctly striated form - Figure 1.

Film stacked composites are either directly formed into their final shape or may be consolidated flat and subsequently reformed using the inherent thermoplasticity of the resin (12). Consolidation is usually carried out at one hundred atmospheres for about two hours in the melt.

3.2   Interpenetration of Reinforcement Fibres with Resin in Film, Fibre or Powder Form

A second approach to producing a woven fabric form is to intermingle or interweave the reinforcement fibres with the resin in the form of film, fibre or powder. These approaches can provide highly drapeable product forms which can be conveniently shaped at ambient temperature before consolidating them by melting.

The product form which follows most directly from film stacking involves co-weaving a tow of carbon fibre with a layer of slit film. Such a product typically has the form

and a product of this form has been commercially developed (13).

An alternative product form replaces the resin film with resin monofilaments which are usually large in diameter by comparison with the reinforcement fibre (14). In both this, and the previous case it will be clear that there remains a considerable amount of work to do to achieve a fully wetted composite although not quite as much as is the case with film stacking.

The principle of fabricating reinforced composites from mixtures of resin and reinforcement fibres (15) includes the hybridisation of resin and reinforcement fibres of comparable diameter so that, in principle, a fully consolidated product form can be achieved by melting alone without the need for further work. However, a simple statistical model reveals that, even if the resin and reinforcement fibres are the same diameter and randomly dispersed, a 50/50 mixture will only produce 50% wetting of the reinforcement fibres leaving typically five fibres in a bundle. This is an excellent starting point for complete wetting, and one which may even be improved by invoking a positive dispersion mechanism, but it is a starting point at which it is difficult to arrive with thermoplastic fibres, and, having arrived there, if the resin fibres are oriented (as such small fibres will usually be), secondary phenomena may complex the melting process. These are formidable obstacles to overcome to which a satisfactory solution is currently being sought.

A further variant replaces the resin fibre by resin in powdered form (16) and sheathes this product in a "sausage skin" of the same, or different, resin to provide a highly flexible "Fibre Impregnated Thermoplastic" (FIT) (17). As with the 'hybridisation' of resin and reinforcement fibres this route requires that the resin particles be very finely divided in order to achieve a high level of impregnation.

We have used the interpenetration route to produce a soft fabric capable of being moulded by hand into complex forms prior to consolidation. However the quality of wetting obtained depends substantially on the pressure and thermal history of the consolidation stage : a satisfactory quality of wetting - Figure 2 (a) - is achieved after one hour at pressures of forty atmospheres.

3.3 Directly Impregnated Woven Fabric

The option of directly impregnating a woven fabric has the attraction of separating the stage of weaving from that of impregnation so allowing the weaver the full freedom of his art in the construction of optimised fabrics and providing a readily recognised fabric to the designer. However the product, once impregnated with thermoplastic, is boardy and the inherent drape characteristics of the weave are only revealed when the resin is melted.

This product, based on carbon fibre reinforced polyether etherketone, has been described in detail in an earlier publication (79). As a fully preimpregnated product total consolidation is achieved under minimum moulding conditions : the recommended conditions for full consolidation (fifteen minutes at 380°C and ten atmospheres) yield a good uniform microstructure - Figure 3.

3.4 Woven Uniaxial Tapes

Uniaxial tapes such as APC-2 based on continuous collimated fibre, can be slit to convenient width and woven or braided into a fabric form. However, the uniformity in thickness, width and a real weight of such slit tape falls some way short of optimum and as a feedstock it is much more desirable to use an impregnated single tow product of controlled width, thickness and fibre content. Research quantities of such material based on 6,000 and 12,000 filament tows are now available and have been braided and woven into fabric forms.

Such narrow tapes have an important role to play in the development of processing technology such as filament winding, and, they can also be woven (18) or braided (19) into structures. Although the tapes themselves are fully impregnated they are thin and flexible so that the weave structure, which may be of carpet width, possesses a measure of dry drapeability which is especially appropriate for large area aerofoil sections. Naturally, since the tapes are impregnated to the quality of APC-2, structures consolidated from the woven tapes demonstrate excellent fibre distribution after short moulding process (fifteen minutes at 380°C and ten atmospheres) - Figure 4.

3.5 Cross-plied Laminate Structures

For completeness, and as control, we include cross-plied laminate structures consolidated from APC-2 preimpregnated tape - Figure 5.

4 COMPARATIVE MECHANICAL PERFORMANCE OF DIFFERENT PRODUCT FORMS

The properties of APC-2 laminate structures have received exhaustive evaluation (20). In Table 1 we compare some of the key mechanical properties for the various ways of producing a balanced (0/90) laminate structure. In this section we relate those properties to the quality of fibre dispersion as judged from the micrographs of Figures 1 - 5.

The reduced tensile strength of the woven structure is typical of the observations made on qualitively similar comparisons for cross-plied laminates and woven cloth structures impregnated with epoxy resins (8). This reduction is usually assumed to result from stress concentrations at the weave cross over combined with the departures from optimum fibre alignment at such points (5,6).

The maintenance of compression strength in the woven fabric composite is particularly satisfying - in such systems the natural kinking of the fibres could have been expected to produce low results.

There is some evidence of reduced interlaminar shear strength (evidenced by the short beam shear flexure test) in those panels made by interpenetrated fibres or film stacking where the uniformity of fibre distribution is reduced. It is unclear whether this is associated with premature yielding in the thicker resin layers or weakness due to incomplete wetting of the fibre bed. There is no loss of property associated with the homogeneous microstructure of woven single tow products or directly impregnated woven fabrics.

The single most significant difference is in the impact resistance where the "Interpenetrated Fibres" and "Film Stacked" systems offer dramatically less energy adsorption than cross-plied laminates, woven single tow products or directly impregnated woven fabrics. This is not associated with the degree of wetting those systems since comparable composites with a reduced level of fibre wetting can be shown to give increased energy absorption under impact (21), further, all other things being equal, structures which delaminate readily also give higher energy absorption to failure (22). An alternative explanation is that consolidation of the interpenetrated fibre systems and film stacked composites involves applying work to the system before the fibres are fully wetted and this involves abrading fibres against one another before they are protected. We have some evidence that such work can damage the fibres and suggest that it is such damage which is responsible for reducing the impact resistance of the product.

5 COMPARATIVE FORMING TECHNOLOGIES

One of the more remarkable features of continuous fibre reinforced thermoplastic is the facility with which inextensible uniaxial tape can be formed into complex double curvature shapes

(22). The same processing strategies can be utilised with woven fabric forms. Highly complex shapes are most conveniently moulded from inter-penetrated fibre systems taking full advantage of the ease with which such fabrics can be draped at room temperature. Particular problems which may be encountered with such systems include shrinkage of the resin when it melts which may lead to local distortion of the reinforcing fibres.

The directly impregnated woven fabric is best suited for thin (1 mm) skin structures of modest curvature. Because of the very high number of fibre tow intersections in this product such thin sheets have an excellent uniform thickness, however, such intersections are also the key areas where excess resin congregates and is squeezed out of the structure and this is particularly noticeable where tight radii of curvature (10 mm) are used. It is also a particularly useful feedstock form for processes such as pultrusion and roll forming, in conjunction with the uniaxial tape product where some degree of transverse strength must be incorporated without undertaking a complex lay up.

6   CONCLUSIONS : COMPLEMENTARY PRODUCT FORMS

With thermoplastic composites we have at least four complementary product forms :

Uniaxial tape will remain the standard form where the greatest possible mechanical property and product definition are required and especially where the full benefits of anisotropic design are to be incorporated.

Woven uniaxial tape will become a standard broadgoods product especially suitable for very large area mouldings with gentle double curvature.

Interpenetrated fibres will be a class of materials especially suited to complex shapes where "hand lay up" into the tool will still be essential and where long moulding times will be acceptable. Because such components are likely to be relatively lightly loaded some compromise in mechanical performance in comparison with totally wetted, well dispersed, fibres will be acceptable.

Directly impregnated woven fabrics will initially be most widely used for thin skin structures and as a feedstock for precision forming technologies such as roll forming. This product form is also likely to be widely used in conjunction with continuous collimated fibres in continuous lamination or pultrusion processes.

All these product forms - Figure 6 - will be mutually compatible, so that complete thermo-plastic structures may frequently contain elements formed from different "prepreg" forms, and this diversity will encourage the development of novel fabrication technology.

7   ACKNOWLEDGEMENTS

We are indebted to N Hayman who prepared the interpenetrated fibre systems, to D C Leach who supervised the mechanical evaluation and to D Coverdale who carried out the micro-scopy studies reported here.

8   REFERENCES

1   SAMPE J. 20, 5 (1984)

2   F N Cogswell and D C Leach, "Continuous Fibre Reinforced Thermoplastics : A Change in the Rules for Composite Technology" Plastics and Rubber Processing and Applications 4,3,271 (1984)

3   P McMahon "Thermoplastic Carbon Fibre Composites" Developments in Reinforced Plastics-4, Applied Science (1984)

4   J B Cattanach and F N Cogswell "Processing with Aromatic Polymer Composites" Developments in Reinforced Plastics-5, Applied Science (in press)

5   T W Chou, B I Shehata and J-M Yang "Analytical and Experimental Studies of Textile Structural Composites" University of Delaware (in press 1985)

6   T Ishikawa, T-W Chou "Stiffness and Strength Behaviour of Woven Fabric Composites" J Materials Sci 17,3211 (1982)

7   U Measuria and F N Cogswell "Aromatic Polymer Composites : Broadening the Range" paper presented at 30th National SAMPE Symposium Anahiem 1985

8   S M Bishop and P T Curtis "An Assessment of the Potential of Woven Fibre Reinforced Plastics for High Performance Applications" Composites 15,4,259 (1984)

9   J T Hartness "Polyether ether ketone Matrix Composites" 14TH National SAMPE Tech Conference 14,26 (1982)

10   D J Murphy and L N Phillips British Patent 1,570,000 (1980)

11   D J Willats "Film Stacked Thermoplastic Composites" SAMPE J. 20,5,6 (1984)

12   L Post and W H M van Dreumel "Continuous Fiber Reinforced Thermoplastics" in Progress in Advanced Materials and Processes : Durability, Reliability and Quality Control, Edited by G Bartelds and R J Schliekelmann, 201, Elsevier, Amsterdam (1985)

13   J McGrath Textile Products Inc.

14   Albany International, product demonstrated at 30th National SAMPE Symposium and Exhibition Anahiem (1985)

15   C J A Slater "Woven Fabrics of Carbon or Glass Fibres with Thermoplastic Polymers Suitable for Fabricating into Reinforced Composites" Research Disclosure 20239 (1981)

16   R V Price "Production of Impregnated Roving" US Patent 3,742,106

17   "ATOCHEM Breaks New Ground in Composites : A Fibre-Impregnated Thermoplastic" Composites at Noveaux Materiaux, 5 May 1984

18   Milliken Ltd, Bury

19   Bentley Harris Inc. 2041 Welsh Post Road Lyonville Pennsylvania 19353

20   APC Data Sheets, ICI PLC, Welwyn Garden City

21   R L Addlemann, I Brewster and F N Cogswell "Continuous Fibre Reinforced Thermoplastics" BPF Reinforced Plastics Congress Brighton 1982

22   J B Cattanach, G Cuff and F N Cogswell "The Processing of Thermoplastics Containing High Loadings of Long and Continuous Reinforcing Fibres" Inaugural Conference of the Polymer Processing Society Akron 1985 (in press).

Table 1   Mechanical properties at 23°C*

Microstructure	Most Uniform		Least Uniform		
Structure / Test	Cross Plied Laminate 3.5	Woven** Single Tow 3.4	Impregnated Woven Fabric 3.3	Interpenetrated Fibres 3.2	Film Stacked 3.1
Tension					
Modulus GPa	72	60	70		
Strength MPa	1100	772	769		
Compression(1)					
Strength MPa	680	580	694		
Flexure					
Modulus GPa	62	50	61	65	57
Strength MPa	907	929	1052	782	680
Shear MPa	76	68	80	60	67
Impact(2)					
Initiation Energy J	5.9	8.5	7.5	3.6	3.4
Failure Energy J	22.6	22.8	29	12.7	9.3
Interlaminar Fracture Toughness (3) J / m^3	2400		2100		

1)   ITRII Compression Test

2)   Instrumented Falling Weight Impact Test on Panels 2 mm Thick

3)   Double Cantilever Beam Method

*All measurements are averaged results of measurements taken in the two directions of the balanced cross plied structure. Such averaging is especially necessary for flexure studies on 0,90 laminates where the surface ply orientation dominates stiffness. The co-efficient of variation on impact measurements is typically 10% that on stiffness and strength is typically 5%. A more detailed comparison of certain of these measurement is available (7).

**The volume fraction of fibre in this composite was measured at 50% compared with other systems which was 60%.

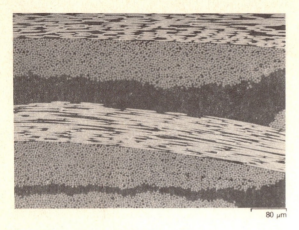

Fig 1   Film stacked composite presized fibre

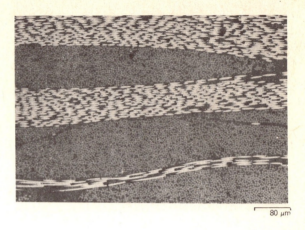

Fig 2   Interpenetrated fibre composites

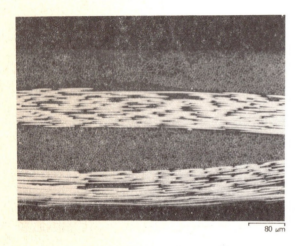

Fig 3   Direct impregnated woven fibre

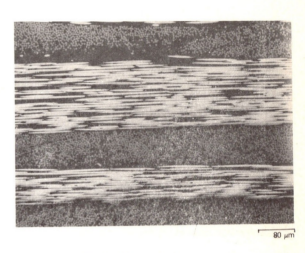

Fig 4   Woven single tow

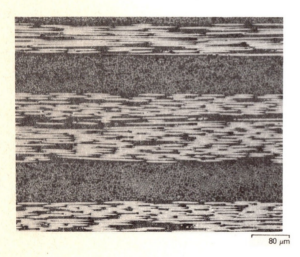

Fig 5   Cross-plied laminate

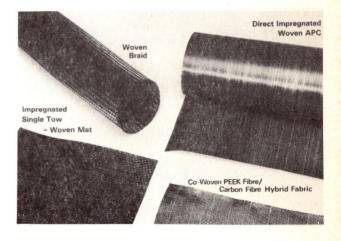

Fig 6   Woven products

The preparation and properties of ultra high modulus polyethylene

I M WARD, MA, DPhil, FRS, FInstP, FPRI
Department of Physics, University of Leeds

SYNOPSIS Comparative data have been obtained for the behaviour of epoxy composites where the reinforcing phase is ultra high modulus polyethylene (UHMPE), carbon, glass or Kevlar fibres, and for hybrid composites incorporating UHMPE and carbon or glass fibres. It has been found that the combination of high strength and high extension to break of the UHMPE fibres leads to high energy absorption of the UHMPE composites. The most useful application of the UHMPE fibres may, however, be in hybrid composites containing glass or carbon fibre. In this way the good compressive properties of glass composites, and the exceptional stiffness and strength of carbon fibre composites can be combined with the high energy absorption and non-shattering capabilities of UHMPE composites.

1 INTRODUCTION

The research described in this paper forms part of an extensive ongoing research programme at Leeds University on highly oriented polymers and fibre-reinforced composites. The present paper is concerned with the behaviour of epoxy composites incorporating ultra high modulus polyethylene (UHMPE) fibres as the reinforcing phase. Basic research at Leeds University in the period 1970-5 established the guidelines for the production of UHMPE fibres by the melt spinning and drawing route [1-4]. A small scale pilot plant was put into operation for the production of kgm quantities of multifilament yarns and monofilaments. The materials were used to make the first UHMPE composites described in this paper, and to find methods for improving fibre-resin adhesion.

More recently, the Leeds technology has been licenced by the British Technology Group to the Celanese Fibres Company USA, so that larger quantities of fibre have become available for development applications trials. The second stage of research on UHMPE composites was therefore based on the commercial route of lamination of pre-pregs in either a compression moulding press or an autoclave. At this stage hybrid composites were also produced with a sandwich structure of glass or carbon fibre layers enclosed symmetrically between UHMPE fibre layers.

In this paper the properties of the UHMPE fibres will be discussed, followed by a brief account of the preparation of the composites and a more detailed consideration of the properties of the composites, especially with regard to a comparison of their behaviour with those of composites incorporating glass, carbon and Kevlar fibres.

2 ULTRA HIGH MODULUS POLYETHYLENE FIBRES

2.1 Comparison of fibre properties

In Table 1 the principal mechanical characteristics of UHMPE fibres presently available from the melt spinning and drawing route are compared with those of carbon, glass and Kevlar fibres. The properties quoted for the UHMPE fibres are conservative, to reflect those which can readily be achieved in practice at economical production rates. Fibres with higher tensile strengths (~ 2.5 GPa) and somewhat higher tensile moduli (~ 100 GPa) can be obtained by gel spinning [5], but these involve a costly solvent recovery process, and are therefore intrinsically more expensive to produce.

In terms of absolute mechanical properties UHMPE fibres are comparable in stiffness to glass fibre, but less stiff and less strong than carbon, glass or Kevlar fibres. The specific properties of UHMPE fibres are, however, more nearly comparable to the other fibres, although the specific modulus of carbon fibres is very much higher than the specific modulus of UHMPE fibres or the other commercial fibres. It is clear from this comparison that UHMPE fibres will challenge both glass and Kevlar fibres in terms of specific properties. Moreover, UHMPE fibres have a significantly greater extension to break than the other fibres. This suggests that UHMPE fibres are likely to be useful in situations where the fracture energy to failure is important i.e. in impact and other ballistic applications. It will be shown that the impact behaviour of the UHMPE composites, and their general damage tolerance reflects the high extensibility and comparatively high strength of the fibres.

2.2 Strength, Creep and Strain Rate Sensitivity

A very important feature of UHMPE fibres which contrasts with the behaviour of the other fibres in Table 1 is their high strain-rate sensitivity[3]. At high strain rates even low molecular weight fibres of comparatively low draw ratio (~ 25) can show tensile strengths of nearly 1.5 GPa, and high molecular weight fibres of much high draw ratio (~ 50) show strengths of at least 2.5 GPa. This is clearly of outstanding interest, especially when associated with a breaking extensibility of 4-5%.

At low strain rates, on the other hand, creep occurs and the effective strengths of the UHMPE fibres falls, depending on molecular weight, to the range 0.1 - 0.5 GPa[7]. The melt spun and drawn fibres can be improved by cross-linking using electron-beam irradiation so that these fibres (but not the gel spun fibres) can retain a substantial proportion of their high strain rate strengths[8]. The present position is that the long term strengths of the UHMPE fibres, irrespective of their origin, is in the range 0.3-0.5 GPa, which should be borne in mind if the UHMPE fibre composites are being considered for permanent load-bearing applications.

3 PREPARATION OF FIBRE COMPOSITES

3.1 Fibre/resin Adhesion

It was appreciated that due to the chemical inertness of polyethylene, and the absence of any polar groups in the molecular chain, there was likely to be a poor bond between the UHMPE fibres and conventional resins, in the absence of some special surface treatment of the fibres. A thorough study was therefore undertaken of the adhesion of the fibres to both epoxy and polyester resins, with particular reference to the effects of speculative surface treatments[9].

These adhesion studies were made easier by the fact that UHMPE fibres can be produced with identical structures in a very wide range of filament diameters. In fact, by varying the process from tensile drawing to die-drawing, products with diameters from ~ 20μ up to several cms can be produced at commercially acceptable production rates. For the adhesion studies it was most convenient to determine the force required to pull one end of a monofilament of diameter 0.26 - 0.55 mm out of a thin disc of resin. The experimental arrangement is shown in Figure 1. A commercial epoxy resin (Ciba-Geigy XD 927), devised for high strength composite structures, was used for most of the comparative tests of surface treatment. This resin is cured at room temperature for ~ 16 hr and post cured for 5 hr at 80°C in an air oven. The pull-out adhesion strength σ_p was defined as the failure load F divided by the surface area of contact between the fibre and the resin

$$\text{i.e.} \quad \sigma_p = \frac{F}{\pi d \ell}$$

where d is the filament diameter and ℓ the immersion length.

Monofilaments were prepared at the three draw ratios of 8:1, 15:1 and 30:1 and subjected to different surface treatments. It was found that immersion in chromic acid at room temperature, and plasma treatment with oxygen as the carrier gas, were both effective in increasing the pull-out strength. Key results are summarised in Table 2. The chromic acid treatment is quite effective for the low draw ratio monofilaments, but the high draw ratio monofilaments are not affected so much. This result is consistent with the known improvements in chemical resistance of the UHMPE fibres with increasing orientation. The plasma treatment, on the other hand, increases in effectiveness with increasing draw ratio.

The examination of the fibre surfaces in a scanning electron microscope (SEM) provided the basis for an explanation of these results. The untreated monofilaments show smooth surfaces, and these are not significantly affected by chromic acid treatment. Pull-out tests show that the failure occurs by sliding the monofilaments out of the resin. The improvement in adhesion can therefore be attributed to surface oxidation and the consequent increase in electrostatic interaction between treated fibre with oxygen containing groups and the resin. Plasma etched monofilaments are quite different in showing a unique cellular surface texture (Figure 2) at the highest draw ratio. This texture is much less developed at low draw ratio. It can therefore be concluded that the plasma treatment with the oxygen gas is effective in improving the adhesion for two reasons. First, at all draw ratios there is an effect similar to that found for chromic acid treatment, which improves the wetting of the monofilament by the resin. Secondly, small pits are found on the monofilament surface. Penetration of these pits by the resin provides a mechanical keying of the monofilament to the resin. This increases the pull-out adhesion strengths from~ 2.5 MPa to~ 5.0 MPa. The SEM photographs show that the surface layer of the monofilament is removed in the pull-out test, suggesting that the limitation in adhesion strength is now determined by the shear strength of UHMPE which is comparatively low, as for all oriented synthetic fibres of this type.

It can be seen from Table 2 that the plasma etching treatment can produce an order of magnitude improvement in the pull-out adhesion strength as determined by this test. Both chromic acid and plasma treatments can also cause a fall in the tensile strength of the monofilaments. In subsequent work, a continuous treatment method was devised for plasma etching multifilament yarns, and care was taken to obtain satisfactory improvements in fibre/resin adhesion without producing any detectable reduction in fibre tensile strength. This is confirmed by the mechanical tests on the composites to be presented.

3.2 Preparation of Fibre Composites

As indicated in the Introduction section fibre composites were prepared in the laboratories at Leeds University prior to adopting a more commercially acceptable procedure.

In the laboratory the composites were prepared by one of two methods

(a) A bundle of fibres was placed in a rectangular loose-fitting mould (the so-called leaky mould). The fibre bundle was fully wetted with liquid resin and then compressed by the smooth fitting top of the mould, which allows excess resin to escape. XD 927 resin was used in the experiments to be discussed in the present paper. The curing conditions were identical to those described above for the adhesion pull-out tests. This procedure produced a rectangular bar containing ~ 55% by volume of UHMPE fibres, the fibre alignment being along the length of the bar.

(b) Layers of square weave UHMPE fabric were laid down in the leaky mould between layers of liquid resin. Excess resin was removed by compressing the layers as in (a) above, and an identical curing procedure was also adopted. The fabric layers were placed in the mould so as to ensure that the warp and weft threads are closely aligned with the axes of the mould. The rectangular bar composites are therefore comparable to those obtained using the fibre bundle, and again contain ~ 55% by volume of UHMPE fibre.

Finally, fibre composites were prepared by lamination of pre-impregnated sheets of fibre. Pre-pregs containing UHMPE, carbon, glass and Kevlar were produced by Rotorways, Bridgewater, UK. In this case CODE 91 epoxy resin was used, and the pre-pregs moulded in a hot press (or more recently in autoclave) to obtain laminated sheets containing either one type of fibre or in a sandwich construction with 2 layers of UHMPE enclosing 3 layers of carbon or glass fibre. In all cases the total fibre content was again ~ 55% by volume. The orientation of the fibres was varied, but in this paper we will discuss lamination where all the fibres are aligned parallel to one chosen direction.

4 MECHANICAL BEHAVIOUR OF FIBRE COMPOSITES

4.1 Test Procedures

The mechanical tests on the fibre composites followed standard test procedures developed at the Royal Aircraft Establishment, Farnborough UK.

In all cases the composite samples were of 2 mm thickness and 10 mm width. To determine the interlaminar shear strength (ILSS) a short beam test was performed. This was a three point bend test with a gauge length of 10 mm. For the determination of flexural modulus (FM) the gauge length was 160 mm, and for the ultimate flexural strength (UFS) the gauge length was 80 mm. Full details of these tests have been given by Sturgeon[10]. The FM was determined for an equivalent tensile strain of 0.03%.

The measurements of tensile modulus (TM) tensile strength (TS) and compressive strength (CS) were undertaken on samples whose ends were sandwiched between soft aluminium alloy plates bonded to the sample surfaces over a 50 mm length at each end. For the TM measurements the gauge length was 50 mm and again the modulus was determined at 0.03% strain. For TS measurements the thickness dimension was waisted with a continuous radius of 1000 mm, to give a minimum thickness of 1.2 mm. For CS measurements the sample gauge length was 10 mm. Full details of these tests are given by Ewins[11].

The UHMPE fibre composites are extremely tough and notch-insensitive. We therefore

followed procedures recommended by Dorey[12] where unnotched specimens are subjected to impact on their broad face, using a custom-built Charpy impact machine with a fixed impact energy of 5.33 kJ.

4.2 Results of Mechanical Tests

The first results were obtained on the leaky mould composites, incorporating fibres or woven yarn and are summarized in Table 3. The major interest at this stage was to evaluate the influence of surface treatments on the interlaminar shear strength. As can be seen the ILSS values for the unidirectional fibre composite are approximately doubled by the plasma treatment of the fibres. Acid treatment produces an intermediate result. The effectiveness of the surface treatments is therefore confirmed, although the magnitude of the effect is less than observed for the single monofilaments. Table 3 indicates that there is no significant reduction in the tensile properties of the UHMPE fibres due to surface treatment. It can be seen that the tensile modulus, flexural modulus and tensile strength of the composites are comparatively low, considering the fibre properties quoted in Table 1. This is because these composites were made using UHMPE fibres produced at Leeds University where fibre properties were not as good as those obtained more recently, as will be apparent from results to be presented below on the composites made using Celanese Fibers Company fibres. However the TM and TS values in Table 3 do reflect the fibre properties very well, taking into account this factor and also the less than perfect fibre alignment achieved in the leaky mould method.

The compressive strength of the UHMPE composites is low, and as we have already mentioned, is consistent with the low compressive strength of the fibres. The ultimate flexural strengths reflect, in part, this low compressive strength but it is important to note that because of the very high ductility of the fibres, specimens can be restraightened after the UFS test, and retested with only a small reduction in measured UFS (typical reduction after three UFS tests is 5-8% in strength).

The most encouraging feature of the results shown in Table 3 is the comparatively high values of impact strength. Moreover the samples retain their integrity and can be straightened and retested, showing a substantial reduction in impact energy after each test. In this respect the effect of surface treatment of the fibres is marginal. Plasma treatment does reduce resin cracking and hence reduce the impact energy, but this is not a major effect and has been found to be comparable to changing the resin formulation to give a ductile rather than a brittle resin. This result is illustrated in Figure 3. These results are consistent with the view that a major source of energy absorption is due to deformation of the UHMPE fibres. This is confirmed by the Charpy impact data shown in Table 3 for the woven fabric composites, where the impact energies are approximately one half of those for the unidirectional fibre composites, exactly in line with the reductions in TM and TS shown in this table.

The results of this preliminary study emphasised the importance of utilizing the UHMPE fibres to improve the impact behaviour of fibre composites and led to the comparative study of UHMPE fibre composites with composites incorporating carbon, glass and Kevlar fibres and with the hybrid composites. The key results of this study are presented in Table 4. At this stage only untreated UHMPE fibres were available.

The results in Table 4 may be commented on as follows. First, the ILSS values for the hybrid composites are intermediate to those obtained with the corresponding single component fibre composites, suggesting that if plasma treated UHMPE fibres had been used acceptable ILSS values would have been obtained. Secondly, the values of TM, FM and TS are much greater than those shown in Table 3, corresponding to the better mechanical properties of the Celanese fibres.

As anticipated on the basis of the fibres properties of Table 1, the absolute values of stiffness and strength for the UHMPE composite are lower than those for the commercial fibre composites, with the exception of the glass fibre composite. The hybrid composites, however, show an interesting range of properties, especially if it is borne in mind that in addition to high impact energies these materials retain their integrity on impact as illustrated in Figure 4. Furthermore, the incorporation of UHMPE fibres produces a marked reduction in density so that a comparison in terms of specific properties is more favourable to the hybrid systems. These specific properties are compared in Table 5 for the carbon and glass fibre composites and the hybrids. It can be seen that excellent impact energies and damage tolerance can be obtained, still retaining good stiffness and strength.

5 CONCLUSIONS

In this paper recent research on the preparation and properties of UHMPE fibre composites has been reviewed. It has been shown that UHMPE fibres can be incorporated into existing resin formulations in a straightforward way, using available techniques. Surface treatments have been developed to give improvements in interlaminar shear strength.

The UHMPE fibres show a potentially valuable combination of properties, namely low specific mass, comparatively high tensile modulus and strength, and a high extension at break. The UHMPE fibre composites therefore offer some possibilities for light weight structures with high energy absorption on impact. It appears that hybrid composites, where the UHMPE fibres are combined with carbon or glass fibres may be of greater interest. The hybrid composites offer the possibility of combining the advantages obtained from the high ductility of UHMPE fibres, in terms of impact energy and damage tolerance, with the outstanding stiffness and strength of carbon fibres, or the good compressive strength of glass fibres.

6 ACKNOWLEDGEMENTS

This research was undertaken with financial support from the Royal Aircraft Establishment, Farnborough UK. We are greatly indebted also for useful discussions with colleagues at RAE, including Mr. L. Phillips, Mr. W. Johnson, Dr. J. Harvey and Dr. G. Dorey.

REFERENCES

(1) CAPACCIO, G.& WARD,I.M. Properties of ultra-high modulus linear polyethylene. Nature Phys. Sci. 1973, 243, 143.

(2) CAPACCIO, G. CROMPTON, T.A. & WARD,I.M. The drawing behaviour of linear polyethylene I. Rate of drawing as a function of polymer molecular weight & initial thermal treatment. J.Polymer Sci., Polymer Phys. Edn. 1976, 14, 1641-1658.

(3) CAPACCIO, G., CROMPTON, T.A. & WARD, I.M. Drawing behaviour of linear polyethylene II. Effect of draw temperature and molecular weight on draw ratio and modulus. J.Polymer Sci., Polymer Phys. Edn. 1980, 18, 301-309.

(4) CAPACCIO, G., GIBSON, A.G. & WARD, I.M. Ultra High Modulus Polymers, A.Ciferri and I.M.Ward eds. Applied Science Publishers, London 1979, 1-76.

(5) SMITH, P. & LEMSTRA, P.J. Ultra-high strength polyethylene filaments by solution spinning/drawing. J.Mater.Sci., 1980, 15, 505-514.

(6) CANSFIELD, D.L.M., WARD,I.M., WOODS, D.W., BUCKLEY, A., PIERCE, J.M. & WESLEY, J.L. Tensile strength of ultra high modulus linear polyethylene filaments. Polymer Comm. 1983, 24, 130-131.

(7) WILDING, M.A. & WARD, I.M. Creep behaviour of ultra-high modulus polyethylene: Influence of draw ratio and polymer composition. J.Polymer Sci., Polymer Phys. Edn., 1984, 22, 561-575.

(8) WOODS, D.W., BUSFIELD, W.K. & WARD, I.M. Improved mechanical behaviour in ultra high modulus polyethylenes by controlled cross-linking. Plast. Rubb.: Process.Appln., 1985, 5, 157-164.

(9) LADIZESKY, N.H. & WARD, I.M. A study of the adhesion of drawn polyethylene fibre/ polymeric resin systems. J.Mater.Sci., 1983, 18, 533-544.

(10) Sturgeon, J.B., Technical Report 71026, Royal Aircraft Establishment, Farnborough UK, 1971.

(11) Ewins, P.D., Technical Report 71217, Royal Aircraft Establishment, Farnborough UK, 1971.

(12) Dorey, G., Royal Aircraft Establishment, Farnborough UK. Personal communication.

Table 1 Properties of reinforcing fibres (room temperature)

Property Fibre	Tensile Modulus (GPa)	Tensile Strength (GPa)	Elong- ation at Break %	Density ρ g/cm³	Specific Modulus GPa/ρ	Specific Strength GPa/ρ	Maximum Working Temperature °C
Carbon	250	3.6	1.5	1.80	139	2.0	>1500
Glass	75	3.0	2.5	2.54	30	1.2	250
Kevlar 49	125	3.0	3.0	1.45	85	2.1	~ 180
Polyethy- lene	40-70*	1-1.5	4-18**	0.96	42-73*	1-1.5	130

* Depending on draw ratio.
** Depending on strain rate.

Table 2 Pull-out adhesion strengths for selected surface treatments

Draw Ratio	Treatment	Pull-out Adhesion (MPa)
8:1	None	0.6
	Chromic acid	2.4
	Plasma (oxygen gas)	2.6
15:1	None	0.5
	Chromic acid	2.2
	Plasma (oxygen gas)	2.7
30:1	None	0.5
	Chromic acid	1.4
	Plasma (oxygen gas)	4.9

Table 3 Mechanical properties of laboratory made ultra high modulus polyethylene fibre composites

Reinforcement	Fibre Treatment	ILSS MPa	TM GPa	FM GPa	TS GPa	CS MPa	UFS MPa	Charpy Energy Absorption kJ/m²
Unidirectional fibres	Untreated	15	19	22	0.31	80	165	160
	Chromic acid	20	20	-	0.33	83	145	-
	Plasma (oxygen)	27	21	19	0.33	85	150	120
Woven cloth	Untreated	17	9	8	0.12	70	85	60
	Plasma (oxygen)	24	9	9	0.13	80	95	45

Table 4 Mechanical properties of pre-preg composite systems

Reinforcement	Density kgm⁻³	ILSS MPa	TM GPa	FM GPa	TS GPa	CS MPa	UFS MPa	Charpy Impact Energy kJm⁻²
UHMPE (untreated)	1.08	15	41	41	0.43	75	165	135
Kevlar	1.35	53	75	66	1.30	277	532	175
Carbon EXAS	1.56	66	137	104	1.95	1050	1600	75
E-Type Glass	1.95	34	56	41	1.56	975	1145	320
UHMPE (untreated) Carbon EXAS	1.28	28	85	48	-	410	495	155
UHMPE (untreated) E-Type Glass	1.45	22	46	42	-	255	245	240

Table 5 Specific mechanical properties of carbon and glass fibre composites and ultra high modulus polyethylene hybrid composites

Reinforcement	TM GPa/ρ ($\times 10^3$)	FM GPa/ρ ($\times 10^3$)	TS GPa/ρ ($\times 10^3$)	CS MPa/ρ ($\times 10^3$)	UFS MPa/ρ ($\times 10^3$)	Charpy Energy kJm^{-2}/ρ ($\times 10^3$)
Carbon EXAS	88	67	1.25	675	1025	50
UHMPE (untreated) Carbon EXAS	66	38	–	320	385	120
E-Type Glass	29	21	0.80	500	585	160
UHMPE (untreated) E-Type Glass	32	29	–	175	170	165

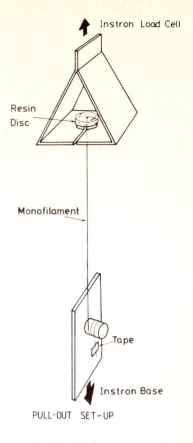

Fig 1   Pull-out test

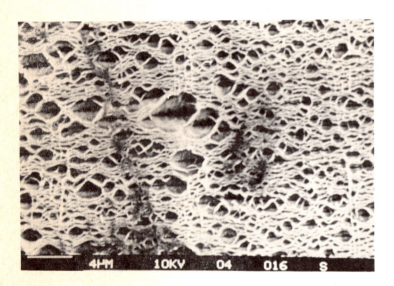

Fig 2   Plasma-treated monofilament, draw ratio 30:1 — x 2850 (photographically reduced by 50 per cent)

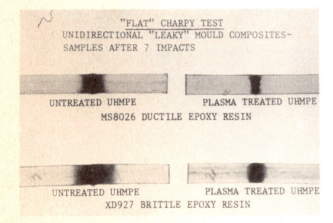

Fig 3   Comparison of impact damage on brittle/ductile resins with surface treated/untreated ultra high modulus polyethylene

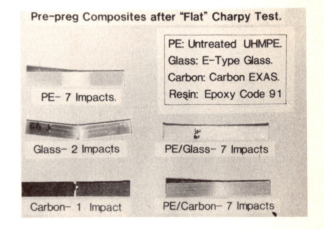

Fig 4   Impact behaviour of hybrid composites

Bismaleimide resins – the properties of compimide BMI-resins

H D STENZENBERGER, Dipl-Ing, PhD and **P KONIG**, Dipl-Chem, PhD
Technochemie GmbH-verfahrenstechnik, Dossenheim, West Germany
K FEAR, S PIERCE and M S CANNING
The Boots Company plc, Nottingham

1. Introduction

Polyimides have attracted a great deal of interest because of their unique properties at elevated temperatures and extreme environments, such as high humidity and fire conditions. However, until recently the history of polyimide resin development has been one of materials difficult to process but with outstanding performance.

Commercially available polyimides may be classified into three distinct groups: condensation, addition and thermoplastic. Condensation polyimides, although amongst the best in elevated temperature performance, can only be processed via the precursor polymer route. Their use is limited to films and fibres. Addition polyimides, generating no volatiles during cure, like bisnadic acidimides and bismaleimides are easier to fabricate and are successfully used for fibre reinforced composites because their processing is similar to epoxy resins. Real thermoplastic polyimides have become available recently, but their processing temperatures are still too high in relation to the temperature capability (Tg = glass transition temperature) achieved.

COMPIMIDE is the general trade mark of a family of thermosetting bismaleimide resins which show an outstanding balance between processability and temperature performance. This paper outlines their use for reinforced composites, as injection moulding compounds, for filament winding and resin injection moulding.

2. Bismaleimide Resin Chemistry

Bismaleimides of the general formula I

$$R = -CH_2-, -O-, -S-$$

are the main building blocks for commercial resin formulations. They are synthesized from maleic acid anhydride and aromatic diamines in high yield. They can be cured thermally to high heat resistant materials. However, they suffer from lack of processability and therefore have to be formulated into resins which can be used in prepregging, low and high pressure cure, in filament winding, resin injection moulding (RIM) and injection moulding. Such formulations may include copolymerisation with other resins (epoxies, phenolics, unsaturated polyesters, cyanates) if necessary in the presence of reactive diluents, like diallylphthalate, triallyl-isocyanurate, triallycyanurate and catalysts like tertiary amines or imidazoles. COMPIMIDE is the general trade mark for a family of bismaleimide resins which are available as unformulated basic resins (COMPIMIDE 353, COMPIMIDE 795) as fully formulated products (COMPIMIDE 800 and COMPIMIDE 183) and as moulding powders. The chemistry involved for all resins except COMPIMIDE 353 is based on the prereaction of bismaleimide with aminobenzoic acid hydrazide (1), as outlined in Fig. 1. All COMPIMIDE resins require the use of a catalyst like DABCO (Diazabicyclooctane) and/or imidazole (2-methylimidazole) to influence the cure kinetics, in practical terms, to allow cure at low temperatures in reasonable time.

3. Processing of COMPIMIDE Resins

3.1 High Pressure Laminate Moulding

The resin designed for this technology is COMPIMIDE 183 (2). The usual technology for the manufacture of fibre reinforced laminates is via prepregs which are prepared by the impregnation of the fabric reinforcement with low viscosity varnish via dip coating techniques. The impregnate is dried in a vertical or horizontal prepregging tower to strip off solvent and the prepreg is simultaneously B-staged, i.e. prereacted to a certain degree, to achieve the appropriate flow properties according to the conditions to be employed for composite moulding. A typical varnish is prepared by dissolving COMPIMIDE 183 resin, supplied as an amorphous resolidified melt, in acetone (or acetone/methylglycol acetate mixtures) to which a solution of the catalyst in the same solvent is added with stirring. At this point it has to be mentioned that the type of solvent used for prepregging also influences the cure kinetics. For some applications, we found it adequate to use mixtures of DMF (Dimethyl formamide) and MPA (methylproxitol-acetate), in which case the catalyst concentration could be reduced to a lower level due to the contribution of the DMF solvent to the reaction kinetics.

3.1.1. Cure Cycles

Laminate moulding is the technology of stacking prepreg material in a given way in a heated platen press between release film, and the conversion of the staple into a dense void free laminate through the application of pressure and temperature. Through this process, the resin is cured i.e. converted into a solid upon gelation follow-ed by vitrification. The cure process and the specific resin states during the cure process can best be described by a generalized time/temperature/transformation state diagram, (3). From a practical point of view, the most important factors which influence the proporties of the cured resin and the moulding operation (cure cycle) are:

- the thickening rate of the resin before gelation
- the gelation time under the specific cure conditions,
- the vitrification conditions,
- the cure time and the postcure time and temperature.

It has been mentioned that the cure temperature and time for COMPIMIDE 183 can be tailored by employing catalysts so the thickening rate (rheology) of the resin during cure can be adjusted to meet the requirements of the specific moulding operations. The following situations can be visualized;

a. Thick laminates, for instance 1-2 inch thick thermal insulation boards have to be mould-ed. The maximum cure temperature is $170^{o}C$, due to the press capacity. In this situation a so-called nett resin prepreg which has a high thickening rate is preferably employed so that the resin loss due to squeeze out is minimized. A simple cure cycle is possible with COMPIMIDE 183 when high concentrations of DABCO catalyst are employed (0.5 - 0.8 %), thus providing the viscosity time profile given in Fig. 2. It is important that the resin minimum viscosity is high enough so that pressure can be applied at the beginning of the cure cycle.

A typical cure cycle for COMPIMIDE 183 reads as follows.

Cure Cycle 1.

- Insert prepreg staple into the press and apply pressure of 10-15 bars.
- Heat to $160^{o}C$ and incease pressure to 30 bars
- Cure for 2-3 hours at $160^{o}C$ and a pressure of 30 bars.
- Cool down to room temperature and demould
- Postcure for 4 hours at $210^{o}C$ plus 4 hours at $250^{o}C$

Comment: the heating and cooling rates within this cycle are dependent on the laminate thickness.

b. Thin laminates have to be moulded and the presses available can operate at $210-220^{o}C$. In this situation a fabricator would be interested in an economical cure cycle

which allows the occupation of the press for the shortest period of time. COMPIMIDE 183 prepregs would have to be prepared using 0.5% DABCO (diazabicyclo-octane) as a catalyst and the following cure cycle:

Cure Cycle 2.

- Insert prepreg staple into the press and apply pressure of 15 bars.
- Heat to $210^{o}C$ and apply full lamination pressure of 30 bars.
- Hold 20 minutes at $210^{o}C$ and 30 bars.
- Cool down to room temperature and demould.
- Postcure 2-3 hours at $200-210^{o}C$.

c. A third situation would be that laminates for printed wiring boards (PWB's) have to be moulded employing prepregs from light glass fabric (US-style 2116) and a high resin content. In this case good flow properties would be required because good wetting of treated copper foil is essential. This is clad onto the laminate during the moulding process. In this situation 2-methylimidazole is preferably used as a catalyst, at a concentration of 0.5 - 0.8 % and the prepregging solvent should be a mixture of acetone and methylglycolacetate. Employing a DMF/MPA (Dimethylformamide/methylproxitol-acetate) solvent mixture would allow the reduction of the 2-methylimidazole catalyst concentration to 0.3 % because the highly polar DMF solvent contributes to the thickening rate. The thickening rate of COMPIMIDE 183 resin pre-pregged from methylglycolacetate in the presence of 2-methylimidazole is given in Fig. 3. The cure cycle for copper clad laminates is as follows:

Cure Cycle 3.

- Insert prepreg with copper foil between release sheets in press.
- Apply pressure of 20 bars and heat up to $180^{o}C$
- Apply full lamination pressure (approx. 40 bars) and cure for 2 hours.
- Cool down to between $50-60^{o}C$ and demould
- Postcure 2 hours at $200^{o}C$.

These various examples of curing flat COMPIMIDE 183 glass fabric laminates show that the cure conditions can be adjusted according to the requirements of the processor which in many cases are dictated by the equipment already established for epoxy and polyester moulding.

3.1.2. Laminate Properties

Typical properties for glass fabric laminates moulded from US-style 2116 prepregs via cure cycle 3 are given in Table 1. The results are self-explanatory, but interestingly the good flexural properties at room temperature and $250^{o}C$ are almost fully developed even without postcure.

3.2 Low Pressure Autoclave Cure

The aerospace industry is interested in carbon fibre composites for large area carbon fibre components. These laminates are usually produced

from hot melt type unidirectional tape.
COMPIMIDE 353, 795 or 795E are meltable, unformulated bismaleimide resins which are in use for hot melt prepreg. COMPIMIDE 800 is a formulated resin which has found attention mainly in Europe and is used by several prepreggers following our recommendations.

Formulation of hot melt resin:

COMPIMIDE 800	100	parts
o-diallylphthalate (DAP)	8	parts
diazabicyclooctane	0.2	parts

The preparation of ready to use material consists of blending a solution of catalyst in DAP with COMPIMIDE 800 at a temperature not exceeding $80^{\circ}C$. The resulting resin melt is applied onto a suitable carrier as a hot melt film into which collimated carbon fibres are pressed employing heated rollers. Proper wet out of the fibres is the key requirement in this operation to achieve satisfactory laminate properties.

3.2.1 Cure Cycle

Controlled bleed lay up and bagging methods are identical to those used for $175^{\circ}C$ curing epoxies. A typical cure cycle is given in Fig. 4. Normally a 12 hour postcure at $230-250^{\circ}C$ is used to fully develop the outstanding high temperature properties.

3.2.2 Laminate Properties

Typical laminate properties for various types of carbon fibres are given in Table 2. The fibre resin interface is of major importance for the development of the expected mechanical properties. Commercially available fibres are designed for compatability with epoxy resins and are therefore not optimised for polyimide matrix resins. According to Table 2, the interlaminar shear strength with all fibre types seem to be sufficiently high. However, the differences in 90° - strength and fracture toughness indicate that optimisation is required. The outstanding hot-wet performance of the COMPIMIDE 800 matrix resin in combination with Celion 6000 fibres has been demonstrated earlier (4).

3.3 Injection Moulding

COMPIMIDE X15 MRK is the resin which has been designed for the injection moulding process. The basic resin is blended with cure catalyst (2-methylimidazole) via solvent/solution techniques and B-staged at elevated temperature to achieve rheological properties to suit the processing requirements of injection moulding. The catalyst level has to be high enough to allow cure at temperatures not exceeding $200^{\circ}C$ within 30 seconds (per mm of wall thickness). The resin is supplied as a powder containing the required catalyst homogeneously dispersed. It has been developed for processing with a wide range of reinforcements such as glass, carbon, aramid and mineral fibres, and with particulate fillers such as PTFE, graphite powder and molybdenum disulphide for friction and wear property modification.

Compounding, blending and kneading may be easily performed using standard compounding equipment.

The provisional data for this moulding resin are given in Table 3. A typical composition for friction part applications consists of 80 parts by weight of COMPIMIDE X15 MRK, 20 parts of PTFE powder plus small amounts (~1%) of a lubricant.

Moulding conditions:

Barrel Temperature, front:	$85 - 100^{\circ}C$
Barrel temperature, back:	$70 - 80^{\circ}C$
Injection pressure:	$100 - 300$ kg/cm²
Mould temperature:	$180 - 200^{\circ}C$
Moulding time:	20 sec/mm thickness

Of course a postcure is required, (5 hours at $210^{\circ}C$, plus 5 hours at $250^{\circ}C$) to achieve the optimum high temperature properties.

3.3.1 Properties of mouldings

Typical properties of COMPIMIDE X15 MRK containing 20% PTFE powder are given in Table 4. The strength property retention up to $250^{\circ}C$ is 40% but the flexural modulus decreases significantly due to the PTFE-filler. The outstanding friction co-efficient and wear rate makes this combination an excellent candidate for friction parts.

3.4 Filament Winding

Filament winding is the technology used to fabricate cylindrical composite configurations. The advantages over other technologies (prepregging and vacuum/pressure bag moulding) include lower cost, reduced manufacturing time, increased repeatability and less labour intensive operation. The majority of current applications use liquid room temperature processable resins, mainly epoxies cured with liquid anhydrides in the presence of catalysts. The major limitation of these resins is their low temperature capability.

High temperature resins in most instances are solids or very high viscosity fluids at room temperature and have to be processed in filament winding at high temperatures. The main requirements for a filament winding resin are a low viscosity (200-500 mPa.s) and reasonable gel time (pot life) at the processing temperature. We have synthesized new co-reactants for bis-maleimides, which when blended with eutectic low melting COMPIMIDE resins (353, 795) provide low viscosity fluids at reasonably low temperatures. Bis-allyl compounds of the general formula II

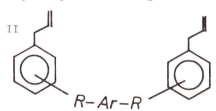

copolymerize with bismaleimides at temperatures between 160 and $240^{\circ}C$ and provide tough cured resins. The bis-allyl compound used for the filament winding bismaleimide resin formulation is called X-120 and has a honey-like consistency at room temperature and a very low viscosity at $110^{\circ}C$.

Viscosity of X-120

at 70°C: 650 - 1800 mPa.s

at 90°C: 260 - 280 mPa.s

at 110°C: 160 - 190 mPa.s

Blends of low melting eutectic bismaleimide (COMPIMIDE 353,795) with X-120 toughening agent provide resin mixtures which can be filamentary wound at temperatures between 90 and 110°C. The mechanical properties of various bismaleimide X-120 blends are provided in Table 5. The optimum properties with respect to fracture toughness are achieved for 35 - 40% X-120 concentrations. These resins are very viscosity stable at 110°C and show gel times of >30 mins at 170°C, which makes them suitable for filament winding.

Typical resin properties:

Viscosity at 100°C: 350 - 550 mPa.s

Viscosity at 100°C:

(After ageing at 100°C for 120 mins)

400 - 650 mPa.s

Gel time at 170°C: >30 mins.

Filament Winding Parameters

Impregnation temperature: 100°C

Mandrel temperature 60 - 70°C

Gel temperature on mandrel: 170°C

Cure temperature: 170 - 190°C

Cure time: 4 - 10 hours

Post cure (after demoulding):

5 hours at 210°C plus 5 hours at 240°C

Boots-Technochemie will shortly be lauching a bismaleimide resin based on the concept described above. The material will be available either as a two component system or as a ready to use resin/diluent mix.

4.   Summary

Bismaleimides are crystalline chemicals which can be modified and formulated into products which meet the processing parameters of solution and hot melt type prepregging, of low pressure autoclave moulding, high pressure platen press moulding, injection moulding and wet filament winding. The filament winding resin is in principle also useful for resin injection moulding (RIM) techniques due to its low viscosity. Due to their properties bismaleimides or laminates based on BMI's find application in the following areas:

* Aircraft structures as carbon fibre laminates for hot sections close to engines or parts requiring hot wet performance.

* Aircraft interior sandwich panels such as floors because of their non-flammability characteristics.

* Electrical applications such as a PCB's

(Printed Circuit Boards) and other dielectrics.

* Thermal insulation boards for injection moulding and press moulding equipment.

* Injection moulded friction parts.

* Filamentary wound structures (e.g. connecting rods) for automotive applications.

COMPIMIDE resins are actually in use or in test for all these applications.

REFERENCES

(1) H.D. Stenzenberger, U.S. Patents 4211861 (1980), 4356227 (1982).

(2) H.D. Stenzenberger, M. Herzog, W. Roemer, K. Fear, M.S. Canning, S. Pierce, SPE, ANTEC 1985, Proceedings p1246 (1985)

(3) J.K. Gillhaun, Polym. Comp. Sci., 19 (10) p676 - 682 (1979)

(4) H.D. Stenzenberger, M. Herzog, W. Roemer, R.Scheiblich, N.J. Reeves, S. Pierce, 29th National SAMPE Symposium No. 29, p1043 - 1059 (1984)

TABLE 1

Mechanical properties of COMPIMIDE 183-glass fabric laminates:

Glass Fabric : US-Styles 2116, Size A1100

Prepreg resin content : 42% by weight

Solvent for prepregging : Methylglycolacetate

Catalyst concentration : 0.35 % (DABCO)

Cure Cycle No. 3

Property	Unit	Test Temperature (°C)			
		23	150	200	250
Fibre content	% by weight	59	–	–	–
Flex. Strength (1)	MPa	474	424	361	288
	ksi	68.7	61.5	52.4	41.8
Flex. Modulus (1)	GPa	17.1	17.6	16.2	15.1
	msi	2.48	2.55	2.35	2.19
Flex. Elongation	%	2.84	2.56	2.39	2.15
Flex. Strength (2)	MPa	481	–	372	299
	ksi	69.8		54.0	43.4
Flex Modulus (2)	GPa	19.6	–	16.2	15.3
	msi	2.84		2.35	2.22

(1) no post cure :   green laminates

(2) post cure   :   2 hours at 200°C,
                    Tg > 250°C

TABLE 2

Mechanical properties of COMPIMIDE 800-carbon fibre laminates

PROPERTY		FD	UNIT	T 300/6000	Celion 6000 PI-finish (1)	IM-6	Besfight HTA-7	Commercial Prepreg Epoxy-sized Besfight HTA-7
Fibre contact			v/o	63	61	63	65,7	64
Flexural Strength	20°C	0	MPa	1892	1884	1839	2253	2069
	250°C	0	MPa	1432	1250	1150	1374	1360
Flexural Modulus	20°C	0	GPa	121	118	142	127	121
	250°C	0	GPa	127	120	144	130	121
Flexural Strength	20°C	90	MPa	85	80	56	73	48
	250°C	90	MPa	–	35	–	33	–
Flexural Modulus	20°C	90	GPa	8,26	7,92	7,17	8,24	8,05
	250°C	90	GPa	–	5,4	–	5,28	–
ILSS	20°C	0	MPa	113	108	92	98	99
	250°C	0	MPa	53	53	46	–	45
ILSS	20°C	0±45	MPa	63	58	43	50	50
	250°C	0±45	MPa	37	47	35	38	36
G_{IC}	20°C	0	J/m²	ca.90 - 100	254	374	–	225

(1) Finish based on NR 150 polyimide resin.
FD = Fibre direction.

TABLE 3

COMPIMIDE X-15 MRK Resin Data

PROPERTY	TEST METHOD	VALUE	COMMENTS
Physical Form			Yellow Powder
Gel Time	TC-TM 26	240±30 secs	At 150°C
Viscosity	x	x	x
Differential Scanning Calorigramme	TC-TM 14	T_B 120±10°C	Heating Rate
		T_{MAX} 199±10°C	10°C/min
Polymerization Energy	TC-TM 14	ΔH 110±20J/g	
Melting Temperature		130 – 145°C	

x Not available when issued

TABLE 4

Properties of COMPIMIDE X-15 MRK/20% PTFE-compound

PROPERTY	UNIT	VALUE		
		23°C	200°C	250°C
Density	g/cm³	1.83		
Flexural Strength	MPa	100	50	40
Flexural Modulus	GPa	3.18	1.62	1.28
Flexural Strain	%	3.4	3.65	3.92
Moulding Shrinkage	%	0.1		
Water Absorption	% 24 hrs cold water immersion	1.25		
Glass Transition Temperature	°C	> 250°C		

TABLE 5

Mechanical properties of toughened bis-maleimide filament winding resins (the system BMI – E/X120)

% BMI-E / % X-120		F.S. (MPa)	F.M. (GPa)	ε (%)	G_{IC} (J/m²)	K_{IC} (kN/m$^{-2/3}$)	WA (x) (%)
80 / 20	(RT)	134	4.23	3.22	150.7	765.5	3.98
	(177°C)	78	2.89	2.80			
	(250°C)	74	2.60	3.89			
70 / 30	(RT)	152	3.90	3.95	272.06	987.7	3.84
	(177°C)	98	2.88	3.67			
	(250°C)	80	2.37	3.80			
65 / 35	(RT)	137	2.83	3.77	301.89	1051.4	
	(177°C)	80	2.53	3.44			
	(250°C)	73	2.10	4.06			
60 / 40	(RT)	123	3.87	3.30	446.07	1267.5	3.15
	(177°C)	97	2.30	5.70			
	(250°C)	50	1.30	6.31			
50 / 50	(RT)	132	4.88	3.20	209.06	883.3	2.07
	(177°C)	16	0.38	8.95			
	(250°C)	–	–	–			
40 / 60	(RT)	43	4.00	1.11	–	–	–
	(177°C)	–	–	–			
	(250°C)	–	–	–			

WA = water absorption, 500 hours at 70°C and 94 % rel. humidity

FS = flexural strength

ε = elongation

G_{IC} = fracture energy

FM = flexural modulus

K_{IC} = stress intensity factor

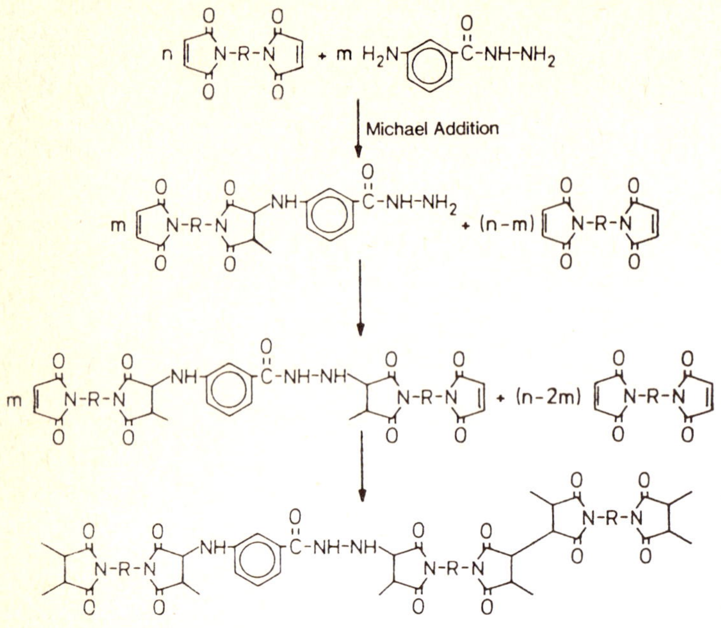

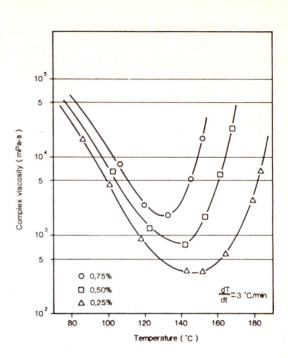

Fig 1 Chemistry of aminobenzoic acid modified bismaleimide resins

Fig 3 Viscosity temperature profile as Compimide 183 resin, (imidazole as a catalyst)

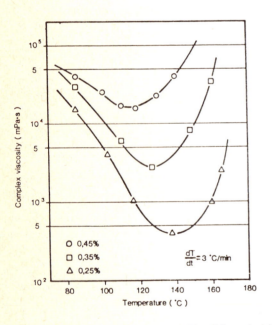

Fig 2 Viscosity temperature profile of Compimide 183 resin, (DABCO as a catalyst)

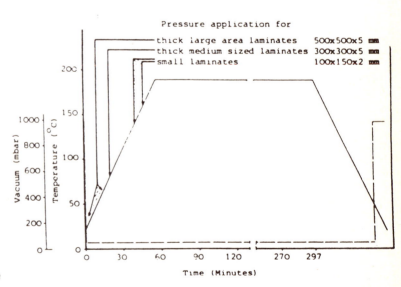

Fig 4 Cure cycle for Compimide 800 laminates

C37/86

Tensile and fatigue behaviour of alumina reinforced aluminium alloys at ambient and elevated temperatures

S J HARRIS, MSc, PhD, CEng, FIM and **T E WILKS**, BSc
Department of Metallurgy and Materials Science, University of Nottingham

SYNOPSIS   The availability of alumina and silicon carbide fibres has rekindled interest in the application and use of metal matrix composites. Short "Saffil" alumina fibres have been incorporated into three aluminium alloys (LM0, LM25 and RR58) of increasing strength by a squeeze casting process. The mechanical properties of these composites has been evaluated at $20^{\circ}C$ and $200^{\circ}C$ and this has shown that short alumina fibres are capable of significantly improving the tensile and fatigue properties of the lower strength and more ductile aluminium alloys. High strength alloy matrices do present a problem because premature failure occurs at transverse fibre-matrix boundaries or through multiple cracking of the fibres. It would appear that control over the fibre-matrix bond strength and residual stresses introduced during composite fabrication are major factors in premature failure.

1   INTRODUCTION

As with polymers metal matrices may be reinforced by continuous fibres. The mechanical properties which have been achieved by these composites depend upon the disposition of the fibres relative to the loading direction. With uniaxially reinforced composites high axial moduli and breaking strengths have been achieved with fibre fractions (V_f) in excess of 0.5, e.g. boron fibres in an aluminium-silicon-magnesium alloy (6061) matrix:225 GPa (modulus) and 1400 MPa (breaking strength)[1]. Perpendicular to the fibre direction these properties are reduced to 150 GPa and 240 GPa. The reduction in properties in the transverse direction (33% in modulus and 83% in strength)[1] are less than those found in unidirectional polymer matrix composites e.g. in Type I carbon fibre-epoxy composites both properties are reduced by 96% in the transverse mode. Transverse strength levels although significantly reduced in metal matrices may be sufficient to deal with intermediate levels of off-axis loading and may be in excess of the strength of the unreinforced matrix material.

In an attempt to reduce the costs as well as to produce fibres which are compatible with metal matrices at elevated temperatures, α-alumina and silicon carbide fibres have been produced in multi-filament tows. The diameter of these fibres is 20μm or less i.e. much less than the expensive 100 or 145 μm boron or silicon carbide coated boron (borsic fibres). The axial and transverse stiffness of alumina reinforced alumininum-lithium alloys and magnesium-zinc-zirconium alloys in the V_f range 0.3-0.6 are the equal of the boron-6061 alloy system[2,3]. Silicon carbide fibres have a lower modulus (200 GPa)[4] than alumina (370 GPa) and hence give axial moduli at $V_f = 0.4$ of 120 GPa in aluminium alloys. Transverse moduli remain at values greater than 75 GPa. Axial strength levels achieved at $V_f \leq 0.3$ of alumina and silicon carbide were close to those predicted by Rule of Mixtures. Above this volume fraction level small increases in strength appear to take place, i.e. most strength levels measured to date lie in the range 600-700 MPa. This observation of poor tensile properties with small diameter brittle fibres in metal matrices has been observed before in carbon fibre reinforced aluminium and copper[5]. Transverse strengths as high as 240 GPa have been measured on well bonded alumina reinforced composites. With increasing test temperatures the axial strength of these composite remains constant until the temperature reaches $400^{\circ}C$[2]. In the transverse direction the properties fall away at test temperatures in excess of $150^{\circ}C$.

Discontinuous fibres and whiskers have been incorporated into aluminium and magnesium alloys to produce net shaped parts. Particular attention has been focussed on silicon carbide whiskers <1μm diameter and up to 50μm long and δ-alumina (Saffil) fibres 3μm diameter and 100μm long. The techniques of composite production have been based on either

(a) powder metallurgy i.e. by mixing powders and fibres and then pressing and extruding (or rolling), or

(b) liquid metal infiltration of fibre preforms.

These techniques are capable of producing composites with fibres arranged with a degree of randomness within the plane at V_f levels ≤ 0.3. In one particular application of the material i.e. in pistons for advanced diesel engines the fibres have been placed in strategic parts of the component. The whole component can be made by injecting liquid metal into a suitably preheated die cavity which contains a fibre preform in the correct location. In association with the liquid metal injection a ram assembly applies a pressure to the die cavity; this process is known as squeeze forming.

The introduction of silicon carbide whiskers into powdered heat treatable aluminium alloys e.g. Al-Cu-Mg (2124) alloy can after suitable ageing treatment (T6) produce moduli of 103 GPa, 0.1% proof stress values of 450 MPa, breaking strengths of 630 MPa and 3% elongation to fracture at ambient temperatures[6]. These properties represent 46%, 12.5% and 20% improvements in modulus, proof and breaking stress above the unreinforced version of the alloy. Reduced properties approaching those of the matrix are found at right angles to the rolling direction because the fibres have been aligned during processing. With increasing test temperatures the properties of the whisker reinforced materials maintain their strengths up to 200°C. Thereafter the breaking strength falls away steadily to 200 MPa at 275°C and 75 MPa at 350°C.

"Saffil" (δ-alumina) fibres have been incorporated into aluminium alloys by squeeze casting[7]. The addition of these fibres ($V_f \geq 0.24$) to an Al-9%Si-3%Cu alloy (in the T6 condition) produced an ambient breaking strength of 300 MPa and a modulus of 110 GPa[8]. Whilst the modulus was attractive and represented a 50% improvement above the unreinforced alloy the breaking strength of the composite was marginally lower than that of the matrix alloy. At elevated test temperatures above 200°C the reinforced material began to achieve strengths in excess of those of the matrix and at 300°C the composite strength was 230 MPa whereas the matrix was reduced to 80 MPa. In excess of 300°C the properties of the composite began to fall away more steeply. Strain to failure of the composites at ambient was less than 1% with increased temperature this parameter increased marginally.

This paper describes work which has been carried out on composites containing "Saffil" fibres ($V_f \geq 0.2 \leq 0.25$) and which have one of three aluminium alloy matrices, i.e. a 99.85% pure aluminium (LMO), an Al-Si-Mg (LM25) casting alloy and an Al-Cu-Mg-Fe-Ni (RR58) alloy which is normally used in a wrought form for elevated temperature purposes. Stress-strain plots have been determined at 20°C and 200°C and limited amount of fatigue data has been obtained on the reinforced LMO alloy at 20°C. From the form of plots, strengths, fracture surfaces etc. some of the factors which control deformation and failure in the three alloys of differing strengths are discussed.

2 EXPERIMENTAL

2.1 Materials

"Saffil" alumina fibres (RF grade) in a discontinuous form were used to reinforce the composites. The mechanical and physical properties of these polycrystalline fibres are given in Table 1. The compositions of the three matrix alloys are given in Table 2.

2.2 Composite Fabrication

A process of "squeeze forming" was used to fabricate the composites. Fibre preforms (or pads) of varying density (0.4-0.8 g/cm^3) were inserted into the heated die prior to injection of the superheated metal and the application of pressure. A silicon binder was used to bind the fibres in the preform. The formed part contained a disc shaped fibre reinforced region (110 mm diamter and 7-15 mm thick). Squeeze cast alloy blanks without fibres were also prepared.

2.3 Mechanical Properties

Flat tensile test samples were cut and shaped in at least two directions in each of the discs. The samples had a gauge length of 25mm and a section of 5mm x 5mm. A pair of strain gauges were mounted on the gauge length of each sample and load-deflection plots were recorded using a conventional tensile testing machine. Fatigue tests were carried out on similar specimens in a Mayes servohydraulic machine operating under load control with a frequency of 10Hz and an R-ratio of 0.1.

2.4 Microstructure and Fractography

Samples were cut in two directions on the gauge length of the samples adjacent to the fracture. These were ground and polished carefully to minimise relief effects prior to examination in either the Optical Microscope or in the Scanning Electron

Microscope (SEM). Fracture surfaces were examined in the SEM.

3 RESULTS

3.1 Tensile Stress-Strain Behaviour at 200°C

Stress-strain plots were constructed for each of the matrix and composite materials, see Fig. 1. At low strains <0.1% all materials behaved in an elastic fashion. Elastic modulus values were determined and average values for at least two tests are given in Table 3. In the 99.85% pure aluminium sample (LMO) the addition of the alumina fibres increased the modulus from 68.0 GPa to 83 GPa (a 22% improvement). A similar quantity of fibres added to the Al-Cu-Mg (RR58) alloy increased the modulus by 32% to 102 GPa, whilst an increased volume fraction (0.25) of fibres added to Al-Si-Mg (LM25) alloy promoted an increase of 29% to 98 GPa.

At low stresses the stress-strain plot for the LMO alloy begins to curve and show evidence of plastic deformation. The introduction of the fibres increases the stress at which the plot becomes non linear; an indication of this is given by the 0.1% proof stress which increases from 27 to 57 MPa as a result of the reinforcement. The LM25 and RR58 alloys show a rather different behaviour, for the stress at which the stress-strain plot becomes non-linear is reduced in each case by the addition of the reinforcement. The observations are supported by the measured reductions in 0.1% proof stress of 100 MPa (from 240 MPa) with the addition of fibres to LM25 and of 106 MPa (from 371 MPa) with the addition of fibres to RR58.

Breaking strength is considerably improved when alumina fibres are added to LMO and this increases the strength by 102 MPa (from 40 MPa). The reinforced versions of LM25 and RR58 alloys show reductions in breaking strength when a comparison is made with the results obtained on their respective non-reinforced matrix alloys, i.e. a reduction of 53 MPa (from 292 MPa) in the case of LM25 and 48 MPa (from 431 MPa) in the case of RR58. Strain to failure of all reinforced alloys is much lower than the unreinforced matrix materials. Recorded values lie between 0.7 and 1.1%. The lower values in this range were recorded with the higher strength matrix materials (LM25 and RR58).

3.2 Tensile Stress-Strain Behaviour at 200°C

The stress-strain plots for each alloy and its reinforced composite are shown in Fig. 2. The plots have the same general form as those shown for tests completed at 20°C. Elastic modulus values for the matrix alloys are approximately 5 MPa lower at 200°C, whilst the reinforced alloys show a marginally larger reduction with the one exception of the LMO which remained almost the same (Table 4). Departure from linear behaviour takes place at lower stress levels in all alloys when comparison is made with the 20°C results. Proof stress values decrease by between 10 and 60 MPa for the unreinforced matrix alloys. Smaller reductions take place in the reinforced alloys e.g. RR58 alloy (50 MPa), LM25 (15 MPa) and LMO (zero). This means that the effective strengthening by the reinforcement at 0.1% strain has been improved in all the alloys at 200°C.

Breaking stress values for unreinforced LM25 and RR58 alloys has been reduced by approximately 90 MPa as a result of the increase in testing temperature. LMO also experienced a reduction in strength but only by one quarter of this value. The introduction of fibres into LM25 and RR58 promoted a smaller reduction in strength due to the temperature change. This would indicate that in both these alloys the reinforcement is beginning to overcome the weakening of the matrix at the elevated temperature. The reinforced version of LMO still maintains a breaking stress above that of the unreinforced matrix, but this level of fibre strengthening had been reduced by the increased test temperature i.e. 83 MPa at 200°C compared with 103 MPa at 20°C. Strains to failure of all materials had increased as a result of testing at 200°C. Values as high as 1.4% were measured on the reinforced LMO alloy.

3.3 Fatigue Tests

Fatigue life plots are shown in Fig. 3 for LMO alloy and its reinforced version with 0.2 V_c alumina. At the high stress-low cycle end of the plot the effect of the fibre reinforcement on the static strength is effectively maintained. As the number of cycles increases to 10^6 the fatigue strength of the reinforced material decreases. Neverthelsss, the ratio of fatigue stress at 10^6 cycles to breaking strength is still close to 0.75 which indicates a significant influence of the reinforcement on fatigue crack initiation and growth. Unreinforced metals generally give ratio values below 0.5. In the case of the unreinforced LMO this kind of value was not obtained because of the material's low strength and ability to creep during cyclic loading.

3.4 Fractography

The fracture surfaces of the tensile samples tested at 20°C and 200°C were examined in the SEM. In the case of all the unreinforced materials, significant amounts of plastic deformation in the form of dimples was found on each of these surfaces. The LMO sample showed evidence of significant amounts of necking down on a macroscopic scale adjacent to its fracture.

The introduction of fibres to the various matrices changed the form of the fracture face markedly. In the case of the reinforced LMO alloy, the fracture surface (see Figs. 4a and b) has the following features:

(a) axial fibres broken and pulled out.

(b) transverse fibres where the fibre-matrix interface has broken down.

(c) limited matrix ductility with cusps of metal between holes where fibres have pulled out.

Raising the temperature to 200°C promotes much more matrix ductility, i.e. deep cusps and less evidence of transverse fibres where the interface has broken down (see Fig. 4c).

With the reinforced LM25 there was less evidence of fibre pull-out, matrix ductility and fibre-matrix interfacial fracture after testing at both 20 and 200°C, see Fig. 5 (a-c). An additional characteristic of fibres cleaving along their length was also found. The fracture faces on the reinforced RR58 alloys were closer to those found on the LMO samples. The level of matrix ductility as represented by height of the cusps was somewhat reduced.

The fracture faces of the high cycle fatigue samples produced on the reinforced LMO material show many of the features found with the samples which were subjected to failure under monotonic tensile testing conditions. (Fig. 4a and b) The major difference between the two fracture surfaces is the reduction in the height of ductile metal cusps which existed between broken fibres in the fatigue sample.

4 DISCUSSION

Elastic modulus results given in Table 3 show significant increases as a result of incorporating discontinuous alumina fibres in all three matrix alloys. The results obtained at ambient temperature are in agreement with those already published for the reinforced version of the Al-Cu-Si alloy[8]. Attempts[9] have been made to account for the effect of random in the plane distribution of fibres and it is clear that contributions to the modulus come from all fibres irrespective of orientation to the loading direction. The somewhat lower increase (22% as opposed to 29-32%) obtained on the LMO alloy might suggest that plastic deformation may be taking place at very low strain levels in this system.

4.1 Yield Behaviour

With the high strength matrix composites (LM25 and RR58) there is clear evidence that departure from linear elastic behaviour occurs at lower stresses than is the case with the unreinforced alloy. This reduction is also shown up by the 0.1% proof stress which indicates a fall of the order of 100 MPa as a result of introducing fibres in the volume fraction range 0.2-0.25. Explanations of this phenomenon are not clear since different factors may be influencing the situation in the different alloys. The following mechanisms may contribute to the increased strain:-

(a) General and local plastic deformation.

(b) Fibre-matrix debonding.

(c) Internal stresses in the composite.

(d) Fibre cracking.

It is possible for local plastic deformation to take place at low composite strains particularly at fibre ends and in matrix regions where fibres are not effectively distributed or where defects or pores exist. Early work by Baker and Cratchley[10] with discontinuous stainless steel wire reinforcement of aluminium indicated that deformation at fibre ends occurred. Fibre-matrix debonding may take place at fibres which are orientated normal to the loading direction. In the current set of tests there is clear evidence of microscopic and fractographic evidence of strong fibre-matrix bonding in the silicon containing LM25 alloy. In which case longitudinal cleaving of the transverse fibres became a more possible cracking mode. Fibre breakage, i.e. axially and in a transverse cleavage mode, is a possible contributing factor, but at low strain values (<0.1%) the number of such breaks will be very low. Arsenault and Taya[11] have shown in the case of silicon carbide whisker reinforced 6061 aluminium alloy that significant levels of residual stress remain in the composite after cooling down from the fabrication process and that further heat treatment does not reduce these stresses. The stresses arise from the fact that silicon carbide has a smaller coefficient of expansion than aluminium and its alloys. As a result the yield stress of a composite containing $0.2\ V_f$ silicon carbide whiskers has a yield stress in compression which is 70 MPa above that measured in tension. The difference in thermal expansion of alumina and aluminium is of the same order and it may be expected that significant residual stress may exist in the composites under study. The unreinforced alloys LM25 and RR58 have different yield (or proof) stress values, i.e. 240 MPa and 371 MPa respectively and yet both are reduced by approximately 100 MPa by the introduction of the alumina fibres. This may suggest that residual tensile stress in the matrix is a common factor in both these cases. With the reinforced LM0 alloy which has an extremely low yield or proof stress (27 MPa) in the unreinforced condition any residual elastic stress held within the system may be small. This material when tested in tension demonstrates its ability to have its proof stress raised by approximately 100% as a result of the introductory alumina fibres.

Raising the test temperature of the composite material from 20 to 200°C on the above basis should

(i) reduce the yield stress of certain alloys to lower levels.

(ii) reduce the degree of internal stress built into the composite material.

In the case of the reinforced LM0 factor (i) may be greater than (ii), hence a reduction in proof stress might be expected, in fact virtually no difference exists between 20 and 200°C test values, which may suggest that any residual stress loss exactly cancels out the reduction in matrix yield stress. With LM25 the proof stress of the composite falls by 15 MPa which is much less than the 41 MPa found for the unreinforced alloy. Whilst the much higher yield stress RR58 showed small differences between composite and unreinforced alloy proof stress values. This would suggest that significant internal stress remained in the reinforced material high strength alloys at 200°C.

4.2 Fracture Behaviour

At 20°C the fracture stress of reinforced LM0 alloy (142 MPa) is three and a half times greater than that achieved on the unreinforced matrix. From the plot given in Fig. 1 it is apparent that the fibres have succeeded in reinforcing material in the region where the fibre is behaving in an elastic manner and the matrix is deforming plastically (Stage II). However, such behaviour leads on to fracture at a strain of 1.1% which is greater than the failure strain of the fibres (0.67%) and much less than that of the matrix (37%). Examination of the fracture face (see Fig. 4) reveals limited amounts of fibre pull-out, fibre-matrix debonding and matrix ductility as being the major activities in the final stages of failure. This would suggest that as increased straining took place, fibre-matrix debonding had taken on a key role. As failure approached, local regions of

debonding will have taken place at numerous places in the cross-section and at final fracture these areas will have linked up by ductile failure of the matrix (cusps) and breaking or pull-out of the fibres which are aligned close to the loading axis.

With the reinforced version of LM25 a somewhat different pattern of events appear to have taken place. Stage II behaviour appears to have been interrupted by fracture at a stress which is still below that found for the unreinforced matrix, i.e. at a failure strain of 0.73%, just in excess of the fibre failure strain. The fracture surface shows (see Fig. 5) evidence of fibre fracture and much less matrix ductility. Fibre fracture is shown to occur in two modes:

(1) fibres close to the load axis have failed with little evidence of pull out.

(2) transverse fibres cleaving.

This type of behaviour indicates that transverse fibre-matrix debonding is not the initial weakening mechanism which initiates fracture. This is because in this silicon containing alloy a strong interfacial bond exists. The alternative modes of fracture initiation appear to be matrix cracking or fibre cracking. The matrix may be more prone to cracking because it has begun to yield at lower stresses, if it is constrained by the fibres or if it contained pores, large intermetallics, or other inclusions. Fig. 5 shows that the matrix is ductile but to a lesser degree than LM0 and although some local problems may exist which allows matrix cracking to occur at low strseses it is unlikely to control the failure process. Cracking of the fibres which are aligned close to the loading axis are probably the major factor in initiating failure. Samples cut from the gauge length of failed test pieces away from the fracture face show clear evidence of multiple fibre cracking, see Fig. 6. Such fractures may take over a range of strains around the average fibre breaking strain (0.67%) depending upon the distribution of defects in the fibre.[9] Previously reported calculations[9] have demonstrated how the incidence of fibre cracking with increased strain may effectively contribute to the reduction in stiffness in stage II behaviour. Eventually such cracking may exist on a wide scale throughout the cross section, this then initiates final matrix failure as well as cleavage of the tranverse fibres.

The reinforcement of RR58 has produced what appears to be a similar situation to that found in LM25, i.e. with failure strain marginally higher at 0.8%. Nevertheless, the failure stress at 383 MPa is 143 MPa greater than that found in LM25. This improvement in breaking strength has probably resulted from the higher stress at which Stage II behaviour began, i.e. the axial fibres would not experience strains close to their average failure strain until much higher stress levels have been achieved. Two further factors may enter the scene; the strength of the fibre-matrix interface and the work hardening characteristics and strength of the matrix. There is evidence from fracture surfaces that the interfacial bond in the RR58 alloy is not as strong as in LM25. Hence, failure may be dependent on fibre-matrix debonding as well as fibre cracking and this is still under investigation.

At elevated temperature (200°C) the breaking strengths of all composite materials are reduced compared with those obtained at 20°C. The strength reduction in the case of the reinforced versions of LM25 ad RR58 are less than those achieved in the unreinforced alloys. No effective change in failure mechanism from those proposed above is expected. What is happening is that the matrix materials are not work hardening (particularly LM25) at anywhere near the same rate and therefore are more tolerant of fibre and interfacial cracking, and this is shown up in the increased levels of composite failure strain. The reduction in breaking strength of the reinforced LM0 alloy at 200°C arises from the weakening of the matrix alloy. Reduced strength is accompanied by increases in strain to failure.

4.3 Fatigue Failure

Previously it has been demonstrated[9] that the introduction of alumina fibres into LM25 alloy produced a poorer fatigue performance when comparisons were made with the unreinforced alloy. The results displayed here clearly indicate that in a low strength ductile matrix (LM0) discontinuous fibres are capable of enhancing fatigue performance. This behaviour relates directly back to the monotonic tensile behaviour which shows that such fibres are capable of reinforcing this alloy. Fibre reinforcement will improve fatigue properties whilst fibre weakening which is effectively taking place in LM25 will produce the reverse.

5 CONCLUSIONS

(1) The introduction of "Saffil" alumina fibres ($V_f \sim 0.20$) will bring about fibre reinforcement of certain aluminium alloys (99.85% Al), i.e. increasing the stress at which plastic matrix deformation begins and raising the breaking strength e.g. from 40 to 142 MPa.

(2) Higher strength aluminium alloys e.g. Al-Si-Mg (LM25) form strong interfacial bonds with "Saffil" fibres and this promotes low strains to failure in which fibre multiple cracking appears to play an initiating role.

(3) Reducing the interfacial bond strength and increasing the alloy strength with an Al-Cu-Mg (RR58) matrix promotes higher composite strengths but retains low breaking strains; this probably occurs because of a combination of fibre cracking, interfacial debonding and matrix work hardening.

(4) Residual stresses from the fabrication stage in high strength matrices may play a significant part in reducing the stress at which plastic matrix deformation begins. This leads to reduced "composite stiffness" and early onset of local failure processes.

(5) Increasing the test temperature from 20 up to 200°C promotes higher strains to failure in all composites. Local failure processes e.g. fibre cracking and interfacial debonding still take place but at temperature the matrix work hardens at a much reduced rate and becomes tolerant of these defects. Eventually as the temperature increases, fibres will reinforce most aluminium alloy matrix materials.

(6) Discontinuous fibres are capable of improving the fatigue performance of aluminium alloy matrix materials provided that they do reinforce the alloy in question.

ACKNOWLEDGEMENTS

The authors are indebted to Dr. J. Johnson, Dr. B. Borradaile and Mr. R. Jones of Rolls-Royce Ltd., Derby; Mr. J. Dinwoodie of ICI plc, Mond Division, Runcorn, and Mr. J. Barlow of GKN Technology Ltd., Wolverhampton, for the provision of materials and general support. They also wish to acknowledge the provision of a studentship (TEW) by the Science and Engineering Research Council.

REFERENCES

(1) Kreider, K.G. and Prewo, K.M. in Composite Materials, vol. 4, edited by K.G. Kreider, Academic Press, New York, 1974, p. 400.

(2) Champion, A.R., Krueger, W.H., Hartman, H.S. and Dhingra, A.K., Fibre F.P. reinforced metal matrix composites, I.C.C.M. 2, Toronto, Metallurgical Society of AIME, 1978, p. 883.

(3) Nunes, J., Chin, E.S.C., Slepetz, J.M. and Tsangarakis, N., Tensile and fatigue behaviour of alumina fibre

reinforced magnesium composites, I.C.C.M. 5, San Diego, <u>Metallurgical Society of AIME</u>, 1985, p. 723.

(4) Tanaka, J., Ichikawa, H., Hayase, T., Okamura, K. and Matsuzawa, T., Mechanical properties of SiC fibre reinforced Al composites, I.C.C.M. 4, Tokyo, <u>Japanese Society for Composite Materials</u>, 1982, p. 1407.

(5) Harris, S.J. and Marsden, A.L., Failure mechanisms in metal matrix composites reinforced with small diameter fibres, <u>Conference on Practical Metallic Composites</u>, Majorca, Institution of Metallurgists, 1974, p. B35.

(6) Nieh, T.G., Rainen, R.A. and Chellman, D.J., Microstructure and fracture in SiC whisker reinforced 2124 aluminium composite, I.C.C.M. 5, San Diego, <u>Metallurgical Society of AIME</u>, 1985, p. 825.

(7) Clyne, T.W. and Bader, M.G., Analysis of a squeeze-infiltration process for fabrication of metal matrix composites, I.C.C.M. 5, San Diego, <u>Metallurgical Society of AIME</u>, 1985, p. 755.

(8) Dinwoodie, J., Moore, E., Langman, C.A.J. and Symes, W.R., The properties and applications of short staple alumina fibre reinforced aluminium alloys, I.C.C.M. 5, San Diego, <u>Metallurgical Society of AIME</u>, 1985.

(9) Harris, S.J. and Wilks, T.E., Tensile and fatigue behaviour of short alumina fibre reinforced aluminium alloys, <u>1st European Conference on Composite Materials</u>, Bordeaux, E.A.C.M., 1985, p. 595.

(10) Baker, A.A. and Cratchley, D., <u>Applied Mat. Res.</u>, 1964, <u>4</u>, 215.

(11) Arsenault, R.J. and Taya, M., The effects of differences in thermal coefficients of expansion in SiC whisker 6062 aluminium composites, I.C.C.M. 5, <u>San Diego, Metallurgical Society of AIME</u>, 1985, p. 21.

Table 1   Properties of "Saffil" Alumina Fibres (RF Grade)

Composition 96–97% Al_2O_3.   Crystal phase δ-alumina
Mean diameter 3μm.   Elastic modulus 300 GPa
Breaking strength 2000 MPa.   Breaking strain 0.67%.

Table 2   Chemical composition of aluminium alloys

Alloy type	%Cu	%Mg	%Si	%Fe	%Mn	%Ti	%Ni	Al
LM25	0.09	0.46	6.96	0.45	0.20	0.17	–	Rem.
RR58	2.00	1.23	0.20	1.23	0.15	0.13	0.92	Rem.
LMO	–	<0.05	<0.05	<0.05	<0.05		–	99.85

Table 3   Properties of Matrix and Composite Materials

Composite/ Alloy type	Matrix Condition	Young's Modulus GPa	0.1% Proof Stress MPa	Breaking Stress MPa	Breaking Strain %
LMO	As-cast	68	27	40	37.0
LMO + 0.2 V_f Al_2O_3	As-cast	83	57	142	1.1
LM25	Overaged	76	240	292	3.7
LM25 + 0.25 V_f Al_2O_3	Overaged	98	140	239	0.7
RR58	T6	77	371	431	2.5
RR58 + 0.2 V_f Al_2O_3	T6	102	265	383	0.8

Table 4 Properties of Matrix and Composite Materials

Composite /Alloy	Matrix Conditions	Young's Modulus Gpa	0.1% Proof Stress Mpa	Breaking Stress Mpa	Breaking Strain %
LM0	As-cast	64	16	19	40.0
LM0 + 0.2 V_f Al$_2$O$_3$	As-cast	84	58	102	1.4
LM25	Overaged	72	199	204	5.5
LM25 + 0.2 V_f Al$_2$O$_3$	Overaged	89	125	190	1.0
RR58	T6	71	310	332	4.5
RR58 + 0.2 V_f Al$_2$O$_3$	T6	92	215	313	1.1

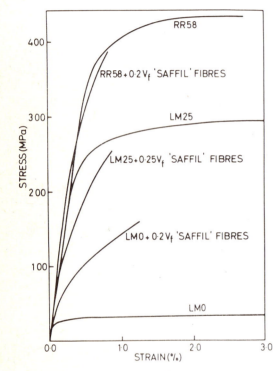

Fig 1   Stress—strain plots obtained at 20°C for matrix alloys and discontinuous fibre composites

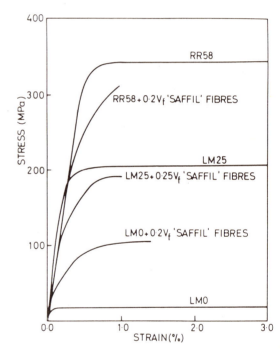

Fig 2   Stress—strain plots obtained at 200°C for matrix alloys and discontinuous fibre composites

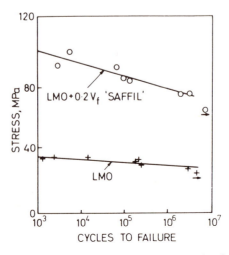

Fig 3   Stress versus number of fatigue cycles plot for the LM0 alloy with and without Saffil fibre reinforcement

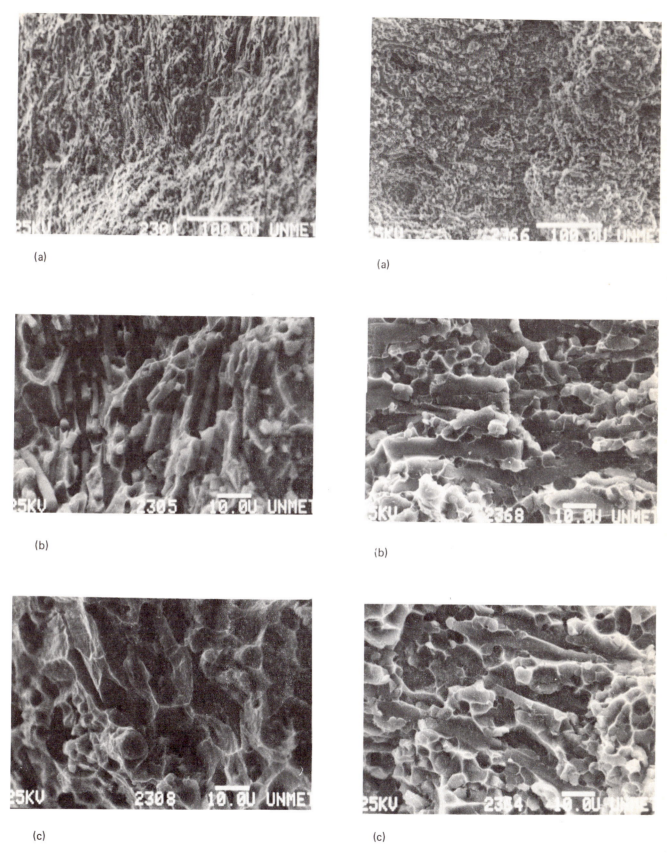

(a)

(a)

(b)

(b)

(c)

(c)

Fig 4   Scanning electron micrograph of the fracture surface
produced with a Saffil reinforced LMO alloy after
(a) and (b) testing at 20°C and (c) testing at 200°C
(all photographs reduced by 80 per cent)

Fig 5   Scanning electron micrograph of the fracture surface
produced with a Saffil reinforced LM25 alloy after
(a) and (b) testing at 20°C and (c) testing at 200°C
(all photographs reduced by 80 per cent)

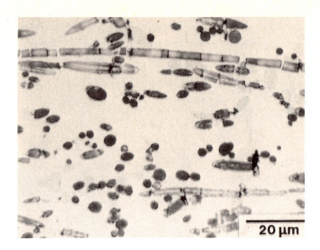

20 µm

Fig 6   Optical micrograph of multiple cracked Saffil fibres
produced in the gauge length of a reinforced LM25
alloy which has been tested and fractured at 20°C
(reduced by 80 per cent)

Buckling of perforated composite plates – an approximate solution

I H MARSHALL, BSc, PhD, CEng, MIMechE, FPRI, **W LITTLE, M M TAYEBY** and **J WILLIAMS**
Department of Mechanical and Production Engineering, Paisley College of Technology, Scotland

ABSTRACT Herein is contained a theoretical investigation into the stability characteristics of un-iaxially compressed, thin, rectangular planform, composite plates with holes or cut-outs.   In particular the effects of positioning such holes offset from the plate centre is considered in detail. Although an approximate method of analysis is employed good correlation with limited experimental evidence from tests on GRP plates with orthotropic material properties is shown.

1   INTRODUCTION

The effects of circular holes, or cut-outs, on the stability of thin rectangular plates has received relatively little attention in the past from either theoretical or experimental points of view.   The relative scarcity of experimental buckling loads combined with a noteworthy divergence between some of the limited existing theoretical models has tended to cloud the issue with more than a modicum of uncertainty. Bearing in mind the inevitability of access holes in a great many instances the problem clearly has practical significance, particularly in the aero space and shipbuilding industries. Consequently, the present problem seemed to be worthy of attention.

Initially a literature survey of previously published work on the equivalent problem for plates manufactured from conventional isotropic materials of construction was instituted. Although by no means exhaustive, references (1) to (9) inclusive appear to realistically cata-logue work to date on perforated isotropic platework systems.

These investigators have largely confined their attention to simply supported plates of square planform ($\lambda=1$) and cite limited experi-mental evidence. Moreover centrally positioned holes of diameter/plate width = 0.5 appear to have been limiting factors in previous studies. The equivalent problem for composite plates has largely escaped attention and, as yet, no papers dealing with the effects of unsymmetrical hole positioning have come to light. Previous work by the authors on the buckling and postbuckling of composite plates with centrally located holes is cited in reference (10) to (12) inclusive.

2   THEORETICAL ANALYSIS

Noting Fig 1 the plate edge boundary conditions consistent with simple supports can be written:

along x = ± $a/2$ along y = ± $b/2$

$\omega = o$ $\omega = o$ Zero Lateral Deflection

$$\frac{\partial^2 \omega}{\partial x^2} + \nu_y \frac{\partial^2 \omega}{\partial y^2} = o \quad \frac{\partial^2 \omega}{\partial y^2} + \nu_x \frac{\partial^2 \omega}{\partial x^2} = o \quad \begin{array}{c}\text{Zero}\\\text{Edge}\\\text{Moment(i)}\end{array}$$

Likewise the boundary conditions prevailing at the periphery of a hole radius r = r_0 can be written:

$$(M_r)_{r=r_0} = o \quad \text{and} \quad \left(Q_r - \frac{\partial M_{r\theta}}{r\partial\theta}\right)_{r=r_0} = o \qquad (2)$$

i.e a stress-free hole boundary.

where:

$$(M_r)_{r=r_0} = \frac{\partial^2 \omega}{\partial x^2}(D_x \cos^2\theta + D_y \nu_x \sin^2\theta)$$

$$+ \frac{\partial^2 \omega}{\partial y^2}(D_y \sin^2\theta + D_x \nu_y \cos^2\theta)$$

$$+ \frac{\partial^2 \omega}{\partial x \partial y}(\frac{t^3}{6} G_{xy} \sin\theta \cos\theta) = 0$$

and $\left(Qr - \frac{\partial M_{r\theta}}{r\partial\theta}\right)_{r=r_0} =$

$$\frac{\partial^3 \omega}{\partial x \partial y^2}(\frac{t}{6} G_{xy} + D_x \nu_y) \cos\theta + \frac{\partial^3 \omega}{\partial x^3} \cdot D_x \cos\theta$$

$$+ \frac{\partial^3 \omega}{\partial y \partial x^2}(\frac{t^3}{6} G_{xy} + D_y \nu_x) \sin\theta + \frac{\partial^3 \omega}{\partial y^3} D_y \sin\theta$$

$$- \frac{\partial^2 \omega}{\partial x^2}(D_x \cos^2\theta + D_y \nu_x \sin^2\theta)$$

$$- \frac{\partial^2 \omega}{\partial y^2}(D_y \sin^2\theta + D_x \nu_y \cos^2\theta)$$

$$- \frac{\partial^2 \omega}{\partial x \partial y}(\frac{t^3}{6} G_{xy} \sin\theta \cos\theta) = 0$$

and   x = $r_0 \cos\theta$

y = $r_0 \sin\theta$ $\qquad\qquad (3)$

Since the present work concerns itself with the buckling load of perforated plates a simple energy-based solution will yield sufficient in-sight into the problem. When an orthotropic (orthogonally anisotropic) plate is subjected to uniaxial compression the total energy is the system (U) will comprise the energy due to bending (U_B) and that due to mid-surface stresses (U_S).

$$U = U_B - U_S \qquad (4)$$

where $U_B = \iint\limits_{A} \left\{ \frac{D_x}{2} \left(\left(\frac{\partial^2 \omega}{\partial x^2}\right)^2 + \nu_y \left(\frac{\partial^2 \omega}{\partial x^2} \frac{\partial^2 \omega}{\partial y^2}\right) \right) + \right.$$

$$\left. \frac{D_y}{2}\left[\left(\frac{\partial^2 \omega}{\partial y^2}\right)^2 + \nu_x \left(\frac{\partial^2 \omega}{\partial x^2}\right)\left(\frac{\partial^2 \omega}{\partial y^2}\right)\right] + \frac{G \times 3}{6} \left(\frac{\partial^2 \omega}{\partial x \partial y}\right)^2 \right\} \, dA \qquad (5)$$

and $U_S = \frac{\sigma_o}{2} \iint\limits_{A} \left\{ \frac{\sigma_x}{\sigma_o} \left(\frac{\partial^2 \omega}{\partial x^2}\right)^2 + 2\frac{\tau_{xy}}{\sigma_o}\left(\frac{\partial \omega}{\partial x}\frac{\partial \omega}{\partial y}\right)^2 + \right.$

$$\left. \frac{\sigma_y}{\sigma_o}\left(\frac{\partial^2 \omega}{\partial y^2}\right)^2 \right\} \, dA \qquad (6)$$

It can also be shown that the in-plane stresses can be adequately described by:

$$\sigma_x = \frac{\sigma_o}{2} \frac{r_o^2}{r^2} \left[\cos 2\theta + (2 - \frac{3 r_o^2}{r^2}) \cos 4\theta \right]$$

$$\sigma_y = \frac{\sigma_o}{2} \left[2 + \frac{r_o^2}{r^2}(3 \cos 2\theta - (2 - \frac{3 r_o^2}{r^2})\cos 4\theta \right]$$

$$\tau_{xy} = \frac{\sigma_o}{2} \frac{r_o^2}{r^2} \left[-\sin 2\theta + (2 - \frac{3 r_o^2}{r^2}) \sin 4\theta \right] \qquad (7)$$

Thus providing a satisfactory deflection function $\omega(x,y)$ can be specified for any given plate geometry the resulting buckling load can be computed using minimum energy principles.

$$\frac{\partial U}{\partial \omega_o} = 0 \qquad (8)$$

In the case of a centrally perforated square plate, as shown in Fig 1, the lateral deflections can be adequately described by:

$$\omega = -\omega_o \left(\cos \frac{\pi x}{a} \cos \frac{\pi y}{b} + L e^{-c\left[\left(\frac{x}{a}\right)^2 + \left(\frac{y}{b}\right)^2\right]} \right) \quad (9)$$

where the first term describes the overall buckling mode and the latter takes account of the localised deformations prevailing around the hole boundaries.

However, a more general approach can be adopted by redefining (9) in the form:

$$\omega = -\omega_o \left(\cos \frac{\pi x}{a} \left(\cos \frac{n\pi(y+y_o)}{b} + \alpha \sin \frac{2n\pi(y+y_o)}{b} \right) + \right.$$

$$\left. + L e^{-c\left[\left(\frac{x}{a}\right)^2 + \left(\frac{y}{b}\right)^2\right]} \right) \qquad (10)$$

This describes the buckling mode of an eccentrically perforated plate as shown in Figure 2.

At this stage it will be noted that, before the plate buckling load can be determined according to equation (8) a number of variables require to be evaluated, i.e (C,L and θ).

Firstly the constants C and L are evaluated as a function of the plate out of plane displacement ω_o, by approximately satisfying the stress free boundary conditions prevailing at the hole periphery - equations (2) and (3). Boundary collocation techniques ($r = r_o$, θ variable) have proved to be an efficient means of achieving a

satisfactory solution of these equations. Essentially this ensures satisfaction of the prescribed boundary conditions at finite points on the hole periphery with the additional proviso that the nett loading on the hole boundary is zero.

Rather curiously, although not explicitly stated by other authors who have employed a similar technique on isotropic centrally perforated plates (1) (6) the collocation technique always yields an optimum value of $\theta = \pi/8$ i.e. exact satisfaction at sixteen points on the hole periphery with small perturbations at intermediate points.

Thereafter, the value of plate buckling load can be determined by substitution of the corrected equation for ω_o into the minimum energy criteria (8).

3 EXPERIMENTAL INVESTIGATION

In view of the dearth of experimental evidence regarding the buckling loads of rectangular planform orthotropic plates it was considered essential to initiate an extensive experimental programme to run in parallel with the theoretical analysis and augment its findings.

3.1 The Test Rig

A novel test rig capable of realisitically applying uniaxial compression to rectangular planform plates was designed and its construction supervised. The salient details of this piece of equipment are given in Fig 3. Essentially the load is applied by a hydraulic cylinder (A) via a pre-calibrated proving ring(B). A set of flat linear bearings (C) along with an adjustable bottom platten allows uniaxial compression to be accurately applied to a variety of plate aspect (length/breadth) ratios (λ). The rig is capable of simulating either simply supported or fully fixed plate edges by the use of interchangeable supports (D).

The test plates were vacumm laminated using uni-directional glass cloth embedded in a polyester resin matrix. After suitable post-curing the plates were accurately machined to a square planform of approximately 250mm by 250mm giving an aspect ratio $\lambda = 1.0$. Each plate was tested in the unpierced condition to determine its critical or buckling load. Thereafter plates with a range of hole diameters and varying degrees of eccentricity were tested. The results of an array of dial gauges were monitored during testing to obtain the respective buckling loads. At each of the aforementioned plate geometries, a minimum of five tests were carried out to verify experimental repeatability.

The test plates of thickness 1.90mm possessed the following material properties:

$$E_x = 3.10 \ \text{GN/m}^2$$
$$E_y = 10.40 \ \text{GN/m}^2$$
$$\nu_y = 0.300$$
$$G_{xy} = 2.142 \text{GN/m}^2$$

4 COMPARISON BETWEEN THEORY AND EXPERIMENT

4.1 Isotropic Centrally Pierced Plates

In order to make comparison with previous experiments the present analysis was simplified for the case of plates with isotropic material properties. Fig 4 compares previously published work on centrally pierced isotropic plates with the present solution. In general, favourable comparison is shown in spite of the relative crudness of the present solution.

4.2 Composite Centrally Pierced Plates

Fig.5 summarises the findings of this part of the investigation. Fuller details may be found in references (10)(11). In general, favourable comparison between theory and experiment is shown for hole diameters/plate widths ≤ 0.4, thereafter some divergence is apparent.

4.3 Composite Eccentrically Pierced Plates

To date the experimental investigation has largely concentrated on centrally pierced composite plates. However, when the limited experimental buckling loads are compared with that predicted by the present theory reasonable comparison has been shown - Fig 6. In general the buckling load is shown to diminish as the hole eccentricity y_0/b increases. Although by no means exhaustive, it is hoped that this on-going study will further corroborate the present theoretical approach.

5 CONCLUDING REMARKS

The present theoretical approach appears to realistically predict the buckling loads of orthotropic plates with centrally positioned holes within the hole diameter/plate width range ≤ 0.40. It has also been shown, for the limited experimental evidence available to date, to similarly model the behaviour of such plates with small degrees of eccentricity.

Although the present method of analysis is a 'hybrid', i.e. a combination of boundary collocation and minimum energy principles, and must therefore be considered as approximate, it has considerable practical potential.

6 ONGOING ASPECTS OF THE INVESTIGATION

Several theoretical avenues of this investigation are presently being actively pursued, e.g. clamped edge supports, post-buckling behaviour, reinforcement of the hole boundaries, etc. Although in many ways in its embryonic state, the investigation has shown considerable promise and will be the subject of future papers.

7 REFERENCES

(1) KUMAI, T. 'Elastic stability of the square plate with a circular hole under edge thrust'. Reports of the Research Institute for Applied Mechanics, Vol.1, No.2, April 1952.

(2) LEVY, S, WOOLLEY, R.M and KROLL, W.A. 'Instability of a simply supported square plate with a reinforced circular hole in edge compression'. Journal of Research National Bureau of Standards, Vol.39, December 1947.

(3) KAWAI, T and OHTSUBO, H. 'A method of solution for the complicated buckling problems of elastic plates with combined use of Rayleight-Ritz's procedure in the finite element method'. Proceedings of the Second Air Force Conference on Matrix Methods in Structure Mechanics, Wright-Patterson AFB, October 1968.

(4) SCHLACK, A.L. 'Experimental critical loads for perforated square plates'. Proceedings of the Society for Experimental Stress Analysis, Vol.25, 1968

(5) PENNINGTON-WANN, W. 'Compressive buckling of perforated plate elements'. Proceedings of the First Speciality Conference on Cold-formed Structures (1971), University of Missouri, Rolla, Missouri, April 1973, pp. 58-64.

(6) YOSHIKI, M, FUJITA, Y, KAWAMURA, A. and ARAI, H. 'Instability of plates with holes (First Report)'. Proceedings of the Society of Naval Architects of Japan, No.122, December 1967.

(7) ROCKEY, K.C., ANDERSON, R.G. and CHUENG,Y.K. 'The behaviour of square shear webs having a circular hole'. Proceedings of Swansea Conference on Thin-Walled Structures, Crosby Lockwood & Sons, London 1967, pp.148-169.

(8) SHANMUGAM, N.E and NARAYANAN, R. 'Elastic buckling of perforated square plates for various loading and edge conditions'. Proc. Int. Conference on Finite Element Methods, Paper No. 103, Shanghai, August 1982.

(9) NARAYANAN, R. and CHOW, F.Y. 'Ultimate capacity of uniaxially compressed perforated plates'. Int. Journal of Thin Walled Structures, Vol.2, No.3, 1984, pp.241-264.

(10) MARSHALL, I.H., LITTLE, W and EL TAYEBY,M.M. 'The stability of composite panels with holes', Reinforced Plastics Congress, Brighton, U.K., November 1984.

(11) MARSHALL, I.H, LITTLE and EL-TAYEBY, M.M. 'The Stability of Composite Panels with Holes' Chapter 17 of "Mechanical Characterisation of Load bearing Fibre Composite Laminates" edited by A.H. Cardon and G. Verchery. Elsevier Applied Science Publishers, 1985.

(12) MARSHALL, I.H., LITTLE, W and EL TAYEBY,M.M. "Post Buckling of Composite Panels with Holes" ICCM/V,San Diego, USA, July 1985.

8 NOMENCLATURE

A	plate surface area
a,b	plate dimensions
D_x,D_y	plate flexural rigidities
G_{xy}	shear modulus
r_0	hole radius
r, θ	polar co-ordinates
t	plate thickness
U_B	Energy due to plate bending
U_S	Energy due to mid-surface bending
U	Total energy in deformed plate
x,y	cartesian co-ordinates
y_0	hole eccentricity
ω	plate lateral deflection coefficient
λ	plate aspect (length/breadth) ratio
$\vartheta_x \vartheta_y$	Poisson's ratios
σ_0	applied axial compressive stress

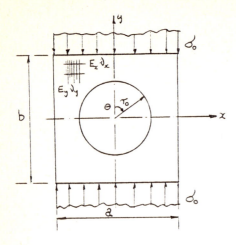

Fig 1   Plate with central hole

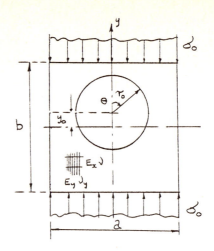

Fig 2   Plate with eccentric hole

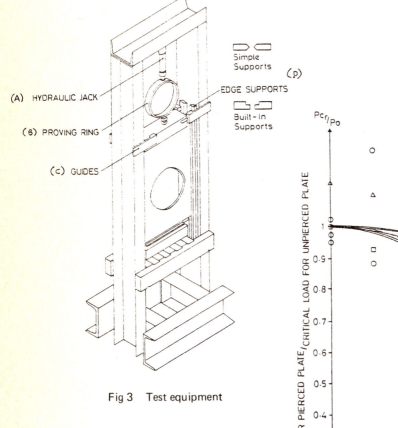

Fig 3   Test equipment

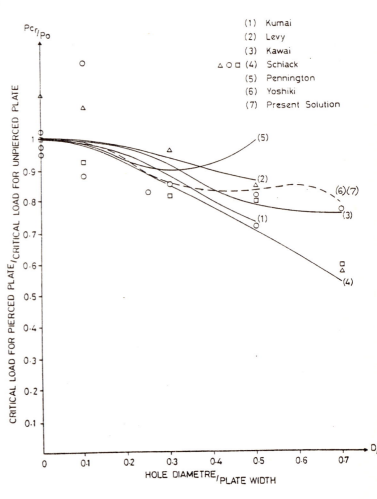

Fig 4   Isotropic plates with central holes

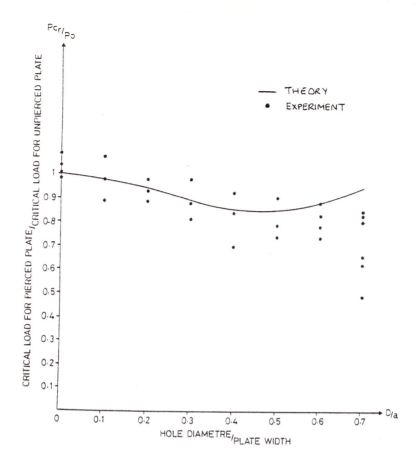

Fig 5   Orthotropic plates with central holes

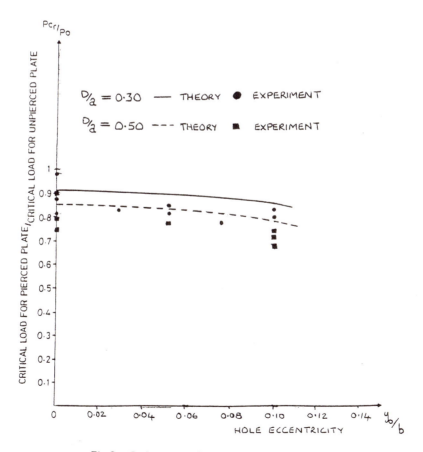

Fig 6   Orthotropic plates with eccentric holes

Stochastic finite element analysis of uncertain in-ply stresses in fibre reinforced plastic laminated plate

S NAKAGIRI, BEng, MEng and T HISADA, DEng
University of Tokyo, Japan
S TANI, MEng,
Mitsubishi Electric Corporation, Amagasaki, Japan
H TAKABATAKE, MEng
Takamatsu Technical College, Takamatsu, Japan

SYNOPSIS A methodology for stochastic finite element analysis is presented on the basis of the second order perturbation technique. Numerical analysis is implemented with regard to the in-ply stress deviation of thin boron/epoxy plate with uncertain stacking sequence and under uniaxial compression.

1 INTRODUCTION

The fiber reinforced plastic (FRP) laminated plate has been extending its front into advanced technology field, taking advantage of the nature with not only high specific strength and stiffness but also anisotropy and heterogeneity. Tailoring design has been made possible by use of the features of anisotropy and heterogeneity of the FRP plate.

The elastic constants of the FRP plate are likely to deviate from the designed ones. Also the fiber orientations and layer thickness fluctuate in the vicinity of the prescribed values depending on the manufacturing process. It is therefore necessary and desired to investigate the effect of the uncertain stacking sequence on the behaviour of the FRP plate.

A methodology is presented to elucidate the relationship between uncertain variation of the stacking sequence and that of the structural response. The methodology makes use of the stochastic finite element method(1). The second order perturbation technique is employed to evaluate the first and second order rates of change of the structural response with respect to the random variables in question. Those rates are combined with the statistical nature of the random variables in order to estimate the expectation and variance of the structural response with the aid of the second order approximation(2).

The numerical examples are related with the uncertainty of static response of thin, rectangular boron/epoxy laminated plate, whose fiber orientations and layer thickness are probabilistic, under axial compression with regard to the input second moments of the random variables.

2 VARIATION OF STIFFNESSES AND DISPLACEMENTS

2.1 Stiffnesses of laminated plate

The mechanical properties of the FRP plate are represented, in a way, by the in-plane stiffness [A], coupling stiffness [B] and bending stiffness [D] in form of 3 by 3 matrix as given below(3).

$$[A] = \sum_{n=1}^{N} [Q]_n (z_{n+1} - z_n) \tag{1}$$

$$[B] = \frac{1}{2} \sum_{n=1}^{N} [Q]_n (z_{n+1}^2 - z_n^2) \tag{2}$$

$$[D] = \frac{1}{3} \sum_{n=1}^{N} [Q]_n (z_{n+1}^3 - z_n^3) \tag{3}$$

In the above, the matrix [Q] is derived from the stress-strain relationship for each layer numbered n under plane stress condition and is determined by the four independent elastic constants E_L, E_T, G_{LT}, n_{LT} and the layer orientation θ(3). The total number of layers is denoted by N, and the coordinate z is set perpendicular to the plate plane, the origin of which is taken at the mid-plane. The Kirchhoff-Love's hypothesis is assumed in the derivation of the coupling and bending stiffnesses.

The resultant forces and resultant moments are connected with the mid-plane strains and curvatures through the following 6 by 6 matrix [R] in the frame of small displacement theory.

$$[R] = \begin{bmatrix} [A] & [B] \\ SYM. & [D] \end{bmatrix} \tag{4}$$

Then the finite element stiffness matrix [k] is given by Eq.(5),

$$[k] = \iint [S]^T [R] [S] \, dxdy \tag{5}$$

where [S] stands for the strain-nodal displacement matrix which is considered deterministic in this study. The non-conforming triangular BCIZ element is employed for the plate bending(4), and the in-plane strains are assumed constant in the element.

2.2 Uncertainties of stacking sequence

The fiber orientation θ and thickness t of each layer are considered uncertain, while the elastic constants are deterministic. The uncertainties are expressed in form of Eqs.(6) and (7).

$$\theta_n = \bar{\theta}_n (1 + \alpha_n) \tag{6}$$

$$t_n = \bar{t}_n (1 + \beta_n) \tag{7}$$

where α and β stand for small random variable, and the upper bar means deterministic term hereafter. The uncertainty of the z coordinate of the r-th layer boundary resulted from the thickness is given by means of converting β to new random variable γ, as follows.

$$z_r = \bar{z}_r + \gamma_r \qquad (8)$$

where

$$\bar{z}_r = \sum_{n=1}^{r-1} \bar{t}_n - \frac{1}{2} \sum_{n=1}^{N} \bar{t}_n \qquad (9)$$

$$\gamma_r = \sum_{n=1}^{r-1} \bar{t}_n \beta_n - \frac{1}{2} \sum_{n=1}^{N} \bar{t}_n \beta_n \qquad (10)$$

Those uncertainties cause uncertain variation of the matrix $[R]$ and further the element stiffness matrix $[k]$. The variation can be expressed by the Taylor series expansion with respect to the random variables. When the Taylor series expansion is truncated at the second order, the overall stiffness matrix $[K]$ is summarized as Eq.(11),

$$[K] = [\bar{K}] + \sum_{i=1}^{M} [K_i^I]\epsilon_i + \frac{1}{2}\sum_i^M\sum_j^M [K_{ij}^{II}]\epsilon_i\epsilon_j \qquad (11)$$

where unified notation ϵ is used for the random variables so that ϵ_i represents α_i for i=1 to N and γ_{i-N} for i=N+1 to M. The total number of the variables is denoted by M=2N+1. The first and second rates of change $[K_i^I]$ and $[K_{ij}^{II}]$ can be evaluated easily by means of differentiating $[R]$ with respect to the pertinent variables, calculating Eq.(5) and merging the result into the overall system. The number of the non-zero constituents of $[K_i^I]$ and $[K_{ij}^{II}]$ is very small compared with that of $[K]$.

The resultant variation of the unknown nodal displacements $\{U\}$ is assumed in the same form of the truncated Taylor series expansion as follows.

$$\{U\} = \{\bar{U}\} + \sum_{i=1}^{M}\{U_i^I\}\epsilon_i + \frac{1}{2}\sum_i^M\sum_j^M\{U_{ij}^{II}\}\epsilon_i\epsilon_j \qquad (12)$$

All the unknowns of $\{\bar{U}\}$, $\{U_i^I\}$ and $\{U_{ij}^{II}\}$ can be determined by substituting Eqs.(11) and (12) into the stiffness equation(13),

$$[K]\{U\} = \{F\} \qquad (13)$$

and applying the second order perturbation technique to the result. The followings are obtained when the force vector $\{F\}$ is unchanged.

$$[\bar{K}]\{\bar{U}\} = \{F\} \qquad (14)$$

$$[\bar{K}]\{U_i^I\} = -[K_i^I]\{\bar{U}\} \qquad (15)$$

$$[\bar{K}]\{U_{ij}^{II}\} = -[K_i^I]\{U_j^I\} - [K_j^I]\{U_i^I\}$$
$$\qquad\qquad -[K_{ij}^{II}]\{\bar{U}\} \qquad (16)$$

All the unknowns are obtained by use of the same solver for $[\bar{K}]^{-1}$ regardless of the random variables, and the right side vectors of Eqs.(15) and (16) are made known by successive substitution of the knowns. It turns out that only short CPU time is required for the determination of all of $\{\bar{U}\}$, $\{U_i^I\}$ and $\{U_{ij}^{II}\}$.

3 VARIATION OF IN-PLY STRESSES

According to the Kirchhoff-Love hypothesis, the strains e at z is given by Eq.(17).

$$e(z) = e_0 + z\kappa \qquad (17)$$

where e_0 stands for the mid-plane strains and κ for the curvatures. The strain-nodal displacement matrix $[S]$ can be divided into two parts of $[S_p]$ for the in-plane displacements and $[S_b]$ for the out-of-plane ones. The strains on the lower boundary of the n-th layer $e(z_n)$ are rewritten as Eq.(18),

$$e(z_n) = ([S_p] + \bar{z}_n[S_b])\{u\} + \gamma_n[S_b]\{u\} \qquad (18)$$

where $\{u\}$ denotes the pertinent nodal displacements for an element. By use of the same expression for $\{u\}$ as Eq.(12), Eq.(18) is rewritten as follows.

$$e(z_n) = [G]\{\bar{u}\} + \sum_{i=1}^{M}([G]\{u_i^I\} + [H_i^I]\{\bar{u}\})\epsilon_i$$
$$+ \frac{1}{2}\sum_i^M\sum_j^M([G]\{u_{ij}^{II}\} + [H_i^I]\{u_j^I\} + [H_j^I]\{u_i^I\})\epsilon_i\epsilon_j \qquad (19)$$

where

$$[G] = [S_p] + \bar{z}_n[S_b] \qquad (20)$$

$$[H_i^I] = \begin{cases} [0] & (\ i \neq N+n\ ) \\ [S_b] & (\ i = N+n\ ) \end{cases} \qquad (21)$$

The plane stresses in the n-th layer are calculated by Eq.(22) based on th $[Q]$ matrix and the strains of Eq.(17).

$$\sigma = [Q]_n e \qquad (22)$$

Taking into account that the $[Q]$ matrix is subject to variation through the fiber orientation, we have the following result for the in-ply stressed at z_n, for example.

$$\sigma(z_n) = \bar{\sigma}(z_n) + \sum_{i=1}^{M}\sigma_i^I(z_n)\epsilon_i + \frac{1}{2}\sum_i^M\sum_j^M\sigma_{ij}^{II}\epsilon_i\epsilon_j \qquad (23)$$

where

$$\bar{\sigma}(z_n) = [Q]_n[G]\{\bar{u}\} \qquad (24)$$

$$\sigma_i^I(z_n) = [\bar{Q}]_n[G]\{u_i^I\}$$
$$\qquad + ([\bar{Q}]_n[H_i^I] + [Q_i^I]_n[G])\{\bar{u}\} \qquad (25)$$

$$\sigma_{ij}^{II}(z_n) = [\bar{Q}]_n[G]\{u_{ij}^{II}\}$$
$$+ ([\bar{Q}]_n[H_i^I] + [Q_i^I]_n[G])\{u_j^I\}$$
$$+ ([\bar{Q}]_n[H_j^I] + [Q_j^I]_n[G])\{u_i^I\}$$
$$+ ([Q_i^I]_n[H_j^I] + [Q_j^I]_n[H_i^I]$$
$$+ [Q_{ij}^{II}]_n[G])\{\bar{u}\} \qquad (26)$$

4 EVALUATION OF RESPONSE STATISTICS

As stated in the preceding sections, any displacement or stress, X, can be approximated in the vicinity of the deterministic term $\overline{X}$ by the following Taylor series expansion with respect to the random variables standing for the uncertainties involved.

$$X = \overline{X} + \sum_{i=1}^{M} X_i^{I} \varepsilon_i + \frac{1}{2} \sum_{i}^{M} \sum_{j}^{M} X_{ij}^{II} \varepsilon_i \varepsilon_j \qquad (27)$$

It should be noted that no distribution function is assumed for the random variables in the above formulation.

Now that the rates of change X_i^{I} and X_{ij}^{II} are determined by the finite element analysis, the expectation $E[X]$ and variance $Var[X]$ of the variable X are obtained in the following form on the basis of the second order approximation(2).

$$E[X] = \overline{X} + \frac{1}{2} \sum_{i=1}^{M} \sum_{j=1}^{M} X_{ij}^{II} E[\varepsilon_i \varepsilon_j] \qquad (28)$$

$$Var[X] = \sum_{i}^{M} \sum_{j}^{M} X_i^{I} X_j^{I} E[\varepsilon_i \varepsilon_j]$$

$$+ \sum_{i}^{M} \sum_{j}^{M} \sum_{k}^{M} X_i^{I} X_{jk}^{II} E[\varepsilon_i \varepsilon_j \varepsilon_k]$$

$$+ \frac{1}{4} \sum_{i}^{M} \sum_{j}^{M} \sum_{k}^{M} \sum_{l}^{M} X_{ij}^{II} X_{kl}^{II} (E[\varepsilon_i \varepsilon_j \varepsilon_k \varepsilon_l]$$

$$- E[\varepsilon_i \varepsilon_j] E[\varepsilon_k \varepsilon_l]) \qquad (29)$$

when the first order approximation is employed, only the first terms in the above equations should be taken into account. $E[\varepsilon_i \varepsilon_j]$ are the second moments of the random variables which are related with the correlation function and further the power spectrum through the Wiener-Khintchine relationship. It is seen that the third and fourth moments of the random variables are required as the input data for such statistics calculation in the case of the second order approximation employed. Those high order moments are prescribed, however, by the second moments if the random variables are Gaussian.

5 NUMERICAL EXAMPLES

Figure 1 illustrates the finite element division of a boron/epoxy laminated plate under uniaxial compression. The edges AB and CD are set free, and the edges AC and BD are simply supported for the bending. The elastic constants used are $E_L = 206.9$GPa, $E_T = 20.69$GPa, $G_{LT} = 6.89$GPa and $\nu_{LT} = 0.3(5)$. Two cases of stacking are analysed by taking the expectation of each layer thickness as 0.1mm. Case 1 is of six layers, whose expectations of the fiber orientation are 120/60/0/0/60/120 degrees. The expectation of the coupling stiffness is set zero in this case. Case 2 is of three layers with the fiber orientations of 135/0/45 degrees. This is an anisotropic case that the coupling between the in-plane and out-of-plane deformations takes place as the expectation.

Figure 2 depicts the deflection at the point P_1 of Fig.1. The circles stand for the deflection calculated by changing deterministically the fiber orientation of the first layer. The broken and solid lines are drawn based upon the first and second order perturbation solutions. The agreement between the deterministic solution and perturbation ones verifies that the first and second order rates of change are calculated correctly.

Figure 3 shows the standard deviations of the deflection of Case 1 along the edges AB and CD in Fig.1. The input standard deviations are 2 degrees for all the six orientations and 0.001mm for the thickness of the six layers. Those variables are assumed as uncorrelated Gaussian. The second moments of the random variables for the z coordinates can be derived easily from those of the thickness by use of the linear relationship between them.

The expectation and three-sigma bounds of the in-ply stresses are illustrated in Fig.4 for Case 1. The input second moments are the same as mentioned above. The distribution of the expectation is uniform all over the plate, but that of the standard deviation varies from point to point. The three-sigma bounds in the figure are evaluated at the point P_1 in Fig.1.

Figure 5 depicts the coefficients of variation of the in-ply stresses, for example, of Case 2 and along the line IJ. It is seen that the in-ply stresses are affected considerably by even such small deviations of the stacking sequence. It has been found that the distribution of the coefficients of variation of the in-ply stresses is sensitive to the stacking sequence.

The effect of the input standard deviations on the output coefficients of variation is shown in Fig.6. The ordinate stands for the in-ply stresses at z_1 on the point P_2 and the abscissa for the magnitude of the input standard deviations of the fiber orientations of Case 1. The standard deviations are increased by the same amount for all the fiber orientations under the assumption that they are uncorrelated and the layer thickness is kept deterministic in this case. As seen by the plot for τ_{xy}, there exists nonlinear relationship. This means that care should be taken to employ the second order approximation, because the relationship is always evaluated as linear by the first order approximation.

6 CONCLUDING REMARKS

The stochastic finite element method, which is based on the second order perturbation technique, is applied to the evaluation of the in-ply stress deviations caused by uncertain stacking sequence.

The formulation is presented which enables us to evaluate efficiently the first and second order rates of change of static response of FRP laminated plate with respect to the random variables taken to represent uncertain stacking sequence.

The variation of the in-ply stresses is estimated quantitatively in regard to the input standard deviations of the fiber orientations and layer thickness. Through the numerical analyses carried out, it has been found that the effect of the fiber orientations is larger than that of the layer thickness, and that the variation of the static response is sensitive when the number of layers is small.

REFERENCES

(1) NAKAGIRI,S. and HISADA,T. _The Stochastic Finite Element Method, An Introduction_, (in Japanese), Baifu-kan, 1985.

(2) ANG,A.H.S. and TANG,W.H. _Probability Concepts in Engineering Planning and Design_, _1_, John Wiley & Sons, 1975, 197.

(3) JONES,R.M. _Mechanics of Composite Materials_, MacGraw-Hill Kogakusha, 1975, 147-238.

(4) ZIENKIEWICZ,O.C. _The Finite Element Method_, 3rd ed., McGraw-Hill, 1977, 226-267.

(5) HIRANO,Y. Optimum Design of Laminated Plates Under Shear, _J.Composite Materials_, 1979, _13_, 329-334.

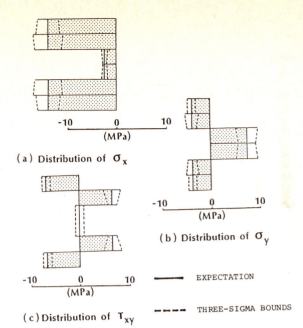

(a) Distribution of σ_x

(b) Distribution of σ_y

(c) Distribution of τ_{xy}

——— EXPECTATION

- - - - - THREE-SIGMA BOUNDS

Fig 4   Distribution of in-ply stresses (Case 1)

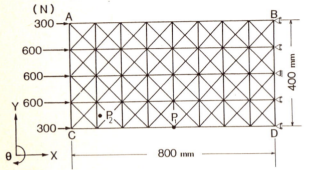

Fig 1   Finite element division of the fibre reinforced plate (FRP)

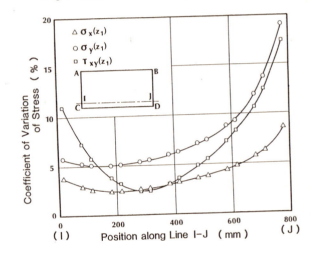

Fig 5   Distribution of stresses along line IJ

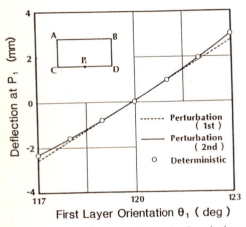

Fig 2   Comparison between determinstic solution and pertubation solutions of deflection

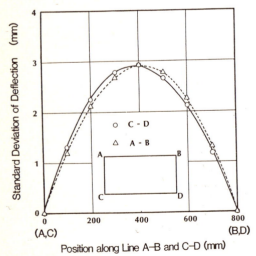

Fig 3   Distribution of standard deviations of deflection along edges AB and CD

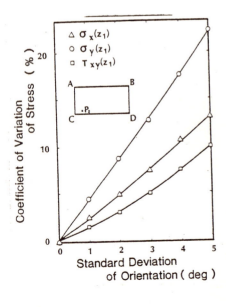

Fig 6   Relationship between output stress variation and input orientation deviation

C17/86

Designing with short fibre reinforced thermoplastics

M W DARLINGTON, BSc, PhD, FPRI
School of Industrial Science, Cranfield Institute of Technology, Bedford
P H UPPERTON, BSc, MSc
Du Pont (UK) Limited, Hemel Hempstead, Hertfordshire

SYNOPSIS The deformation and strength behaviour of a number of components injection moulded in short fibre reinforced thermoplastics have been studied under static loading. The experimental results have been compared with predictions. Experimentally determined modulus and strength data representing the 'bounds' of behaviour of these anisotropic materials, and data appropriate to a random-in-the-plane fibre orientation distribution (RITP FOD) have been used in the calculations. For the stiffness predictions, the use of modulus data appropriate to a RITP FOD in a linear elastic analysis gave good agreement with experiment. For strength, the use of tensile strength data appropriate to a RITP FOD in a plastic collapse analysis gave predicted failure forces which were in good agreement with experiment. Reasonable predictions of component deformation at failure were also obtained using failure strain data in a linear elastic analysis.

1 INTRODUCTION

The mechanical properties of thermoplastics vary in a complex manner with the applied stress (or strain) level, the time under load, temperature, relative humidity (some thermoplastics), the method and conditions of fabrication and even the storage conditions and time between component manufacture and use in service (1). General approaches to data presentation and design with thermoplastics, which allow to varying degrees for this complex behaviour have been in existence for some time (2,3,4).

The stiffness of a thermoplastic at room temperature is typically some two orders of magnitude lower than that of metals. Nevertheless, the attractions of the injection moulding process for thermoplastics have led to their successful use in a wide range of load bearing applications.

The incorporation of short rigid fibres into a thermoplastic is now well established as a means of obtaining increased rigidity in injection moulded components. Unfortunately, the flow of the fibre filled melt into the mould gives rise to a complex fibre orientation distribution (FOD) in the injection moulded component (5,6). In general, the fibre orientation pattern changes through the thickness of the moulding at any chosen place, and this pattern itself varies with position in the moulding. The resulting inhomogeneity and anisotropy of mechanical properties leads to uncertainty in design for stiffness and strength with these materials.

Mould geometry plays an important role in establishing the fibre orientation pattern, but it is important to note that different base polymers (e.g. polypropylene and nylon 66) may give different fibre orientation patterns in the same mould geometry (5). In relatively thin section mouldings (say up to 4mm), the fibres tend to lie in planes parallel to the mould surfaces. As moulding wall thickness increases, the tendency for planar alignment decreases, particularly for those fibres in the central 'core' layer. Voiding may also be a problem with the thicker sections.

The injection moulding process may cause fibre breakage. The degree to which this occurs depends on the injection moulding conditions and the method of composite manufacture. Hence the mechanical properties of the material in a moulded component may differ from those determined in the laboratory. The extrusion compounding method of composite production causes fibre breakage as the fibres are dispersed in the thermoplastic and the resulting product usually contains a broad distribution of fibre lengths. With reasonable care, further fibre breakage during moulding can be minimised and the properties of the material in the component will then be similar to those measured on injection moulded test specimens in the laboratory. This paper is limited to such materials.

At the present time, the fibre orientation in a moulded component can not be predicted with any certainty at the design stage. Even if it were possible to predict the exact FOD throughout the moulding, the complexity of any stress analysis required to handle the resulting anisotropy and inhomogeneity would be impracticable in most cases. Furthermore it must be remembered that the rewards for such effort are nowhere near as great as those obtained with continuous fibre reinforced composites. There is therefore a need to establish design procedures, for use with short fibre reinforced thermoplastics, which offer a realistic compromise between accuracy and ease of use.

The purpose of this paper is to describe recent work in this area. Both design for stiffness and design for strength are considered; the latter being limited to behaviour under 'static' loading conditions.

2 DATA PRESENTATION

Initially, tensile modulus and strength data for short fibre reinforced thermoplastics (SFRTP) were derived from tests on standard injection moulded dumb-bells. However, in such bars, fibre alignment is high in the direction of the tensile axis. Consequently, the data obtained approach the best (i.e. the highest tensile stiffness and strength) that can be expected of the material. Such data are unrepresentative of the behaviour of a typical SFRTP injection moulding in which either the fibres may be less well aligned or the stresses experienced in service may be transverse to the direction of fibre alignment. Hence, a 'bounds' approach to data presentation was suggested (7).

The basic idea of the bounds approach is to provide data that are representative of the possible extremes of behaviour for the material in a moulded part. The above-mentioned tensile dumb-bell data provide the 'Upper Bound'. It is apparent that tensile tests in which the applied stress is transverse to the usual tensile axis of the dumb-bell will give stiffness and strength values approaching the lowest that would normally be expected of the material. Such data are therefore referred to as the 'Lower Bound'. In practice, alternative sources of lower bound specimens are obviously needed, which offer a high degree of fibre alignment and the opportunity to extract tensile test-pieces having their longitudinal axis transverse to the fibre alignment direction.

The bounds values of tensile modulus and tensile strength are normally widely separated and the use of such data in standard isotropic stress analysis equations could lead to significant over- or under-design. The selection of appropriate modulus and strength values for use in design calculations is an important part of the overall procedure.

In contrast, the in-plane shear modulus shows far less variation with direction of test, even for mouldings in which there is a high degree of fibre alignment (8). There would therefore appear to be less uncertainty associated with the use of this parameter.

3 DESIGN FOR STIFFNESS

In a complex, three-dimensional component, a high degree of fibre alignment in one direction throughout the component is unlikely. Thus, a number of studies have shown that layers of high fibre alignment in one direction will usually be balanced either by changes in the preferred alignment direction through thickness or from place to place in the moulding. The general deformation behaviour of the component will be controlled by some average of the response of these regions of differing fibre orientation (and hence, differing moduli) over the area of interest. It was therefore proposed that, in many instances, it would be realistic to use creep modulus data corresponding to a fibre orientation distribution in which the fibres are orientated randomly in the plane of the moulding

and to carry out design calculations assuming the material to be isotropic (9).

The above approach will be referred to as the 'random-in-the-plane fibre orientation distribution', or RITP FOD, approach. This approach has been applied to a number of components or parts of larger components (referred to as sub-components) and the predictions compared with experimental results. Predictions based on the use of alternative stiffness values (e.g. Upper Bound) in the same stress analysis have also been produced for comparison. One example from this study is given below.

3.1 Stiffness study on a sub-component from a tray

The tray shown in Fig. 1 was injection moulded in unfilled polypropylene (PP) and polypropylene reinforced with 30% by weight of short glass fibre (GFPP). The tray was sprue fed at the centre of the base. The 400mm side wall of the tray was removed to form a sub-component which could be tested in three point bend (with a span of 255mm).

Stiffness data for use in the predictions were obtained using highly accurate tensile creep rigs. A 100 second isochronous stress – strain procedure was used to obtain low strain modulus values. Injection moulded bars of the GFPP were used to obtain the Upper Bound modulus. A true Lower Bound specimen is extremely difficult to obtain in GFPP. The tray base was found to be moderately anisotropic and a specimen cut with its tensile axis transverse to the main fibre alignment direction was therefore used. (The modulus of this 'transverse' specimen will be a little higher than that of a true Lower Bound specimen). A modulus applicable to a RITP FOD was estimated from these and other measurements.

A standard linear elastic analysis was used to predict the force-deformation behaviour. Modulus values corresponding to the unfilled polypropylene and the Upper Bound, RITP FOD and the 'transverse' specimen for the GFPP were used in the calculations.

The experimental results for the PP and GFPP sub-component are compared with the predicted force-deformation behaviour in Fig. 2.

Good agreement between prediction and experiment was obtained for the unfilled polypropylene sub-component. This suggests that the analysis used was satisfactory.

For the GFPP sub-component, it is apparent that use of Upper Bound data leads to significant under-prediction of the deformation for a given force whilst use of true Lower Bound data would lead to a very pessimistic prediction. The prediction based on the use of a modulus applicable to a RITP FOD is in reasonable agreement with experiment.

3.2 Additional stiffness studies

Similar studies have been carried out on a range of components embracing a variety of static loading situations. For example, the tray sub-component has also been subjected to a distributed load and the resulting central deflection measured (10). The entire tray has also been studied in several different loading modes. A gear selector component has been moulded in a number of different materials and loaded in two different modes (11).

In all cases, the comparison of experiment and prediction shows similar trends to those apparent in Fig. 2 and the deformation response predicted using the RITP FOD approach has given reasonable agreement with experiment.

4   DESIGN FOR STRENGTH

The random-in-the-plane FOD approach to design for stiffness with short fibre reinforced thermoplastics gives the designer a simple and effective design method for low strain deformation. However, in the case of design to failure, the use of tensile strength data applicable to a random-in-the-plane FOD may lead to unsafe predictions of component performance. For example, failure may be initiated at a point in the moulding where the fibres happen to be highly aligned in a direction tranverse to a major component of tensile stress. This suggests that, in critical areas, the possibility of failure should be examined using lower bound tensile strength data in conjuction with the appropriate formula (unless the FOD can be assumed with certainty to be more favourable).

Having selected the appropriate strength data, the problem of the selection of the appropriate analysis/failure criterion must be considered. For cases in which the dominant deformation mode is uniaxial tension, this should be straightforward. However, these cases are usually also those for which a sensible estimate of FOD can be made and hence evaluation of the Lower Bound proposal on such components or sub-components is of little value.

In typical moulded components, the stress system is usually multiaxial and/or the dominant deformation mode is flexure. In these circumstances, the choice of analysis/failure criterion is less obvious. Unfortunately, in design for strength, tests on the components moulded in the unreinforced polymer can not be relied on to remove any uncertainty in the approach adopted since the failure mode of the fibre reinforced polymer will frequently differ from that of the base polymer. Thus, in evaluating proposals for strength design, the problem of data selection (i,e. Lower Bound, Upper Bound etc.) can not be separated from the problem of the choice of analysis/failure criterion.

A number of components, or parts (sub-components) of larger mouldings, have now been tested to failure and the results compared with predictions obtained using the design approaches outlined in section 4.1 below.

In one study, the problem of design for strength when the component is deformed in flexure was examined using sub-components of T and U section taken from an injection moulded ribbed plaque (12). Mouldings produced in GFPP and GFPA66 were examined. The data used in the design calculations were obtained on tensile specimens cut from the plaques. The results of

studies on two further components are presented in sections 4.2 and 4.3. The results are discussed in section 4.4.

4.1   Failure criteria

The value of the applied force at which failure occurs has been calculated using the following approaches:

(a) Maximum Stress Criterion

This criterion assumes that the component fails when the maximum tensile stress within the component reaches the tensile strength of the material.

(b) Total Plastic Collapse

When an elastic-perfectly plastic material is tested in flexure, failure does not occur when the maximum outer fibre stress reaches the yield stress. Instead, the material at the outer fibre yields. As the applied force is increased, material further below the surface also yields and failure is deemed to have occurred when the entire section has yielded (3).

The materials used in this study do not exhibit elastic-perfectly plastic behaviour. However, the stress-strain behaviour is highly non-linear and the stress at failure is taken as the yield stress in the analysis.

(c) Maximum Strain Criterion

This criterion assumes that the component fails when the maximum strain within the component reaches the tensile failure strain of the material. Using this criterion, it is possible to predict the maximum deflection that the component can withstand before failure occurs.

In all three cases, the material is assumed to be homogeneous.

4.2   Failure study on a subcomponent from a milk crate

The milk crate was injection moulded in polypropylene reinforced with 30% by weight of short glass fibres. Specimens cut from the base of the crate were used as a simple sub-component. The specimen positions and nominal dimensions are shown in Fig. 3. Three crates were used in order to obtain three specimens for each position.

The specimens were tested in three point bend (with the rib in tension) on a constant extension rate test machine using a span of 130mm, crosshead speed of 5mm/min and a test temperature of 23°C. The force and deflection when total failure occurred was recorded for each specimen. The mean values for each specimen position are presented in Table 1. The highest spread recorded for the three values at any one site was 20N for the failure force and 0.9mm for the failure deflection. It is apparent from Table 1 that the variation between sites exceeds these limits. The variation between sites could be due to small differences in the fibre orientation distribution resulting from differing flow geometries.

Table 1 Force and deflection at failure for the
sub-component from the milk crate

Specimen position	Failure force (N)	Failure Deflection (mm)
1	264	5.6
3	249	5.2
4	299	6.0
6	301	6.3
7	289	4.9
9	298	5.1
10	298	5.9
12	298	5.8
13	256	5.3
15	256	5.3

The overall mean force and deflection at
failure are 280 N and 5.5 mm respectively.

The tensile strength data used in the
predictions have been obtained from two sources.
Firstly, injection moulded ASTM tensile dumb-
bells were used to provide Upper Bound data.
Secondly, tensile specimens of the form of ASTM
tensile dumb-bells were machined from the base
of the injection moulded tray shown in Fig. 1.
The specimens were cut at 0° and 90° to the main
flow direction in order that the degree of
anisotropy in the tray base could be assessed.
All the specimens were tested to failure in
uniaxial tension at a constant extension rate of
2 mm/min and a test temperature of 23°C. The
mean values of tensile strength and failure
strain from tests on ten specimens of each type
are shown in Table 2.

The time between moulding and testing of
the specimens and the sub-components from the
crates was in excess of 15 months. Any differences in the properties of the material in the
various mouldings due to differences in storage
time are not therefore expected to be
significant.

Table 2 Tensile failure data for GFPP

Specimen	Tensile strength (MPa)	Tensile strain (%)
ASTM bar	60	3.1
Tray, 0	42	3.6
Tray, 90	32	2.6

The predictions of component failure have
been carried out as follows:

The Maximum Stress Criterion has been used
together with the equation relating applied
force to maximum outer fibre stress given by
linear elastic theory for the deflection of a
beam in three point bend. Thus the failure
force, F, has been calculated using the equation

$$F = \frac{4 \ S \ I}{L \ C}$$

where S is the tensile strength of the material
       I is the second moment of area
       L is the span (0.13m)
       C is the distance to outer fibre from
         neutral axis (7.4mm).

For the case studied here, this gives F =
3.34 S.

The force at which total plastic collapse
occurs has been calculated using standard stress
analysis techniques. For the particular geometry and loading studied, this gave F = 5.93 S.

For the above two approaches, values of F
obtained using the tensile strength for the
Upper Bound (moulded ASTM bar) and the 0° and
90° specimens from the tray base are given in
Table 3.

The Maximum Strain Criterion has been used
in conjunction with the standard linear elastic
analysis. Thus the deflection at failure, y,
has been calculated using the equation

$$y = \frac{L^2 \ e}{12 \ C}$$

where e is the tensile failure strain of the
material. Values of y obtained using the
tensile failure strain for the Upper Bound and
the 0° and 90° specimens from the tray base are
included in Table 3.

Table 3 Predicted force and deflection at
failure for the sub-component from the
milk crate

Source of failure data	Predicted failure force using :		Predicted failure deflection (mm)
	MSC (N)	TPC (N)	
ASTM bar	200	355	5.9
Tray, 0	140	250	6.8
Tray, 90	106	190	4.9

MSC - Maximum stress criterion. TPC - Total
plastic collapse.

4.3 Failure study on a gear selector

The component, shown in Fig. 4, was injection
moulded in glass fibre reinforced grades of
polypropylene (GFPP), polybutylene terephthalate
(GFPBT), nylon 6 (GFPA6) and nylon 66 (GFPA66).
All contained nominally 30% by weight of glass
fibre.

The component was tested to failure on a
tensile test machine at a constant extension
rate of 5 mm/min. and a temperature of 23°C.
The loading mode is shown in Fig. 4. The nylon
based components were tested in the dry condition and after having been stored for approximately 6 months in air at 20 °C and 65 % r.h.

Table 4 Tensile failure data for the materials used in the gear selector study

Material	Storage condition	Stress at failure: ASTM bar (MPa)	Plaque 0° (MPa)	Plaque 90° (MPa)	Strain at failure: ASTM bar (%)	Plaque 0° (%)	Plaque 90° (%)
GFPBT	–	140	101	98	2.95	3.4	2.7
GFPA66	dry	184	136	136	3.7	3.2	3.6
GFPA66	50% rh	165	132	128	3.8	3.9	3.6
GFPA6	dry	164	122	116	–	–	–
GFPA6	50% rh	141	112	106	4.05	4.8	3.2
GFPP	–	60	–	–	3.05	–	–

Five of the gear selectors were tested for each material and condition. The mean values of force and deflection at failure are included in Table 5.

The tensile strength data used in the predictions have been obtained from two sources. Directly moulded ASTM tensile dumb-bells were used to obtain Upper Bound data for each material. Corner-gated square plaques of 4 mm thickness were also moulded in all but GFPP. Specimens of the form of ASTM tensile dumb-bells were machined from the plaques at 0° and 90° to the main direction of melt flow (i.e. the tensile axis of the 0° specimen lies on the diagonal from the gate corner to the opposite corner of the plaque). All the specimens were tested to failure in uniaxial tension at a constant extension rate of 5 mm/min. and 23°C. The storage time and condition of the specimens matched those of the gear selectors. Five specimens of each type were tested for each material and condition. The mean values of tensile strength and failure strain are given in Table 4.

The predictions of component failure have made use of a linear elastic finite element stress analysis of the gear selector. This analysis had already been used successfully to predict the low strain force-deflection behaviour (11).

The finite element programme predicts that at an applied force of 14.7N the maximum stress created in the component is 8.14MPa. The Maximum Stress Criterion has been applied to this relationship in order to predict the force at failure (i.e. it is assumed that the relationship stays constant up to failure). Values of predicted failure force calculated using only the Upper Bound tensile strength data are shown in Table 5.

The predicted failure plane for the gear selector can be determined from the stress distribution given by the finite element analysis. (This predicted failure plane was found to coincide with that observed experimentally). From the position of the failure plane and the component geometry, the bending moment created at this plane could be calculated. From this bending moment and standard stress analysis procedures it is possible to determine the relationship between the 'yield' stress of the material and the value of the applied force at which Total Plastic Collapse will occur (3). Values of failure force obtained using this approach are included in Table 5. Only data appropriate to a random-in-the-plane FOD have been used in the predictions. (The RITP value was estimated from the plaque data of Table 4).

The Maximum Strain Criterion has been used with the relationship between deflection (at the loading point), y, and maximum outer fibre strain in the gear selector, e, derived from the finite element analysis. Thus, on the assumption that the relation between y and e remains linear up to failure, this approach enables the maximum deflection that the component can withstand before failure to be calculated using

Table 5 Comparison of predicted and experimental failure results for the gear selector

Material	Storage condition	Force at failure: Exp. (N)	Pred. MSC/UB (N)	Pred. TPC/RITP (N)	Deflection at failure: Exp. (mm)	Predicted MeC (mm)
GFPBT	–	331	254	314	16.9	14.3
GFPA66	dry	411	333	428	17.3	16.8
GFPA66	50% rh	393	299	411	24.9	17.9
GFPA6	dry	386	296	375	18.9	–
GFPA6	50% rh	356	255	345	26.0	19.1
GFPP	–	140	108	–	15.1	14.5 (*)

Exp. = Experimental. Pred. = Predicted. MSC/UB = Maximum stress criterion with Upper Bound strength data. TPC/RITP = Total plastic collapse analysis with strength data approximated to a random-in-the-plane FOD. MeC = Maximum strain criterion with failure averaged from data in Table 4. (*) - Only ASTM bar data was available for GFPP.

the tensile failure strain of the material. The predicted deflections are shown in Table 5. The failure strain used for each material was the average for the Upper Bound and the plaque 0° and 90° specimens, from Table 4.

4.4 Discussion

In both the examples given above, the use of the maximum stress criterion in a linear elastic analysis predicts that the component will fail at a lower force than that found by experiment. The degree of under-prediction of the failure force is substantial for all four materials studied, even when Upper Bound data is used in the predictions. A similar trend was found in the study on the sub-components from the ribbed plaque (12).

It might be considered that the above trend arises because the strength of the material in the failure region of each component is superior to that found in the Upper Bound test specimens. This would require that the degree of fibre alignment in the major stress direction in the component was superior to that found in the bar axis direction of the Upper Bound test specimens. Fibre orientation distribution studies have shown that this is not the case. Instead, these studies, and stiffness studies, have shown that the fibres are less well aligned in the failure region of each component, when compared with the Upper Bound specimens.

It is apparent that, if Lower Bound data were used with the maximum stress criterion, the experimental values of failure force would exceed the predicted values by nearly a factor of three.

The use of an Upper Bound strength value in the total plastic collapse analysis predicts a failure force that is higher than that found experimentally for the sub-component from the GFPP milk crate. On the basis of the data for the tray specimens in Table 3, a strength value applicable to a RITP FOD would probably lead to a predicted failure force of 215 N for the plastic collapse analysis. This is some 15% lower than the lowest experimental failure force given in Table 1.

Strength values applicable to a RITP FOD were used in the total plastic collapse predictions for the gear selector (Table 5). The predicted failure forces agree very well with the experimental values. (The use of Upper Bound data in this analysis would therefore obviously lead to predicted failure forces which were much higher than those found experimentally).

The study on the ribbed plaque showed a similar improvement in accuracy of prediction when the plastic collapse analysis was used instead of the maximum stress criterion (12). (In both cases, data obtained on specimens cut from the ribs were used in the calculations).

The above predictions have assumed that the material in each component is homogeneous. A significant difference in modulus was found between tensile specimens cut from the base and ribs of the ribbed plaque. Additional predictions were therefore carried out using the maxi-

mum stress criterion but treating each sub-component as a sandwich structure. Good agreement with experiment was obtained (12). However it is apparent that this approach could not at present be used at the design stage of a complex component.

A significant difference in modulus between the 'top' and 'bottom' surfaces in the failure region of the crate and gear selector components is less likely. The necessary measurements have not been carried out and a sandwich analysis has not been attempted. It is not therefore possible to tell whether or not the good agreement between the sandwich analysis and experiment, found for the ribbed plaque, was fortuitous.

The deflection at failure for the sub-component from the GFPP milk crate, predicted using the maximum strain criterion, is shown in Table 3. The lowest predicted deflection (obtained using the failure strain for the tray 90° specimen) agrees with the lowest value found experimentally (see Table 1). The highest predicted value is some 8% above the highest value of failure deflection given in Table 1. The mean of the three predicted deflection values is 5.9mm which compares with a mean experimental value, over all the specimen positions, of 5.5mm.

For each material in Table 4 for which three values of failure strain were available, the mean value was used in the prediction of failure deflection for the gear selector. The predicted values are in reasonable agreement with the experimental values of failure deflection. The nylon materials conditioned in a 20°C/65% rh environment do not show quite such good agreement as the other examples in Table 5. This could be associated with the fact that the conditioning time was not sufficient to ensure complete equilibrium of the test specimens with the environment.

5 CONCLUDING REMARKS

The studies on component stiffness have embraced a range of short fibre reinforced thermoplastics, and a number of different component geometries and loading patterns. In all cases, the use of modulus data applicable to a random-in-the-plane fibre orientation distribution in 'isotropic' design procedures has been found to give predictions of component deformation behaviour that are in reasonable agreement with experiment. It is considered that this approach offers a reasonable compromise between accuracy and ease of use for the designer.

This approach may be less reliable in those instances where the component (or a citical part of a larger component) and the stress field are of simple form. For such cases, the accuracy of the predictions could be improved if the fibre orientation distribution were known at the design stage. Fortunately, current understanding of the influence of mould geometry and base polymer on FOD is probably adequate for reasonable estimates of FOD to be made in these simple cases. Modulus values, appropriate to the estimated fibre orientation and the stress field, can then be used in the design calculations.

The strength studies described above were limited to flexural deformation under 'static' loading conditions. The results indicate that, in such instances, the use of lower bound tensile strength data with a maximum stress failure criterion in a linear elastic analysis will lead to very significant over-design in many cases. Examination of the stress - strain curves of the tensile test pieces and the components suggests that this could be due to the occurrence of some degree of plastic collapse prior to component failure. Some preliminary experiments described elsewhere also support this suggestion (10).

The predictions based on the total plastic collapse approach and the use of data appropriate to a random-in-the-plane FOD agreed well with experiment. However, in all three cases examined, the fibre orientation was either random-in-the-plane or biased towards the direction of the major stress component in the failure region. Less favourable fibre orientation distributions could occur in components. This suggests that the use of lower bound tensile strength data in a plastic collapse analysis will lead to realistic, but safe, predictions of component performance. However, further study is needed before reliable guidelines can be given on the range of applicability of this approach.

It is apparent that the above approach should only be used when the tensile behaviour of the material (under the appropriate test conditions) shows some similarity to that of an elastic-perfectly plastic material. It must also be remembered that the behaviour of a material can be changed from ductile to brittle by factors such as chemical environment and stress concentration.

The use of averaged failure strain data with the maximum strain criterion in a linear elastic analysis gave predictions of component deflection at failure which were in good agreement with experiment. If this procedure were adopted in a design situation, the component could be designed on a stiffness basis, and the failure strain analysis used to ensure that any deformations do not produce strains in the failure region for the material.

Finally, it should be noted that the stiffness and strength studies described above have been carried out on mouldings of thickness up to 6mm. No significant voiding was encountered in the mouldings and, in general, the fibres tended to align in the plane of the moulding. Further consideration must be given to these factors as moulding wall thickness increases.

ACKNOWLEDGMENTS

Grateful acknowledgment is made to the Polymer Engineering Directorate of the Science and Engineering Research Council for financial support of the programme. We also wish to thank Courtaulds PLC for the loan of the gear selector tool and access to the finite element programme. Grateful acknowledgment is also made to Akzo Plastics bv, GPG International and ICI PLC for the supply of materials and/or moulded components.

REFERENCES

(1) DARLINGTON, M. W. and TURNER, S. Creep of thermoplastics. Chapter 11 in Creep of engineering thermoplastics, edited by C. D. POMEROY, Mechanical Engineering Publications (London), 1978.

(2) CRAWFORD, R. J. Plastics engineering, Pergammon Press, 1981.

(3) WILLIAMS, J. G. Stress analysis of polymers (2nd edition), Longman, 1980.

(4) POWELL, P. C. Engineering with polymers, Chapman and Hall, 1983.

(5) BRIGHT, P. F. and DARLINGTON, M. W. Factors influencing fibre orientation and mechanical properties in fibre reinforced thermoplastics injection mouldings. Plastics and Rubber Processing and Applications, 1981, 1, 139-147.

(6) FOLKES, M. J. Short fibre reinforced thermoplastics, Research Studies Press, 1982.

(7) DUNN, C. M. R. and TURNER, S. The characterisation of reinforced thermoplastics for industrial and engineering uses. Composites - Standards, Testing and Design, National Physical Laborarory, IPC Science and Technology Press, 1974.

(8) CHRISTIE, M. A., DARLINGTON, M. W., McCAMMOND, D. and SMITH G. R. Shear anisotropy of short glass fibre reinforced thermoplastics injection mouldings. Fibre Science and Technology, 1979, 12, 167-186.

(9) CHRISTIE, M. A., DARLINGTON, M. W. and SMITH, G. R. Reinforced thermoplastics: a comparison of mechanical properties which considers the effects of processing and product end use. Reinforced plastics congress, Brighton, British Plastics Federation, 1978.

(10) DARLINGTON, M. W. and UPPERTON, P. H. Procedures for engineering design with short fibre reinforced thermplastics. Chapter in The mechanical properties of reinforced thermoplastics, edited by D. W. Clegg and A. A. Collyer, Applied Science, 1986.

(11) DARLINGTON, M. W. and UPPERTON, P. H. Designing with short fibre reinforced thermoplastics. Plastics and Rubber International, 1985, 10, 35-39.

(12) DARLINGTON, M. W. and UPPERTON, P. H. Design with short fibre reinforced thermoplastics. Short fibre reinforced thermoplastics, Brunel University, Plastics and Rubber Institute, 1985.

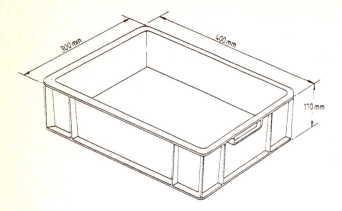

Fig 1   The injection moulded tray

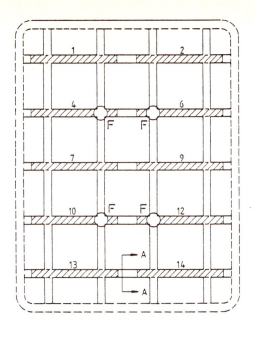

SECTION A·A

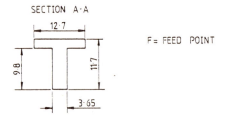

F = FEED POINT

Fig 3   Base of the milk crate moulding showing feed points. The numbered, shaded areas indicate the location of the sub-components cut from the moulding

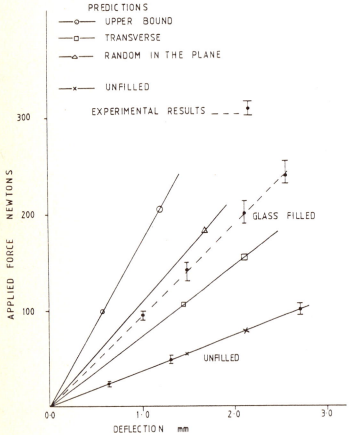

Fig 2   Force—deflection behaviour of the sub-component from the tray. Comparison of prediction with experiment for polypropylene (PP) and short glass fibre reinforced polypropylene (GFPP)

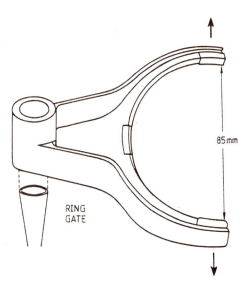

Fig 4   Schematic representation of the gear selector. The arrows show the direction of application of the applied force

Reliability assessment of damage tolerance of unidirectional fibre reinforced composites based on failure process simulation

I KIMPARA, DEng and **T OZAKI**, MSc
University of Tokyo, Japan

SYNOPSIS   The present paper proposes a new failure simulation model considering the effect of matrix shear failure as well as fibre breaks based on a shear-lag theory in which failure can occur randomly not only in fibre elements but also in matrix elements.   A stochastic tensile failure process is simulated by means of a Monte Carlo method based on a repeated increment scheme using a finite difference technique.   The statistical nature is characterized on the effect of accumulated failure in fibres and matrix on the ultimate tensile strength of unidirectional fibre reinforced composites based on the simulation technique, which enables one to assess the effect of fibre break, void and interfacial debonding on the damage tolerance level of unidirectional composites.

1   INTRODUCTION

A general judgement method to evaluate material properties of advanced composites such as carbon, aramid and hybrid fibre reinforced plastics has not been well established on account of a much broader range of properties and a much stronger anisotropy than conventional glass fibre reinforced plastics.   The number of factors to affect a scatter of properties is generally larger in composite materials, which are composed of many different phases of constituents than in conventional monolithic materials: a variation in defects mixed in and dispersion state of constituents during a composition process tends to make the amount of scatter in mechanical properties of composites larger, in addition to a variation in constituent material properties themselves.

Furthermore, as the failure process of fibre reinforced composites is a very complicated accumulation process of damage due to random failure of fibres, matrix and interface, which leads to a catastrophic fracture, it should be necessary to introduce a reliability assessment system to evaluate efffectively a decrease in strength due to cumulative damage and defects, taking every aspect of variation into consideration in order to understand thoroughly the statistical nature of strength properties of composite materials.

For this purpose, the present paper aims at establishing a general assessment system to predict a damage tolerance of composites.   A new failure simulation model of composite materials has been introduced considering the effect of matrix shear failure as well as fibre breaks, based on a repeated increment scheme using a finite difference technique.   Then a Weibull analysis technique has been applied to determine design allowable properties, and the effect of some patterns of defects on a damage tolerance level of composites has been systematically examined.

2   NOTATION

u_i^j	nodal displacement
$F(i,j)$	fibre element
$M(i,j)$	matrix element
$\sigma(i,j)$	tensile strength of fibre element
$\tau(i,j)$	shear strength of matrix element
$\bar{\sigma}$	mean strength of fibre
τ_{max}	shear strength of matrix
E	Young's modulus of fibre
A	cross-sectional area of fibre
d	distance between parallel fibres
G	shear modulus of matrix
h	thickness of lamina
u	forced displacement
δ	ineffective length
k	stress concentration factor
m	shape parameter
σ_0	scale parameter
σ_u	location parameter
σ_{max}	ideal strength of fibre

3   PROCEDURE

3.1   Basic Failure Model

A Monte Carlo simulation is one of the most effective methods to analyze such a complicated probabilistic phenomenon as a failure process of composites, and several investigations have so far been carried out.(1,2,3,4)   In most of the past investigations, however, a simple failure model has been applied in which only random fibre breaks and stress concentration in the nearest fibre to the broken one are taken into account, leading to no more than a flat cleavage plane of a specimen.   In an actual tensile failure process of unidirectional composites, not only fibre break, but also interfacial debonding between fibres and matrix and pull-out of fibres are frequently observed.   Such a complex failure mechanism often leads to a complicated zigzag cleavage plane of a specimen.

Although a few investigations taking interfacial debonding into account have been carried out, the interfacial failure criterion has not been clearly defined.(5)   Therefore, in order to analyze the effect of various kinds of defects on the strength of composites, as mentioned above statistically, it is important to take interfacial strength into consideration.

Consider a lamina of unidirectional composite as shown in Fig.1. Not only each reinforcing fibre arranged parallel to the direction of load has a varied fibre strength, but also every part of a fibre has a different strength. So each of k reinforcing fibres arranged at the same intervals is assumed to be composed of n links of 'ineffective length', δ, where ineffective length means the minimum one in which a fibre does not break at more than two places. And it is assumed that each fibre should break at the centre of a fibre element. Then nodal displacement, u_i^{j-1}, u_i^j, are set up above or below fibre element, F(i,j), which is the i-th fibre from the leftside and the j-th element from the top as shown in Fig.1. And as for the matrix, the matrix element, M(i,j), is introduced between nodes u_i^j and u_{i+1}^j, which is characteristic of this model.

Considering a stress transmission mechanism in the tensile failure of a unidirectional composite, it is approximated by the shear-lag theory, which means that fibres transmit only axial force and matrix only shear force. Therefore tensile strength is allocated at n × k fibre elements and shear strength is alloted at (n-1) × (k-1) matrix elements. Namely, random numbers assumed by a statistical distribution are generated about fibres and matrix. $\sigma(i,j)$ is allocated at the fibre element, F(i,j), and $\tau(i,j)$ is alloted at the matrix element, M(i,j), for the same number of elements.

As for the strength distribution of reinforcing fibres, several experiments and analyses have been carried out from the point of reliability engineering. For example, Miki et al. manufactured a trial chain-load tensile testing machine and showed that the strength of glass fibre followed the Weibull distribution and the normal distribution.(6) As for carbon fibre, Manders et al. made a similar study by means of a statistical analysis on two kinds of samples which have different fibre length, and showed that the strength followed the Weibull distribution with a size effect term.(7)

Therefore, in this study, the strength distribution of fibre elements is given by the random number, $\sigma(i,j)$, based on the Weibull distribution in the same form as Oh(3):

$$F(\sigma) = \begin{cases} 0 & (\sigma \leqslant \sigma_u) \\ \left(1 - \exp\left\{-\left(\frac{\sigma - \sigma_u}{\sigma_0}\right)^m \delta\right\}\right)/Fx & (\sigma_u \leqslant \sigma \leqslant \sigma_{max}) \\ 1 & (\sigma_{max} \leqslant \sigma) \end{cases} \quad (1)$$

where, δ: ineffective length, m: shape parameter, σ_0: scale parameter, σ_u: location parameter. σ_0 is calculated from fibre strength $\bar{\sigma}$.

$$\sigma_0 = \delta^{1/m} \bar{\sigma}/\Gamma(1+1/m) \quad (2)$$

and

$$F_x = 1 - \exp\left\{-\left(\frac{\sigma_{max} - \sigma_u}{\sigma_0}\right)^m \delta\right\} \quad (3)$$

where, σ_{max}: ideal strength of fibre.

As for the matrix elements, the same distribution of shear strength should be assumed, but there are almost no measured data. So, supposing the effect of fibre strength distribution to be the most important factor to determining the strength of composites, a uniform shear strength, τ_{max}, is assumed in this study.

From a shear-lag theory the equilibrium equation of force is given by:

$$EA\frac{d^2u_i}{dx^2} + \frac{Gh}{d}(u_{i-1} - 2u_i + u_{i+1}) = 0$$
$$(1 \leqslant i \leqslant k)$$

$$EA\frac{d^2u_i}{dx^2} + \frac{Gh}{d}(u_2 - u_1) = 0 \quad (i = 1) \quad (4)$$

$$EA\frac{d^2u_i}{dx^2} + \frac{Gh}{d}(u_k - u_{k-1}) = 0 \quad (i = k)$$

where, E: Young's modulus of fibre, A: cross-sectional area of fibre, G: shear modulus of matrix, d: distance between fibres, h: thickness of lamina.

Boundary conditions are as follows:

$$u_i^0 = 0 , \quad u_i^n = u \quad (5)$$

where, u: forced displacement.

The second-order differential equations such as Eq.(4) are approximated by the following equation in the same way as (3):

$$\frac{d^2u_i}{dx^2} = \frac{u_i^{j-1} - 2u_i^j + u_i^{j+1}}{\delta^2} \quad (6)$$

The tensile stress caused in the fibre element, F(i,j), is calculated from the difference of nodal displacements, u_i^j, u_i^{j-1}, by $E(u_i^j - u_i^{j-1})/\delta$ and compared with the given strength $\sigma(i,j)$. If $E(u_i^j - u_i^{j-1})/\delta \leqslant \sigma(i,j)$, then the fibre element, F(i,j), is supposed to be broken.

When a fibre element is broken, the stress field disorder takes place around the broken element. As it is assumed that a fibre always breaks at the centre of an element, when the fibre element, F(i,j), breaks, the nodal displacement of the broken point is expressed by $u_i^{j-1/2}$. Therefore, when F(i,j) breaks as shown in Fig.1, as for u_i^{j-1},(3)

$$\frac{d^2u_i^{j-1}}{dx^2} = \frac{4}{3\delta}\left(\frac{u_i^{j-2} - u_i^{j-1}}{\delta} - \frac{u_i^{j-1} - u_i^{j-1/2}}{\delta/2}\right) \quad (7)$$

and since $u_i^{j-1/2} = u_i^{j-1}$,

$$\frac{d^2u_i^{j-1}}{dx^2} = \frac{4}{3\delta^2}(u_i^{j-2} - u_i^{j-1}) \quad (8)$$

similarly, as for u_i^j,

$$\frac{d^2 u_i^j}{dx^2} = \frac{4}{3\delta^2}(u_i^{j+1} - u_i^j) \qquad (9)$$

so that u_i^{j-1} and u_i^j are replaced by the equations as shown above.

On the other hand, shear stress caused in the matrix element, $M(i,j)$, is calculated from the difference between nodal displacements, u_i^j, u_{i+1}^j, by $\left| G\tan^{-1}\{(u_{i+1}^j - u_i^j)/d\} \right|$, and compared with the shear strength of matrix, $\tau(i,j)$. If $\left| G\tan^{-1}\{(u_{i+1}^j - u_i^j)/d\} \right| \geqslant \tau(i,j)$, the matrix element, $M(i,j)$ is supposed to break. In this case, the equilibrium equations become as follows:

$$EA\frac{d^2 u_i^j}{dx^2} + \frac{Gh}{d}(u_{i-1}^j - u_i^j) = 0$$

$$\qquad\qquad\qquad\qquad\qquad\qquad (10)$$

$$EA\frac{d^2 u_i^j}{dx^2} + \frac{Gh}{d}(u_{i+2}^j - u_{i+1}^j) = 0$$

In the tensile failure process simulation, a repeated increment scheme is used. Figure 2 shows a flow chart of the simulation procedure.

Prior to carrying out a tensile failure simulation by the above modeling, stress concentration factors due to fibre break are examined. A simulation model having the number of fibres, $k=25$, and stress concentration factors, $k_r (r=1$ to 3), is calculated when 1 to 3 fibres break in a unidirectional CFRP lamina. As the result, $k_1=1.340$, $k_2=1.584$, $k_3=1.768$ are obtained, which are in good accordance with the analytical solutions by Hedgepeth[8]. Therefore the accuracy of calculating stress redistributions in this simulation procedure are shown to be adequate.

3.2 Reliability Assessment System

A reliability assessment system incorporating the tensile failure simulation as described above is made in order to examine the damage tolerance of composites. Figure 3 shows a general flow chart of the system. The conditions to be input are as follows: (1) volume fraction of fibre, (2) the number of trial simulations, (3) selection of defect types, (4) required strength level. Defect types are such as defects of fibres, notches, matrix voids, and so on. This system gives an allowable defect level which satisfies the required strength level.

First, under given conditions, a tensile failure simulation is carried out, and outputs are stored in a datafile. The outputs consist of strength, failure strain, number of broken elements of fibre and matrix. The strength is defined as the maximum stress when a composite supports no more load. Then a statistical analysis is carried out based on the derived datafile. From these data, parameters of statistical distributions such as normal distribution, log normal distribution and Weibull distribution are estimated by means of a least squares method. These distributions are often available to express variations in mechanical properties of composites: normal distribution is used to express variations in static strength or modulus, while

log normal distribution is available for life of composites. On the other hand, as Weibull distribution is applicable for both static strength and life of composites, it is widely used to assess the reliability of composites. When parameters are estimated, a test of significance is carried out, in which the Kolmogorov-Smirnov's test is used here because of its availability not only for a large sample but for a small sample.

Finally design allowables are calculated. The reliability in material design can be quantified in terms of probability of failure or probability of survival. So from a group of data whose distribution is known, the design allowables can be derived. It should be noted, however, that the reliability level is influenced by the number of data. From these points of view, A and B design allowables are proposed by the MIL standards in USA.[9] The A allowable is defined as the value above which at least 99 per cent of the population survived with 95 per cent confidence and the B allowable as the value above which at least 90 per cent of the population survived with 95 per cent confidence.

In this system, these allowables are compared with the required strength, and a chain of calculations are repeated by increasing the number of defects until the allowables are less than the required strength, which gives an allowable defect level.

4 RESULTS

The effects of some types of defects of fibre or matrix on the tensile strength of unidirectional CFRP are systematically examined based on the assessment system. The model consists of 25 fibres, 15 units of ineffective length of fibre, and the volume fraction(V_f) is 60 per cent. 50 simulations are carried out for every type of defect. The numerical values used in the simulation are as follows:

a) Cross-sectional area of fibre: $A=3.85 \times 10^{-11}$ m^2
b) Young's modulus of fibre: $E=230.3$ GPa
c) Mean tensile strength of fibre: $\bar{\sigma}=2.74$ GPa
d) Shear modulus of matrix: $G=1.25$ GPa
e) Shear strength of matrix: $\tau_{max}=0.098$ GPa
The Weibull parameters of the fibre are assumed or estimated as follows:
f) Location parameter: $\sigma_u=0$
g) Scale parameter: $\sigma_0=2.80$ GPa
h) Shape parameter: $m=6.98$ (according to (7))
i) Ideal strength: $\sigma_{max}=23.03$ GPa (according to (3), taken as E/10)

4.1 Effect of Fibre Break

Among defects of fibre, three types are examined, which are (1) centre-notched type, that is, r defects of fibre are arranged continuously in the middle of a lamina, (2) edge-notched type, that is, r/2 defects of fibre are arranged continuously from each side of a lamina, (3) dispersed-defects type, that is, r defects of fibre are dispersed over a lamina.

Figure 4 shows typical simulated stress-strain curves, and Fig.5 shows typical simulated failure patterns for these types of defects.

A failed element of fibre and matrix is expressed by an asterisk in Fig.5. As far as a unidirectional CFRP with no defects is concerned, it is a general tendency that the stress increases straight up to the maximum level, when fibres and matrix break one after another leading to a final fracture at one time.

In case of both centre-notched and edge-notched types, it is shown that a longitudinal crack grows gradually along the fibre at the notch root with the increase of strain. This corresponds to some small drops in stress in the range $\varepsilon \leqslant 0.8$ per cent as shown in Fig.4. When some fibre elements which are not failed previously begin to break, a lamina tends to lose its load carrying capacity rapidly. In case of centre-notched type, however, two areas divided by the two londitudinal cracks at the notch root fail one by one, so that such a stress-strain curve as shown in Fig.4 is often observed. That is, the decrease in stress at $\varepsilon = 1.0$ per cent is due to the failure of the left-hand area, and the final decrease in stress at $\varepsilon = 1.4$ per cent is due to that of the right-hand area in Fig.5 (b).

In case of dispersed-defects type, the stress increases straight up with a little smaller slope than that of a no-defect lamina, in the range $\varepsilon \leqslant 0.8$ per cent, where there is neither fibre break nor matrix crack. When the strain gets to 0.8 per cent, a few broken elements are connected to each other, resulting in a decrease in stress. The stress increases, however, again until the strain arrives at 1.0 per cent, where the final failure of a lamina takes place.

Mean strength, A basis allowable and B basis allowable of each type of defects are shown in Fig.6. Figure 6(b) shows the values for centre- and edge-notched types. It can be seen that the strength level decreases gradually with the increasing number of fibre defects(r). Although there is little difference in the value between these two types, examining the design allowables, especially in A value, however, the value of centre-notched is less than that of edge-notched type, which is due to the larger variation in strength of the former type. Figure 6(a) shows the values for dispersed-defects type. This also shows that the strength level decreases with r, but little by little. So as r increases, the strength level of dispersed-defects type becomes higher than that of notched-type. This indicates that the succession of defects is significant for the decrease in strength level. Referring to Fig.6(b) again, it is also supported by the fact that there is a rather large difference in strength levels between r=1 and r=2. Moreover, in case of dispersed-defects type, as some previous defects have often nothing to do with the formation of the cleavage plane of a lamina, there is not so much decrease in strength level in relation to the increase of r.

Figure 7 shows cumulative probability distributions for dispersed-defects type, which are derived by the Weibull analysis. Suppose the strength level as 1.1 GPa, the allowable defect should be no more than one based on the probability of survival level 90 per cent. If its level is, however, decreased to be 80 per cent, the defects up to four can be accepted as shown in Fig.7.

The results for such damage tolerance levels are summarized in Table 1.

4.2 Effect of Matrix Voids

As for defects of matrix, consider a matrix void, which is simulated by the model containing randomly arranged void elements in the matrix region. Figure 8 shows the results. As far as the shear-lag theory is concerned, matrix voids are considered to have nothing to do with the londitudinal tensile strength of a unidirectional lamina directly. It is, however, shown that the strength of a lamina decreases little by little with the increase of void content, and that A and B values become considerably lower on account of the increase in variation.

The decrease in strength due to matrix void itself is not so significant. However, it is noteworthy that succession of voids such as poor adhesion between fibre and matrix or successive defects of both fibre and matrix can be a critical factor to reduce the reliability level of a lamina.

5 CONCLUSIONS

In this study, a reliability assessment system is made based on the tensile failure process simulation for unidirectional lamina models in order to examine the damage tolerance level of composite materials. A successful simulation is carried out tracing faithfully an actual failure process by considering interfacial debonding between fibres and matrix as well as fibre breaks, and the reliability is estimated on CFRP which includes some defects previously defined by means of statistical analyses.

As an example, three types of models with fibre defects and one model with matrix voids are examined, and the results are summarized as follows:

(1) Continuous fibre defects have more significant influence on a decrease in reliability of composite materials than dispersed fibre defects.
(2) Matrix voids have small influence on the londitudinal tensile strength of unidirectional composites, and decrease the reliability of composites with an increase of variance.

REFERENCES

(1) KIMPARA, I., WATANABE, I., OKATSU, K., UEDA, T. A simulation of failure process of fibre-reinforced materials. The Seventh Symp. of Composite Mater., JUSE, 1974, 169-174.

(2) FUKUDA, H. and KAWATA, K. On the strength distribution of unidirectional fibre composites. Fibre Sci. Tech., 1977, 10, 53-63.

(3) OH, K. P. The strength of unidirectional fibre-reinforced composites. Jour. Composite Mater., 1979, 13, 311-328.

(4) FUKUDA, H. and CHOU, T. W. A statistical

approach to the strength of hybrid composites. <u>Proc. ICCM-4</u>, 1982, <u>2</u>, 1145-1151.

(5) OKUNO, S. and MIURA, I. Analysis of fracture process and strength in fibre reinforced alloys by Monte Carlo simulation. <u>Jour. Japan Soc. Metals</u>, 1978, <u>42</u>, 736-742.

(6) MIKI, M. and YOSHIDA, H. New chain-loading fibre tensometer and the strength of glass fibres. <u>Jour. Soc. Mater. Sci.</u>, Japan, 1979, <u>315</u>, 1204-1210.

(7) MANDERS, P.W. and BADER, M. G. The strength of hybrid glass/carbon fibre composites. <u>Jour. Mater. Sci.</u>, 1981, <u>16</u>, 2246-2256.

(8) HEDGEPETH, J. M. Stress concentrations in filamentary structures. NASA TN D-882, 1961.

(9) <u>MIL-HDBK-5D</u>, U.S.A., Depertment of Defense, 1983.

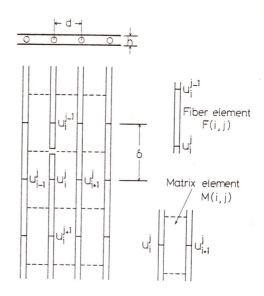

Fig 1   Fibre and matrix elements, nodal displacements

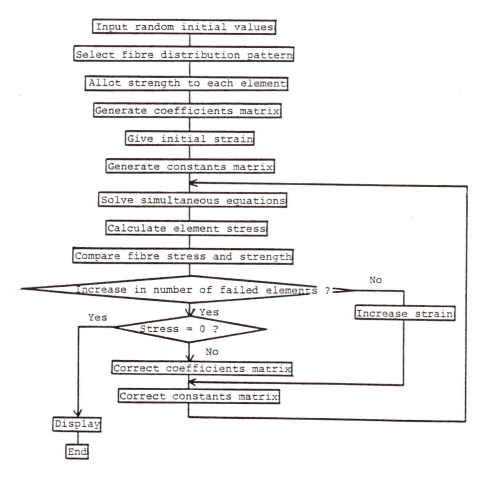

Fig 2   Flow chart of failure process simulation

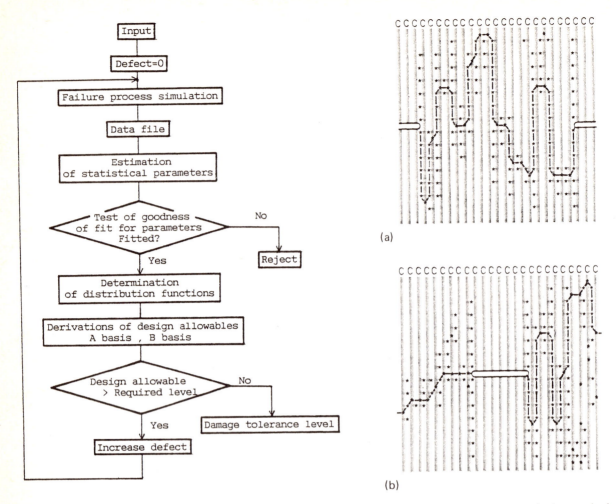

Fig 3   Flow chart of reliability assessment system

Fig 5   Failure patterns for centre- and edge-notched composites:
(a)   edge-notched
(b)   centre-notched

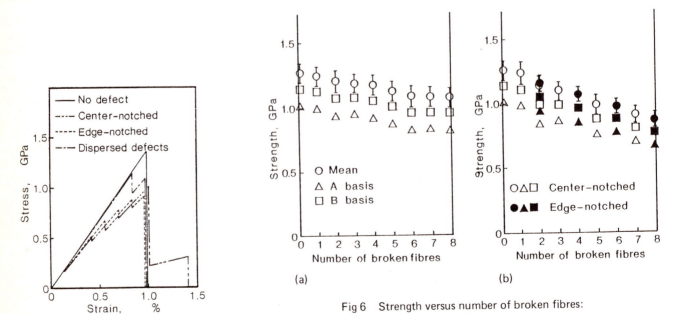

Fig 4   Stress-strain curves for undirectional composites with and without defects

Fig 6   Strength versus number of broken fibres:
(a)   dispersed defects
(b)   centre- and edge-notched

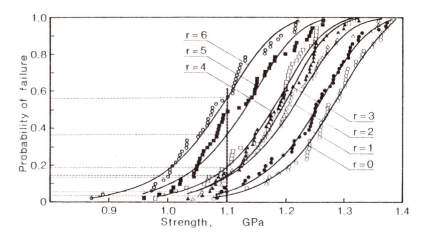

Fig 7 Probability of failure versus strength for varied numbers
        of broken fibres

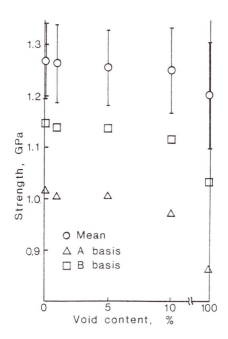

Fig 8 Strength versus void content

Table 1 Maximum allowable number of broken fibres to support required strength level

Required strength GPa	Center-notched		Edge-notched		Dispersed-defects	
	A	B	A	B	A	B
0.70	7	7	6	8	8	8
0.75	5	7	6	8	8	8
0.80	3	7	4	6	8	8
0.85	3	5	4	6	5	8
0.90	1	3	2	4	4	8
0.95	1	3	0	4	1	7
1.00	0	1	0	2	0	4
1.05	–	1	–	0	–	1

C25/86

The role of diffusion in the micromechanisms of stress corrosion cracking of E-glass/polyester composites

B D CADDOCK, CChem, MRSC, and **K E EVANS**, BSc, PhD, ARCS, MInstP, CPhys, AFIMA
Department of Metallurgy and Materials Science, University of Liverpool
D HULL, BSc, MA, PhD, DSc, DEng, FIM, FPRI
Department of Metallurgy and Materials Science, University of Cambridge

SYNOPSIS A radioactive tracer method has been used to study the solubility and diffusion of aqueous hydrochloric acid and water in polyester resins. Water diffuses freely into the resin at room temperature to establish a saturation concentration, but HCl exhibits a negligible solubility. The implications of this result are discussed.

1 INTRODUCTION

Composite materials, made from polyester resins reinforced with E-glass fibres, have found application in the construction of storage tanks, chemical plant and pipework, in situations where resistance to attack by solvents or other aggresive media is of importance. Experience has shown that such composites may be subject to stress corrosion cracking when they are exposed to the simultaneous effects of stress and dilute mineral acids (1). In a few instances catastrophic failure has been reported (1,2). It has been established (3) that the principal cause of the stress corrosion problem is the weakening of the glass reinforcement by the acid to an extent where the local stress exceeds the strength of the weakened fibres. Cracks propagate through the composite by mechanisms which depend on the nature of the environment and on the magnitude of the applied load.

Work by Hull et al (1,4) has identified three regimes of specimen failure by examination of the fracture surfaces. At high loads the surface exhibits extensive fibre pullout and is similar to that found when a dry specimen fails under a tensile load. At lower loads the surfaces appear stepped but at very low loads the fracture surfaces are remarkably smooth. This very low load regime is of special interest because of the difficulties of developing satisfactory design criteria for tanks and vessels subject to very slow crack growth. In some cases cracks may grow for several years before complete failure occurs.

It is known (3,5,6) that failure of the glass reinforcement will only occur if the aggressive agent gains access to the fibres. Thus, the use of a barrier layer of unreinforced polymer (e.g. gelcoat) is in general an effective way of avoiding or reducing stress corrosion problems. If, however, cracks develop in the barrier layer or if the layer is or becomes permeable to the corrosive agent, stress corrosion is likely to occur in regions of the component exposed to local stresses even though these may be well below the dry strength of the composite.

Extensive studies have been made of the strengths of bare E-glass fibres in various aqueous environments (3,5,7,8,9). Under these conditions a given fibre is most likely to break under load at its weakest point, usually at a flaw in the fibre. The strengths of single fibres and fibre bundles depend critically on the size and distribution at these flaws (3,10). For a composite undergoing stress corrosion cracking, however, the location of the point of fracture for a given fibre is restricted to the relatively small region ahead of the growing crack and co-planar with it. This region may or may not contain a flaw large enough for the fibre to break. Flaw growth or the nucleation and growth of a new flaw in this localised region may occur by the demineralisation of the surface layers of the E-glass fibres by contact with the acid solution. This possibility was discussed by Hogg and Hull (1) and by Lhymn and Schultz (11) and involves the same chemical mechanism as that which leads to the 'core-sheath' appearance of bare fibres exposed to aqueous mineral acids (3, 12).

Whether or not fibre weakening is a consequence of stress-assisted flaw growth or results simply from the overall depletion of calcium and aluminium from the outer layers of the locally exposed fibre it is likely that the extent of weakening will depend on both the exposure time and on the acid strength. To be able to model crack propagation in a composite in an acid environment it is therefore necessary to know what combinations of acid strength and exposure are needed for significant fibre weakening to occur. The possibility of achieving such acid concentrations by diffusion through the resin webs separating the fibres in a composite can then by examined.

1.1 Matrix cracking and diffusion

These are the two processes by means of which the liquid environment can reach the fibres. Situations in which matrix cracking can occur in composites, even in the absence of an applied load, have been described by Jones, Mulheron and Bailey (13). It has been shown that crossply laminates are particularly susceptible to transverse cracking because of the thermal stresses that develop in the intralaminar regions of the laminate during curing of the resin. The presence of these cracks is likely to stimulate stress corrosion if the composite is subsequently subjected to a load in an acid environment. Alternatively it has been suggested that initiation of stress corrosion cracking results from the diffusion of the aqueous acid through the resin in the outer layers of the composite (1,8).

Various mechanisms for crack propagation have been proposed but at low loads it has been suggested (1,4) that at the crack front fibre breakage is followed by a waiting period. During this period acid may diffuse through the resin web separating the fibre from its nearest neighbour until a sufficient amount has accumulated for the fibre to be weakened and to break so that the cycle can start again.

Water is known to permeate polymers and it is typical for polyesters to have a maximum solubility for water in the range 0.5%wt. - 2.5%wt. at room temperature (14-17). Various techniques have been applied for the measurement of water solubilities and diffusion coefficients, including the direct measurement of weight increase on exposure to water (16,17) and the use of a radioactive tracer method to determine concentration profiles within an exposed specimen (14,15). In an early study by Regester (18) an attempt was made to follow the penetration of both water and hydrochloric acid into an E-glass/polyester composite material with a gelcoat barrier layer. Regester showed that the composite was readily permeable to water but there was no detectable penetration of the hydrochloric acid through the barrier layer. A limitation to these measurements, however, was imposed by the relative insensitivity of the X-ray fluorescence method used for the chlorine determinations, which restricted the amount detectable to about 0.1%wt.. Sensitivity and sample handling problems also restrict the application of the EDAX technique with the scanning electron microscope (SEM) for chlorine measurements in polyester resins.

In the present work a modified version of a radioactive tracer method (15) has been used to examine the diffusion and solubility of molar aqueous hydrochloric acid in a range of resins. The objective was to establish whether or not diffusion through the resin can create a sufficiently aggressive environment for fibre weakening to occur. If the results show that this is not the case it must be assumed that matrix cracking is the principal mode of acid transport to the fibre surface during stress corrosion crack propagation.

2 EXPERIMENTAL METHODS

2.1 Radioisotopes used and their measurement

The radioactive tracer method was selected for use in these studies for the following reasons. Firstly, it is possible to achieve very high sensitivity. Because the radioisotopes used do not occur naturally this sensitivity can be used to the full without interference from contaminants in the samples or the environment. This means that low diffusion rates can be measured without the need for excessively long exposure times and that small specimens that are convenient to handle can be used. Secondly, since the source of the diffusing material is uniquely

tagged it provides unambiguous evidence of the movement of material from the diffusant solution into the polymer. Finally, it enables the diffusion of different molecular species to be followed simultaneously. This is of interest because the diffusion and solubility characteristics of aqueous solutions of hydrochloric acid in resins are complicated by the fact that the water and the hydrochloric acid are almost certain to have different solubilities and diffusion rates. If penetration of the polymer structure is a thermally activated process occurring at the molecular level the diffusion rates would be expected to be significantly temperature dependent and the relative proportions of acid and water would vary with diffusion depth. On the other hand the permeation of the resin by an acid solution through cracks or pores in the material could lead to the transport of the acid solution through the resin without change in concentration and with only a relatively small effect of temperature on the rate of movement of the acid through the material. In seeking to explore these different possibilities it was therefore of interest to consider, in some of our experiments, a method in which the movement of both the water and the hydrochloric acid through the resins could be followed simultaneously. This was found to be possible by using two different beta-emitting isotopes to label the two compounds. The water was traced with tritium (^{3}H, $t_{\frac{1}{2}}$ = 12.3 years, E_{max} = 0.018 MeV) whilst the hydrochloric acid was traced with ^{36}Cl ($t_{\frac{1}{2}}$ = 300 000 years, E_{max} = 0.714 MeV) where $t_{\frac{1}{2}}$ is the half-life of the radioactive isotope, E_{max} is the maximum energy of the emitted beta particle. The energy difference between the two isotopes is very large and it is easily possible to resolve the counts from the two isotopes by beta spectrometry, using a liquid scintillation counter.

In the present application it was necessary to bring the polymer samples, after exposure to the diffusant, into suspension or solution in a liquid phosphor (e.g. toluene containing fluorescent solutes) such that the photons generated by each individual beta particle emission are detected by a photomultiplier and amplified for pulse counting. In practice, 15 ml. samples of the liquid phosphor containing the dissolved or suspended radioisotopes were used. It was straightforward to find pulse discrimination settings that enabled ^{3}H to be counted in one channel of the counter and ^{36}Cl in the other. Quenching control and quantitative estimations of the amount of radioisotope present in each sample was achieved by counting matched liquid phosphor solutions containing known amounts of the radioisotope in question.

2.2 Specific activities of radioisotopes used

Results from recent published work (15,16,17) shows that water diffuses through polyester resins in accordance with Fick's law at room temperature with diffusion coefficients of the order of $10^{-9} cm^2 sec^{-1}$. The specific activity of the tritiated water used for the present experiments was about 10 Ci ml^{-1} and that of the ^{36}Cl-labelled hydrochloric acid about 1 Ci ml^{-1}.

The diffusant solution was molar hydrochloric acid and it was estimated that at this activity level the presence of 0.001 M HCl in the fluid dissolved in 40 mg polymer samples saturated with diffusant could be detected and measured.

2.3 Solubility and diffusion measurements

The method used was required to provide accurate values of both solubilities and diffusion coefficients. The total uptake method (19) was found to be best suited to meet these requirements, since it gives a direct measure of both the solubility and the diffusion coefficient of each diffusant in the polymer. In using this method a number of nominally identical polymer specimens were prepared, weighed and exposed to the radioactive diffusant for various times. After exposure the uptake of diffusant in each specimen was measured by radioactive assay and the results were used to construct uptake vs time profiles. The flat portions of these profiles represent the situation where the specimen is saturated with diffusant and are a direct measure of diffusant solubility. Diffusion coefficients are calculated from the rise time of the curve as discussed by Ellis and Found (16), who have shown that for thin flat plates the half-saturation time, $T_{\frac{1}{2}}$, is related to the diffusion coefficient D by the equation $D = .049L^2/T_{\frac{1}{2}}$ where L is the plate thickness. For cylindrical specimens, which we have used to minimise edge effects, $D = .063 a^2/T_{\frac{1}{2}}$ where a is the cylinder radius.

The two types of specimen used in the present work were (a) thin plates, approximately 36mm x 8mm with thicknesses varying from .08mm to .25mm (b) cylindrical rods 25mm long with radius 0.6mm. All the specimens were prepared by casting catalysed liquid resins which were allowed to cure overnight at room temperature prior to postcuring in the usual way. Exposure to the radioactive diffusant was achieved by immersing the specimens in 2ml of radioactive molar hydrochloric acid.

2.4 Radioactivity measurements

On completion of the exposure to the radioactive diffusant it was imperative to wash the excess liquid from the specimen surface after removal from the solution. This is potentially a serious source of error since the amount of this surface material can exceed that which has diffused into the specimen. A suitable procedure was developed in which the specimen was washed by momentarily dipping it in three separate baths of inactive .01 M HCl. Tests on blank specimens showed that this procedure freed them from surface contamination.

The measurement of the ^{36}Cl activity of the specimens was relatively easy since they were sufficiently thin to avoid self-absorption problems. However, the very low energy of the ^{3}H beta particles causes very severe problems with self-absorption and the only satisfactory method of measurement by liquid scintillation counting is one in which the radioactive material is uniformly dissolved in the liquid phosphor. This means that the water that has diffused into the polymer must be extracted and dissolved in the counting medium. Three different ways of achieving this were explored and the results were self-consistent: (a) Combustion of the washed specimens in a stream of air in a quartz tube containing a section packed with quartz chips and maintained at 1000°C. This treatment drove the diffusants out of the polymer, which was oxidized to $CO_2 + H_2O$. The radioactive water and hydrochloric acid was consensed in a cold trap together with the water of combustion and was subsequently dissolved in 20ml of liquid phosphor. (b) Slow extraction of the diffusants from the polymer specimens into 15ml of liquid phosphor containing 0.2 ml of .01M aqueous HCl. By carrying out this extraction directly in a counting vial the ^{36}Cl assay could be made immediately since virtually all the high energy beta particles escape from the specimen into the scintillation medium. The ^{3}H counts slowly increase with time as the radioactive water in the specimen equilibrates with that in the liquid phosphor. The massive excess of inactive water in the phosphor leads to essentially complete recovery of the ^{3}H into the phosphor after about one week at room temperature. (c) Reverse diffusion by immersing the washed resin samples in 2ml inactive aqueous 0.1 M HCl. After 1 week the specimen was removed and placed in a second 2ml of 0.1 M HCl. Each 2ml of solution produced in this way was dissolved in 15ml of liquid phosphor and counted. The counts from two or three extractions on each specimen were combined to give the total specimen counting rate, although in fact 95% of the activity was recovered in the first extract.

From the recorded counts on each specimen and the measured specific activities of the diffusant solutions it was easily possible to calculate the total uptake per unit weight for each specimen, and these data were used to construct the required uptake vs time profiles.

2.5 Measurement of demineralisation of the glass fibres

The glass fibre reinforcement can be weakened by the corrosive action of the acid environment at the crack tip by the local depletion of calcium and aluminium from the fibre surface (8). To find what combinations of hydrochloric acid concentration and exposure time lead to measurable depletion some tests were carried out in which short lengths of desized E-glass roving were exposed to acid of various concentrations for various times at 25°C and 50°C. After exposure the strands were washed with water, then with acetone and dried at room temperature. The dried fibres were mounted in casting resin and polished to reveal the cross sections of the fibres. The polished specimens were coated with carbon to enable them to be examined in a scanning electron microscope (SEM) both visually and by the EDAX technique.

3 RESULTS AND DISCUSSION

3.1 Hydrochloric acid solubility measurements with thin plate specimens

Solubility and diffusion data were aquired on the following polyester resins: (a) Scott-Bader Crystic 272 (b) blends of Scott-Bader Crystic 272 and Crystic 586 (c) Beetle 870.

Measurements were made first on thin plate specimens of Crystic 272 weighing about 50 mg. These were exposed to aqueous molar HCl labelled with [36]Cl for periods varying from 1 hour to three months. In all a total of 50 specimens were tested and were counted by the combustion and slow extraction methods described earlier. Examination of the results obtained showed that the uptake of [36]Cl by the resin was very low in relation to the expected solubility of water in the resin. If the molar HCl had permeated the resin without change of concentration it would be expected that the chlorine content at saturation would be about .06%wt.. The results obtained had a mean value of .0018%wt.. Moreover, the results showed much scatter, did not gradually increase with time and did not show a trend to level out at an equilibrium level representing the saturation solubility. It is believed that the acid may not be diffusing into the resin at all but is becoming absorbed on the specimen surface or entering surface microcracks, particularly at the cut edges of the specimens. Further experiments were made with Crystic 272 plate specimens of varying thickness which were exposed to the [36]Cl labelled diffusant. In a parallel experiment a similar set of thin plate specimens of varying thickness were exposed to tritium labelled aqueous molar hydrochloric acid for 310 hrs. These latter specimens were counted by the reverse diffusion method. From the radioactivity measurements on all these specimens the HCl and H_2O uptakes were calculated and the results obtained are given in Table 1.

best correlation appeared to be with the surface areas of the specimens.

3.2 Solubility and diffusion measurements with rod specimens using aqueous molar HCl labelled with ^{3}H and [36]Cl.

To demonstrate further the difference in the solubility and diffusivity of water and HCl in polyester resins dual labelling with ^{3}H and [36]Cl was used. Polymer rod specimens were used to minimise end effects and the results obtained are shown as a cumulative uptake curve in Figure 1. These data were obtained without thermostatic control at a laboratory temperature of 20°C and the points represent measurements made by the combustion and slow extraction procedures. The curve through the water uptake points is that for Fickian diffusion. It can be seen that the water is diffusing into the polymer according to Fick's law and that the saturation solubility is 1.82%wt. The [36]Cl data reflect very low or zero bulk diffusion into the polymer.

Complete diffusion profiles for the other polymers of interest were not measured at 20°C. It was considered sufficient to show that on exposure to molar HCl water diffused into the polymers to reach a saturation level whereas a negligible uptake of HCl occurred. Results for other polymers are given in Table 2 and relate to experiments with polymer rods exposed to the dual-tagged molar HCl for 2 weeks.

Although these data show a trend for HCl content to correlate with water content the measured levels are still extremely low in relat-

Table 1 Crystic 272 solubility experiments at 20°C with thin specimens

Plate No.	Exposure Time hrs.	wt mg	Thickness mm	HCl uptake, %wt from [36]Cl assay	H_2O uptake, %wt from ^{3}H assay
1	2280	.09858	0.25	.00136	
2	2280	.09817	0.25	.00129	
3	310	.09673	0.25	.000752	
4	2280	.04928	0.12	.00282	
5	2280	.04782	0.12	.00339	
6	310	.02884	0.08	.00300	
7	1968	.02592	0.08	.00410	
8	310	.09904	0.25		1.79
9	310	.04652	0.12		1.86
10	310	.03080	0.08		1.84

The water uptake by the resin can be seen to lead to saturation within the exposure time of 310 hrs and the measured solubility is independent of sample thickness. The water uptake is reversible, since the counts were obtained by the reverse diffusion method, and the results are therefore consistent with the known Fickian diffusion of water into polyester resins.

In contrast to this the HCl uptake, as measured by the [36]Cl uptake of the resin samples was extremely low and erratic. This clearly indicates that the samples do not saturate within the timescale of the experiments. In fact the

ion to the value of .07%, which would be obtained from Crystic 272 at saturation if the acid diffused without change of concentration.

The effect of temperature on the solubility and diffusion of water and HCl into these polymers is being studied further and in Figs 2 and 3 curves for water and HCl uptake from molar HCl into Crystic 272 and Beetle at 25°C are presented. In these experiments exposures were carried out in a thermostatically controlled water bath. The results show that for water both the saturation solubility of and the Fickian diffusion coefficients for these two resins are quite strongly influenced by temperature, but that again the up-take of HCl is either very low or is confined to surface occlusion or adsorption.

Table 2 Uptake of water and HCl by polymer rod specimens.

Polymer	Water content, %wt after 345 hrs exposure	HCl content, %wt after 345 hrs exposure
Crystic 272	1.87	.00063
25% Crystic 272 75% Crystic 586	2.41	.00105
50% Crystic 272 50% Crystic 586	3.21	.00204
Beetle 870	1.30	.00050

3.3 Acid concentration and exposure time needed for fibre demineralisation

Strands of desized roving were exposed to aqueous hydrochloric acid solutions of various concentrations for various times and were subsequently examined in the SEM. Figures 4, 5 and 6 are SEM photographs of fibre cross sections showing how the core-sheath structure progresses from a regular annulus to an irregular shape as the time of exposure to molar hydrochloric acid is increased. In these experiments no demineralisation or surface cracking was found for acid concentrations less than 1 molar.

3.4 Interpretation of results

The results of the radioactive tracer experiments strongly indicate that whilst water is able to diffuse freely through the polymer matrix in a polyester composite, the material is impermeable to hydrochloric acid. Strictly speaking the experimental data shows that chloride ions will not diffuse through the polymer network and it is assumed that the need to overcome electrostatic attraction forces will preclude the diffusion of solvated protons into the polymer. We consider this to be a reasonable assumption since it would be necessary to provide energy to drive a charge separation process and it is unlikely that this would occur spontaneously. Regester [18] attempted to measure hydrogen ion diffusion experimentally through polyester composites exposed to 5% H_2SO_4, 25% H_2SO_4 and 15% HCl at $100^{\circ}C$. No evidence of such diffusion was found after an exposure period of six months and Regester concluded that H^+ ion mobility is restricted by the anion mobility to preserve electrical neutrality within the laminate.

The results from the fibre demineralisation experiments show that relatively high acid concentrations and long exposure times are needed before detectable losses of calcium and aluminium occur and the formation of a core-sheath structure becomes evident. It is clear that the achievement of a sufficient HCl concentration to promote fibre demineralisation can not be achieved by diffusion alone within time taken [20] for sequential fibre breakage during typical stress corrosion crack propagation events. It is of course possible that high stresses at the crack tip are capable of opening up the polymer network for HCl diffusion or that at higher temperatures the acid

becomes significantly soluble in the polymer. It is also possible that the weakening of the fibres at, or just ahead of, the crack tip involves processes other than the demineralisation of the fibre surface. However the weight of evidence at the present time points to matrix cracking as the dominant process by which the aggresive agent reaches the fibres. If this is the case and if the demineralisation mechanism is the source of fibre weakening, one possible way in which diffusion and solubility can influence crack propagation is via the local increase in acid concentration in matrix cracks as a consequence of the diffusion of water into the surrounding polymer, leaving behind an acid solution of gradually increasing strength. All these possibilities are under current examination

CONCLUSIONS

A radioactive tracer method capable of measuring simultaneously the uptake of water and hydrochloric acid into polymers exposed to aqueous molar HCl has been developed. Results obtained with this method show that water diffused into polyester resins in accordance with Fick's Law to reach a saturation solubility in the range 0.5%-3% at or near room temperature. The diffusion process appears to be thermally activated. Hydrochloric acid, measured by chloride ion mobility, does not appear to diffuse into or dissolve in unstressed polyester resin at or near room temperature to a sufficient extent to account for the fibre weakening that leads to stress corrosion crack propagation. The effect of applied stress on HCl mobility in these resins, however, needs examination. In the stress corrosion of E-glass/polyester composites matrix cracking appears to offer a more likely route for the transport of the aggresive medium to the fibre surface.

The demineralisation of the surface layers of fibres exposed to the aqueous acid may be responsible for their eventual breakage under load. This demineralisation requires contact with relatively strong acid (0.5M - 1.0M) for the effect to be detectable within the known timescale of stress corrosion crack propagation. If demineralisation of the fibres by acid penetrating through matrix cracks is the dominant fibre weakening Process the selective diffusion of water out of the cracks may be of importance as a way of achieving a high enough acid concentration to promote the demineralisation.

ACKNOWLEDGEMENT

The radioactive tracer measurements described above were carried out at Manchester Polytechnic. Helpful advice and encouragement together with access to radioisotope facilities were provided by Dr. J.M. Marshall and this assistance is gratefully acknowledged.

REFERENCES

1. HOGG, P.J. and HULL, D. Developments in grp technology - 1, Edited by B. Harris, Applied Science Publishers, Barking, 1983, pp 37-90.

2. NORWOOD, L.S. and HOGG, P.J. Grp in contact with acid environments - a case study. Composite Structures, 1984, 2, pp 1-22.

3. METCALFE, A.G. and SCHMITZ, G.K., Mechanism

of stress corrosion in E-glass filaments. Glass Technology 1972, 13, (1) pp 5.

4. PRICE, J.N. and HULL, D. Propagation of stress corrosion cracks in aligned glass fibre composite materials. J. Materials Sci. 1983, 18, pp 2798-2810.

5. SCRIMSHAW, G. The effect of the environment on the properties of grp. Paper 5 in PipeCon Proceedings June 1980, published by Fibre-glass Ltd. and Amoco Chemicals SA.

6. PROCTOR, B.A. The long term behaviour of glass fibre reinforced composites. Glass - current issues NATO Advanced Study Institute April, 1984.

7. AVESTON, J. and SILLWOOD, J.M. Long term strength of glass reinforced plastics in dilute sulphuric acid. J. Material Sci, 1982, 17, pp 3490-3498.

8. NOBLE, B., HARRIS, S.J. and OWEN, M.J. Stress corrosion cracking of grp pultruded rods in acid environments. J. Material Sci 1983, 18, pp 1244-1245.

9. JONES, F.R., ROCK, J.W., and BAILEY, J.E. The environmental stress corrosion cracking of glass fibre-reinforced laminates and single E-glass filaments. J.Material Sci. 1983, 18, pp 1059-1072.

10. KELLY, A. and McCARTNEY, L.N., Failure by stress corrosion of bundles of fibres. Proc. Roy.Soc.London 1981, 17, A377, pp 475-489.

11. LHYMN, C and SCHULTZ, J.M., Chemically assisted fracture of thermoplastic PET reinforced with short E-glass fibre. J.Material Sci. 1983, 18, pp 2923-2938.

12. BARKER, H.A., BAIRD-SMITH, I.G. and JONES, F.R., Large diameter grp pipes in civil engineering - the influence of corrosive environments Symposium on reinforced plastics in anti-corrosion application, 1979, NEL East Kilbridge Paper 12.

13. JONES, F.R., MULHERON, M., and BAILEY, J.E. Generation of thermal strains in grp. J. Materials Sci., 1983, 18, pp 1522-1532.

14. MARSHALL, J.M. Ph.D. thesis CNAA/Manchester Polytechnic, 1981.

15. MARSHALL, J.M. MARSHALL, G.P. and PINZELLI, R.F. 37th annual SPI Conference, 1982, Washington DC.

16. ELLIS, B, and FOUND, M.S. The effects of water absorption on a polyester/chopped strand mat laminate composites. 1983, 14, 3, pp 237.

17. SHEN, C.H. and SPRINGER, G.S. Moisture absorption and desorption of composite materials. J. Composite Materials 1976, 10, pp 2.

18. REGESTER, R.F. Behaviour of fibre reinforced plastic materials in chemical service corrosion. 1969, 25, pp 157.

19. CRANK, J. The Mathematics of Diffusion. Oxford University Press, 1985.

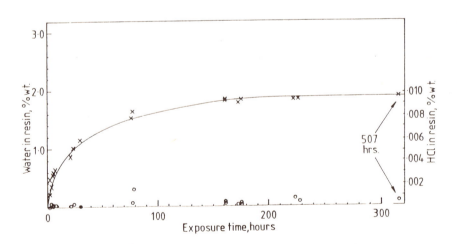

Fig 1   Diffusion of water and HCl into Crystic 272 at 20°C
x water uptake by resin, o HCl uptake by resin (expanded scale)

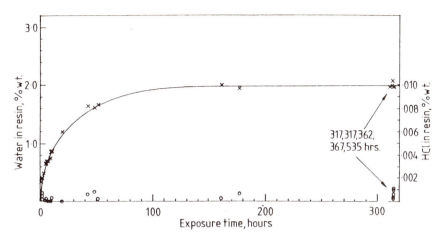

Fig 2   Diffusion of water and HCl into Crystic 272 at 25° C
x water uptake by resin, o HCl uptake by resin (expanded scale)

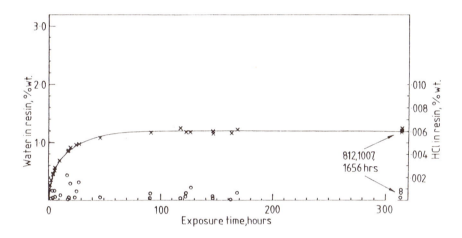

Fig 3   Diffusion of water and HCl into Beetle 870 at 25° C
x water uptake by resin, o HCl uptake by resin (expanded scale)

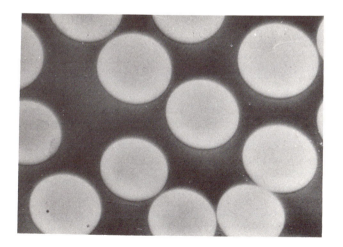

Fig 4   Original fibres

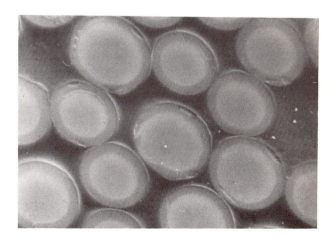

Fig 5   Fibres exposed to M HCl for 16 hours at 25° C

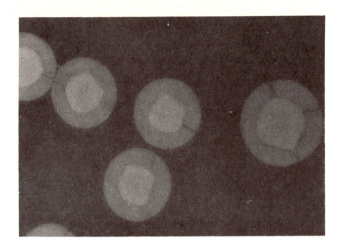

Fig 6   Fibres exposed to M HC1 for 64 hours at 25°C

C38/86

The effect of environment on stress corrosion of single E-glass filaments and the nucleation of damage in glass fibre reinforced plastics

P A SHEARD, BSc
Department of Materials, Science and Engineering, University of Surrey, Guildford
F R JONES
Department of Ceramics, Glasses and Polymers, University of Sheffield

SYNOPSIS   The stress corrosion of E-glass fibres has been investigated by measuring their times-to-failure in dilute acidic and basic environments. Microprobe analysis of the fractured fibres has shown that an ion exchange process occurs. A preliminary comparison of the corrosive damage which occurs in the form of transverse cracking of both plies in $0^o/90^o/0^o$ epoxy laminates, has been made. The rate of fibre failure appears to have little correlation with the rate of damage accumulation in the laminates which is considered to be more dependent upon the solubility of the corrosion products.

INTRODUCTION

The stress corrosion cracking of GRP has recently received considerable interest because of the increased use of this material for transporting and storing corrosive liquids. Good resistance can be achieved by using relatively tough matrix resins, at the expense of their overall permeability and chemical resistance (Hull and Hogg, 1982), or by replacing the E-glass with the more resistant ECR type of glass fibre (Aveston and Sillwood, 1982). Of particular importance to the long-term durability is the mechanism of initiation of the catastrophic failure crack. This was evident in the so-called Type II failure of 0° and $0^o/90^o/0^o$ epoxy and polyester composites which occurred in the non-immersed part of the coupons at relatively low strains (Jones et al, 1983). They further observed the formation of damage in partially immersed non-externally stressed composites. These phenomena were attributed to the precipitation of insoluble calcium salts at fibre/resin interfaces within the regions exposed to a moisture gradient. The crystallisation pressure was considered to nucleate a sharp flaw which would lead to failure when the fibres were placed in tension. The thermal strains placed the 0° fibres in cross ply laminates into compression so that the unstressed coupons became only damaged (Type III). Assessing durability on the basis of short term stress-rupture tests, however, has indicated that in the long term the mechanisms of failure could change to one in which damage accumulation occurs or a Type II crack is nucleated. Previous experiments (Jones et al, 1983) have suggested that the latter can occur sooner at equivalent applied strains. Since the former Type III damage accumulation may dominate the long-term failure, it follows that environments in which stress-corrosion cracking is slower may still cause stress induced failures.

The aim of this study was to analyse the stress corrosion behaviour of single glass fibres and the Type III damage accumulation of reinforced laminates in selected environments. Since the insolubility of the glass degradation products is believed to be an important contribution to the laminate stress corrosion the following five aqueous environments were chosen because their calcium salts had varying degrees of insolubility: sulphuric acid, orthophosphoric acid, sodium bisulphate, hydrofluoric acid, sodium hydroxide. Although orthophosphoric acid and sodium hydroxide have little effect on single fibres the unstressed laminate coupons corroded rapidly. In contrast hydrofluoric acid corroded single glass fibres extremely rapidly but caused only limited localised damage to the laminate.

EXPERIMENTAL

Materials

All chemicals used were analytical grade reagents diluted to 0.5 molar concentrations with distilled water. The glass fibre was supplied by Silenka UK Limited, 084-1200 TEX coated with a silane coupling agent. The epoxy resin system chosen for the laminate coupons was Shell Epikote 828. This was hardened with NMA and catalysed by BDMA in the ratios of 1:0.8:0.017 by weight respectively.

Single Fibres

Single fibre strands were carefully separated from the tows and adhered to test cards with a punched out gauge length of 50mm. The fibre diameters were measured accurately using a laser beam which refracted on passing through the glass. The number of fringes between two specified angles can be related to the fibre diameter by the following equation (Smithgall et al., 1977):

$$N = \frac{2r}{\lambda}\left(\sin\frac{\theta_2}{2} + \{\mu^2 + 1 - 2\mu\cos\frac{\theta_2}{2}\}^{\frac{1}{2}}\right)$$

$$-\left(\sin\frac{\theta_1}{2} + \{\mu^2 + 1 - 2\mu\cos\frac{\theta_1}{2}\}^{\frac{1}{2}}\right)$$

where N = No. of fringes within the angle $\theta_2 - \theta_1$
    μ = Refractive index of glass fibre
    λ = Wavelength of laser light used
    r = Fibre radius

A hooked glass rod and a rubber loop were adhered to the fibre strand which was cut from the test card when the glue had hardened. A beaker containing the chosen environment was placed into a constant temperature water tank at 20°C. The fibre was carefully lowered into the environment and hooked onto a pivoted sensor clamped to the side of the water tank. A glass weight was hung onto the rubber loop in order to apply the required tensile strain to the fibre. A magnetically operated switch was positioned at the end of the pivot such that on failure of the fibre the electric circuit would be interrupted. The fibre failure data was recorded on a Commodore Pet computer which was multiplexed to scan 64 fibres simultaneously.

Laminates

0°/90°/0° cross ply laminates were fabricated by winding glass tows onto a metal frame in the required orientations. The lay up was vacuum pre-impregnated with de-gassed resin and cured in an autoclave for 3 hours at 100°C. This method produced laminate sheets of 2.5mm thickness with a fibre volume fraction of around 55%. Coupons measuring 230 by 10mm were cut from the sheets using a diamond saw and subjected to a postcure treatment of 1½ hours at 150°C.

Plastic cells of 100mm in length with 2ml capacities were clamped onto the surface of the coupons and sealed using a high vacuum grease. The specimens were hung on a free standing frame and kept at a constant temperature of 21°C at 32% relative humidity. The tensile thermal strain in the transverse ply was calculated to be in the region of 0.3%, no external strain was applied to the coupons.

The environment was introduced via the open upper face of the cell and topped up daily. The absorption of the environment by the laminate coupon could then be calculated after taking into account the evaporation which was being monitored at the same time. The time to first visible damage was recorded and the rate of damage propagation followed photographically.

RESULTS AND DISCUSSION

Stress Corrosion of Single E-glass filaments

The stress corrosion effects of the fibres has been studied and the statistics of fracture analysed. Since the fibre surface was also exposed to the environment the microprobe has been used to examine any changes in elemental composition.

Fig 1 gives the fracture surface of a glass fibre exposed to dilute sulphuric acid, where the microprobe line scans for calcium and

aluminium across the central portion of the fibre have been superimposed. A core sheath structure is clearly visible and the analysis shows a decrease in both calcium and aluminium concentrations at the surface of the fibre. The sheath is ≈1.5μm after 9 hours immersion.

Similar results were obtained with sodium bisulphate solution. The analysis of the degradation products which are formed on the surface of the glass fibres is given in Fig 2, which clearly shows the formation of calcium sulphate. A silica trace is still detected due to the fibre beneath but the aluminium content is very low and assumed to have dissolved into the acidic environment as aluminium sulphate.

The attack on the fibres by dilute orthophosphoric acid was found to be virtually non-existent. Microprobe analysis of the surface of these fibres revealed a depletion in the levels of calcium and aluminium (see Fig 3) and the incorporation of a phosphorus containing specie. After two months exposure most of the fibres remained intact and examination of these, under a scanning electron microscope (SEM) and microprobe analysis, showed no evidence for the ingress of a leached layer. Accurate measurement of the fibre diameters revealed that these were also unchanged. Fox, 1977 and Ray, 1970, have suggested that a protective layer of silicon phosphate can form on the fibre surface which would prevent any further removal of the alkali metal cations from the lattice. This type of behaviour could well explain the observed passivity.

Hydrofluoric acid attacks the fibres by rapidly dissolving the silica network. SEM analysis of the fracture surface showed a pronounced core sheath structure. The sheath became detached after fibre failure, in the form of a porous tube. This is shown to be composed of calcium and aluminium fluoride in Fig 4.

Fibre failures were scattered over quite a wide range in time. This was primarily due to the statistical distribution of the inherent flaw size. The data has been analysed by Weibull statistics (Weibull, 1951) according to the following equation;

$$p(Ft) = 1 - \exp\left(-\frac{t}{t_o}\right)^W$$

where p(Ft) = Probability of failure at time t
    t_o = Characteristic failure time
    w = Weibull parameter
    t = Failure time

The characteristic failure time is the time which corresponds to 0 on the Ln Ln (1/(1-p_F)) axis. The nature of the single fibre tests render this value as rather meaningless. The Weibull parameter, sometimes called shape parameter or Weibull modulus, indicates the shape of the distribution to which the data has been fitted. Table 1 gives this value and other sampling data for three of the environments considered. Significant changes in the Weibull parameter can be indicative of a mechanistic change.

Fig 5 shows the results for fibres immersed in dilute sulphuric acid over a range of applied strain values. Results for fibres immersed in sodium bisulphate were virtually identical and these have not been presented separately. An increase in the applied strain is seen to produce more rapid fibre failures and the similarity between the two environments attack rate would suggest the same mechanism.

Fig 6 compares the effect of sulphuric acid, sodium hydroxide and orthophosphoric acid on the glass fibre failure rate. Sodium hydroxide is observed to attack far slower than sulphuric acid. In this case the dominant mechanism is known to be the hydrolysis of the siloxane bonds. The difference between the Weibull parameters of the two plots would support this change in mechanism. Work published by Cockram, 1981, has indicated that alkaline solutions with the same pH values as those used in the present study cause less severe corrosive damage than currently observed, even at elevated temperatures. This increased severity of attack coupled with observations of only limited reductions in fibre diameters would suggest that stress corrosion can also occur in alkaline environments.

Table 2 gives a summary of the single fibre test results. The characteristic failure times are taken from the mean of the population to the maximum failure time recorded. This is in order to remove the effect of a possible bias introduced by particularly weak fibres which would fail after similar time periods regardless of the test environment.

Laminate corrosion damage

The time to initiate Type III damage in unstressed coupons with attached face-cells has been recorded using time-lapse photography as described in the experimental section.

Damage to the laminate usually occurred by means of a three stage process. Stage one involved the initiation of some minor damage, normally a small longitudinal split in the 0° ply about 5mm above the level of the environmental cell. During stage two slow growth of this crack and the development of associated transverse cracks in the 90° ply occurs. Finally in stage three these cracks propagated rapidly until the whole coupon was covered with multiple cracking. Table 3 gives the results for the various environments in terms of the average time taken to initiate stage one and stage three damage for five coupons. These are presented graphically in Fig 7 and compared with the single fibre results.

Aqueous sodium hydroxide is shown to rapidly damage the laminate severely. This is due in part to the hydrolysis of the resin which causes significant weakening and also since the hydroxyl ion is very mobile it is likely that the fibre/matrix interface is also quickly attacked. This could cause early precipitation of insoluble corrosion products within the laminate contributing to the increased rate of damage propagation.

The acidic environments on the other hand regardless of acidity cause damage over similar time periods. This contrasts greatly with their behaviour towards single fibres.

Hydrofluoric acid attacks the single fibres very rapidly but has much less effect on the laminate coupons. Time to stage one damage is comparable to the other acids, however, the extent of the cracking was limited to a narrow band immediately above the environmental cell. This could be explained in terms of the silica network in the fibres being removed so rapidly that the formation of calcium and aluminium fluoride within the laminate is unable to exert any substantial internal stress.

The severity of the attack by orthophosphoric acid is not so easy to comprehend. Both calcium and aluminium phosphate, however, are extremely insoluble ($<2 \times 10^{-3}$g/ℓ). It is possible that even minute quantities of these being formed, before the protective silicon phosphate coating, as proposed by Ray, 1970, is completed, could be sufficient to cause significant crystallisation pressures. Another consideration is that the passivating reaction becomes inhibited within the laminate. Fox, 1977, speculated on the importance of the point of zero charge (pzc) on the reaction occurring at the surface of the glass fibre. This would be affected by the presence of a contacting matrix material.

Dilute sodium bisulphate and sulphuric acid have similar times to the initiation of stage one damage. Stage three is, however, markedly delayed in the case of the former environment. This difference presumably results from the lower acidity and the complex solution chemistry in the vicinity of the fibre.

CONCLUSIONS

The effect of several corrosive environments on the stress corrosion of single E-glass fibres has been evaluated. A strong dependence on the applied strain has been observed and a good fit to a Weibull distribution for failures in aqueous H_2SO_4, $NaHSO_4$ and $NaOH$ solutions has been obtained. The failure times in orthophosphoric acid do not fit the Weibull analysis because the fibres are found to be especially resistant. This is believed to result from the formation of a passive phosphate layer on the surface of the glass fibres.

The fracture of the fibres by stress corrosion is known to be responsible for the premature brittle failure of stressed composites in acidic environments. The nucleation of predamage by corrosion of the glass fibres does not correlate with single filament failure times indicating that the mechanisms which determine long-term durability in corrosive environments may differ. For example, aqueous H_3PO_4 causes rapid damage in the form of microcracking of the laminate but the fibres alone were particularly resistant.

REFERENCES

HULL, D., HOGG, P.J., Proc. 13th Reinf. Plas. Cong., Brighton, Paper 29 (1982), p115.

AVESTON, J., SILLWOOD, J.M., J. Mat. Sci., 17, (1982), 3491.

JONES, F.R., ROCK, J.W., BAILEY, J.E., J. Mat. Sci., 18, (1983), 1059.

FOX, P.G., Met. Soc. Conf. Proc., London, (1977), p298.

RAY, J., Non Cryst. Solids, 5, (1970), 96.

WEIBULL, W., J. Appl. Mech., Sept, (1951), 293.

COCKRAM, D.R., Glass Tech., 22, (1981), 211.

SMITHGALL, D.H., WATKINS, L.S., FRAZEE, R.E.,
Appl. Optics, 16, (1977), 2395.

Table 3   Laminate damage times

ENVIRONMENT	FIRST DAMAGE s×10³		RAPID PROPAGATION s×10³	
	x	sd	x	sd
H₂SO₄	750	500	1300	600
NaOH	20	10	60	20
H₃PO₄	1000	600	1600	700
NaHSO₄	1300	600	2500	900
HF	700	100	4000	+

+ Denotes damage did not propagate rapidly during test

Table 1   Weibull data

ENVIRONMENT	STRAIN %	No. FAILURES	WEIBULL PARAMETER	CORRELATION COEFFICIENT
H₂SO₄	0.3	98	0.79	0.98
	0.4	156	0.79	0.99
	0.5	41	0.77	0.93
NaOH	0.5	15	1.24	0.95
H₃PO₄	0.6	8	0.37	0.95

Table 2   Single fibre failure times

ENVIRONMENT	STRAIN %	FAILURE TIMES CHARACTERISTIC RANGE s×10³	
H₂SO₄	0.3	40	250
	0.4	20	120
	0.5	5	30
NaOH	0.4	350	670
	0.5	150	320
H₃PO₄	0.6	1000	4000 +
NaHSO₄	0.3	55	300
	0.4	35	140
	0.5	8	25
HF	0.175	0.6	0.9
	0.25	0.5	0.7
	0.325	0.3	0.5

+   Denotes most fibres survived test

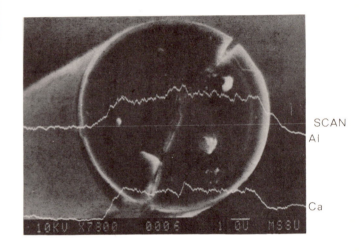

Fig 1

Sample : Glass Corrosion Debris

KeV : 10   Scan Time : 100 seconds

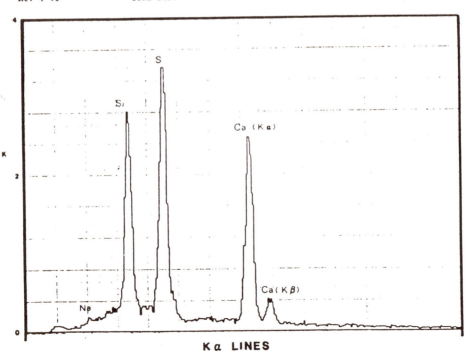

K α LINES

Fig 2   Microprobe analysis of material deposited on the fracture surface of a
glass fibre after immersion in dilute NaHSO₄ for 55 x 10³ seconds at
0.4 per cent strain

Sample : Glass Fibre Surface

KeV : 10   Scar. Time : 100 seconds

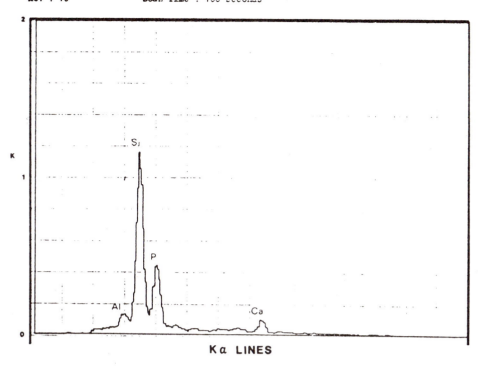

Fig 3   Microprobe analysis of a glass fibre surface after immersion in dilute H_3PO_4 for 1×10^6 seconds at 0.6 per cent strain

Sample : Glass Fibre Sheath

KeV : 10   Scan Time : 100 seconds

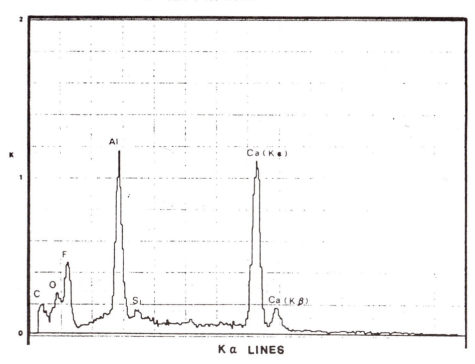

Fig 4   Microprobe analysis of sheath material that became detached after immersion in dilute HF for 500 seconds at 0.25 per cent strain

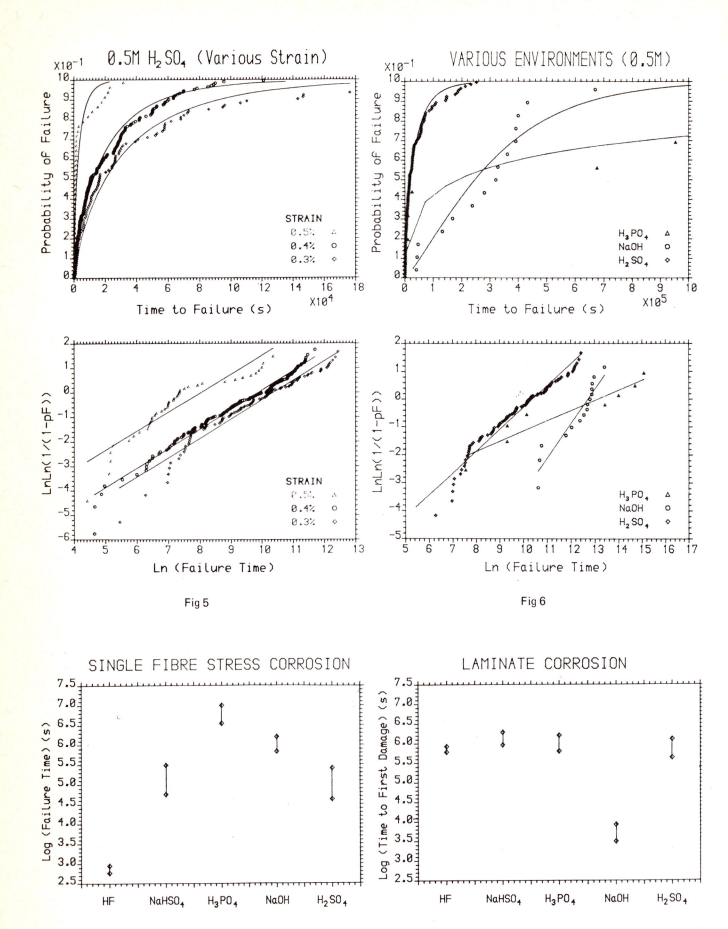

Fig 5

Fig 6

Fig 7

C15/86

Postbuckling behaviour of cylindrically curved composite plates with clamped edges, subject to uniaxial and shear loading with geometric imperfections

T J CRAIG, BSc, PhD and F L MATTHEWS, BSc(Eng), ACGI, CEng, MRAeS, FPRI
Department of Aeronautics, Imperial College of Science and Technology, London, SW7

SYNOPSIS The Galerkin method is used to solve the nonlinear governing equations of an arbitrarily laminated, composite, curved plate. Results are presented for the postbuckling of such a plate under uniaxial and shear loading, with different combinations of simply supported and clamped edge conditions. The effects of geometric imperfections are also considered.

1   INTRODUCTION

Composite materials have several advantages over their metal counterparts, the most commonly recognised being the increased strength-to-weight ratio and the 'designability' of composites. The price to pay for these advantages is increased complexity in the analysis of composite structures. This is due to their laminated nature, and the variation of material properties with direction. In order to gain the greatest benefit from composites, it is important to understand the response of these materials in the geometrically nonlinear region of structural analysis, of which postbuckling behaviour is a major consideration. Until recently the postbuckling analysis of composite plates has been restricted to flat plates under direct loading, Refs (1) to (6), and the majority of these results have been restricted to simple supports. Chia in his text (7) provides a useful summary of non-finite element-type solutions to the postbuckling of plates, under a range of boundary conditions and loadings. The early work of Chia and Prabhakara (1),(2), has been considerably extended and modified by Zhang and Matthews (8),(9),(10), to include completely general lay-ups, curved plates, shear loading and geometric imperfections ; however the edge conditions have been restricted to those that represent simple supports.

In the present paper the earlier work of Chia and Prabhakara (1),(2), and Zhang and Matthews (8),(9),(10), is extended to include clamped edge conditions. Perfect and imperfect cylindrically curved panels of completely general lay-up subjected to either uniaxial or shear loading will be considered.

The analysis has been incorporated into the computer program BUCCAL (Buckling of Curved Composite, Anisotropic Laminates), and some results generated using this program are presented in this paper.

2   GOVERNING EQUATIONS

A typical cylindrical plate is shown in Fig 1, with length a, thickness h, mid-surface radius R, and arc length b. The plate is assumed to be made up of layers of fibre-reinforced plastic, in each of which the material axes are arbitrarily oriented to the principal geometric axes.

In the following analysis the Kirchhoff hypothesis is adopted so that the plates under consideration must be assumed to be 'thin'. This classical plate theory approach leads to the following constitutive equations,

$$\left\{ \begin{array}{c} \underset{\sim}{N} \\ \underset{\sim}{M} \end{array} \right\} = \left[\begin{array}{cc} \underset{\sim}{A} & \underset{\sim}{B} \\ \underset{\sim}{B} & \underset{\sim}{D} \end{array} \right] \left\{ \begin{array}{c} \underset{\sim}{\varepsilon} \\ \underset{\sim}{K} \end{array} \right\} \qquad (1)$$

where

$$\underset{\sim}{N} = \left\{ N_x \quad N_y \quad N_{xy} \right\}^{\dagger}$$
$$\underset{\sim}{M} = \left\{ M_x \quad M_y \quad M_{xy} \right\}$$
$$\underset{\sim}{\varepsilon} = \left\{ \varepsilon_x \quad \varepsilon_y \quad \gamma_{xy} \right\}$$
$$\underset{\sim}{K} = \left\{ -\frac{\partial^2 w^0}{\partial x^2} \quad -\frac{\partial^2 w^0}{\partial y^2} \quad -2\frac{\partial^2 w^0}{\partial x \partial y} \right\}$$

The quantities in the matrices $\underset{\sim}{N}$ and $\underset{\sim}{M}$ are all per unit length. The strain quantities in $\underset{\sim}{\varepsilon}$ must all be composed of the effects of in-plane membrane displacements and further effects due to the out-of-plane deflection. In this work the equations are expressed in the von Karman sense, that is:

$$\varepsilon_x = \frac{\partial u}{\partial x} + \frac{1}{2}\left(\frac{\partial w^0}{\partial x}\right)^2 + \frac{\partial w^0}{\partial x}\frac{\partial \bar{w}^0}{\partial x} $$

$$\varepsilon_y = \frac{\partial v}{\partial y} + \frac{1}{2}\left(\frac{\partial w^0}{\partial y}\right)^2 + \frac{\partial w^0}{\partial y}\frac{\partial \bar{w}^0}{\partial y} - \frac{w}{R} \Bigg\}$$

$$\gamma_{xy} = \frac{\partial u}{\partial y} + \frac{\partial v}{\partial x} + \frac{\partial w^0}{\partial x}\frac{\partial w^0}{\partial y} + \frac{\partial \bar{w}^0}{\partial x}\frac{\partial w^0}{\partial y} + \frac{\partial w^0}{\partial x}\frac{\partial \bar{w}^0}{\partial y}$$

$$(2)$$

where u and v are the in-plane membrane displacements in the x and y directions respectively, w^0 is the lateral deflection, and $\bar{w}^0$ is the initial imperfection. The coefficients in Eqn (1) are defined in the usual manner of classical lamination theory.

In deriving the governing equations it is more convenient to use a partially inverted form of Eqn (1) as follows,

$$\left\{ \begin{array}{c} \underset{\sim}{\varepsilon} \\ \underset{\sim}{M} \end{array} \right\} = \left[\begin{array}{cc} \underset{\sim}{A}^* & \underset{\sim}{B}^* \\ -(\underset{\sim}{B}^*)^t & \underset{\sim}{D}^* \end{array} \right] \left\{ \begin{array}{c} \underset{\sim}{N} \\ \underset{\sim}{K} \end{array} \right\} \qquad (3)$$

in which

$$\underset{\sim}{A}^* = \underset{\sim}{A}^{-1}, \quad \underset{\sim}{B}^* = -\underset{\sim}{A}^{-1}\underset{\sim}{B}, \quad \underset{\sim}{D}^* = \underset{\sim}{D} - \underset{\sim}{B}\,\underset{\sim}{A}^{-1}\underset{\sim}{B} \quad (4)$$

In deriving the governing equations, using the

---

† Curly { } brackets always denote a column matrix.

principle of virtual displacements and virtual forces, use is made of an Airy stress function defined by,

$$N_x = \frac{\partial^2 \phi}{\partial y^2}, \quad N_y = \frac{\partial^2 \phi}{\partial x^2}, \quad N_{xy} = -\frac{\partial^2 \phi}{\partial x \partial y} \quad (5)$$

Use is also made of the following non-dimensional parameters,

$$\xi = \frac{x}{a} \quad \eta = \frac{y}{b} \quad \beta = \frac{a}{b} \quad w = \frac{w^0}{h}$$

$$F = \frac{\phi}{A_{22}h^2} \quad K_R = \frac{b^2}{Rh} \quad \bar{w} = \frac{\bar{w}^0}{h}$$

The governing equations themselves are long and complex, and are given in Appendix 1.

3   SOLUTION

Solutions to the governing equations have been obtained for plates with simple supports (8),(9),(10), it is now proposed to adopt the following conditions,

at $\xi = 0,1$

$$w = M_\xi = 0 \quad \frac{\partial^2 F}{\partial \eta^2} = -\bar{N}_x, \frac{\partial^2 F}{\partial \xi \partial \eta} = -\beta \bar{N}_{xy} \quad (6)$$

or

$$w = \frac{\partial w}{\partial \xi} = 0 \quad \frac{\partial^2 F}{\partial \eta^2} = -\bar{N}_x, \frac{\partial^2 F}{\partial \xi \partial \eta} = -\beta \bar{N}_{xy} \quad (7)$$

and at $\eta = 0,1$

$$w = M_\eta = \frac{\partial^2 F}{\partial \xi^2} = 0 \quad \frac{\partial^2 F}{\partial \xi \partial \eta} = -\beta \bar{N}_{xy} \quad (8)$$

or

$$w = \frac{\partial w}{\partial \xi} = \frac{\partial^2 F}{\partial \xi^2} = 0 \quad \frac{\partial^2 F}{\partial \xi \partial \eta} = -\beta \bar{N}_{xy} \quad (9)$$

Eqns (6) and (8) correspond to simply supported edges, and Eqns (7) and (9) relate to clamped conditions. $\bar{N}_x$ and $\bar{N}_{xy}$ are non-dimensional edge compression and edge shear respectively, and are defined as follows,

$$\bar{N}_x = \frac{P_x b^2}{A_{22}h^2} \quad \text{and} \quad \bar{N}_{xy} = \frac{P_{xy} b^2}{A_{22}h^2} \quad (10)$$

where P_x and P_{xy} are the applied edge stresses per unit length.

Solutions can be obtained to the governing equations by applying the Galerkin method, with suitable stress and displacement functions. The stress function is,

$$F = \frac{\bar{N}_x \eta^2}{2} - \beta \bar{N}_{xy} \xi \eta + \sum_{m=1}^{r} \sum_{n=1}^{r} F_{mn} X_m(\xi) Y_n(\eta) \quad (11)$$

The lateral deflection takes the following form,

$$w = \sum_{i=1}^{r} \sum_{j=1}^{r} W_{ij} S_i(\xi) S_j(\eta) \quad (12)$$

and the initial imperfection is defined as follows

$$\bar{w} = \sum_{i=1}^{r} \sum_{j=1}^{r} \bar{W}_{ij} S_i(\xi) S_j(\eta) \quad (13)$$

In the above equations r specifies the number of terms used in the series, F_{mn}, W_{ij} and $\bar{W}_{ij}$ are constants, the first two of which are to be determined, and the latter is specified for the initial imperfection. $X_m(\xi)$, $Y_n(\eta)$, $S_i(\xi)$ and $S_j(\eta)$ are all known functions each of which must satisfy the appropriate boundary conditions. For all the boundary conditions detailed in Eqns (6) to (9), the functions $X_m(\xi)$ and $Y_n(\eta)$ can be taken to be,

$$X_m(\xi) = \cosh \alpha_m \xi - \cos \alpha_m \xi - \gamma_m (\sinh \alpha_m \xi - \sin \alpha_m \eta) \Big)$$

$$Y_n(\eta) = \cosh \alpha_n \eta - \cos \alpha_n \eta - \gamma_n (\sinh \alpha_n \eta - \sin \alpha_n \eta) \Big)$$

$$(14)$$

The above equations are, in fact, the eigenmodes of a vibrating Euler beam with clamped ends. The constants α_m and γ_m introduced in the above equations are obtained from the characteristic equation of the vibrating Euler beam. The values of these constants are given to fifteen significant figures in Ref (8). In other work (1),(2), only six significant figures have been used. It has been found however (11), that a larger number of significant figures is required to ensure the orthogonality of the functions given above, therefore the values given in Ref (8) are adopted here.

The functions $S_i(\xi)$ and $S_j(\eta)$ used in Eqns (12) and (13) depend on the type of geometric boundary conditions imposed at the plate edges. Plates can be defined by a nomenclature which represents the boundary conditions on the edges of a plate. For example, an SCSC plate has the edge $\xi = 0$ simply supported and then, proceeding clockwise around the plate succeeding edges are clamped, simply supported and clamped. Using this nomenclature to define the plate edge boundary conditions, the functions required in Eqns (12) and (13) are as follows:

SSSS: $S_i(\xi) = \sin i\pi\xi$ SCSC: $S_i(\xi) = \sin i\pi\xi$

$\quad\quad S_j(\eta) = \sin j\pi\eta$ $\quad\quad\quad\quad S_j(\eta) = Y_j(\eta)$

CSCS: $S_i(\xi) = X_i(\xi)$ CCCC: $S_i(\xi) = X_i(\xi)$

$\quad\quad S_j(\eta) = \sin j\pi\eta$ $\quad\quad\quad\quad S_j(\eta) = Y_j(\eta)$

$$(15)$$

Having decided on the correct functions for a particular analysis, Eqns (11),(12) and (13) can be substituted into the governing equations and the Galerkin method applied, resulting in two simultaneous nonlinear equations in unknowns F_{mn} and W_{ij}. These equations can be written in matrix form as follows,

$$(\underline{k}_1 + \mu_1 \underline{k}_2 + \mu_2 \underline{k}_3)\underline{W} + (\underline{k}_4 + \underline{W}^T \underline{k}_5)\underline{F} + \underline{k}_6 = \underline{0} \quad (16)$$

$$\underline{k}_7 \underline{F} + (\underline{k}_8 + \underline{W}^T \underline{k}_9)\underline{W} = \underline{0} \quad (17)$$

and here $\mu_1 = \bar{N}_x/\pi^2 D_{11}$ and $\mu_2 = \bar{N}_{xy}/\pi^2 D_{11}$. $\underline{W}$ and $\underline{F}$ are column matrices containing the unknowns W_{ij} and F_{mn} respectively.

A linearised form of the above equations for initial buckling analysis can be obtained by dropping the quadratic terms $\underline{k}_5$ and $\underline{k}_9$ and the constant term $\underline{k}_6$ (this results from unsymmetric lay-ups for which there can be no initial buckling). $\underline{F}$ is eliminated between Eqns (16) and (17) forming a single set of linear equations in $\underline{W}$. This can be solved using standard eigenvalue techniques to yield the initial buckling load, and its mode shape.

For nonlinear analysis a similar procedure is adopted, retaining all the terms in Eqns (16) and (17); eliminating $\underline{F}$ between these two equations results in a single equation, quadratic in $\underline{W}$. In order to obtain a postbuckled path history one entry in $\underline{W}$ is incremented by a small amount (the 'step length'), and a modified Newton-Raphson method is used to update the remaining entries in $\underline{W}$, reducing the residuals of the equations to zero. This may, of course, involve a number of iterations. A suitable step length must be specified so that the iterative method does not fail. Values ranging from $W_{11} = 0.05$ to $W_{11} = 0.4$ have been used in the present work.

The starting point for the postbuckling path is determined in one of three ways. For problems involving bifurcation-type behaviour (e.g. symmetric or antisymmetric, perfect plates), the initial buckling problem is solved first, and the eigenvector (suitably scaled) is used as an initial guess of the postbuckled shape of the plate. In the iterative solution, the largest term in the eigenvector is incremented and for all the problems described herein this happens to be the first term, W_{11}. For problems of a non-bifurcational nature, the starting point is provided by either the initial imperfection, or just zero displacements.

The above analysis is incorporated into the computer program BUCCAL.

4   RESULTS

This section divides into two parts. The first is concerned with the comparison between the current work and earlier work by Chia and Prabhakara (2) for the postbuckling of flat plates. The second part is concerned with a study of a particular curved plate, and the influence of boundary conditions on its postbuckling response.

In Ref (2) Chia and Prabhakara detail results for the postbuckling of fully clamped, square, flat, isotropic and antisymmetrically laminated composite plates. Results for the fully clamped isotropic plate (with Poisson's ratio $\nu = 0.3$) are given in Fig 2. The BUCCAL results have been obtained in two ways. The two sets of results differ in the size of the step length, both sets of results were generated with $r = 3$ in Eqns (11) and (12). The two step lengths have been taken to be $0.15\,h$ and $0.05\,h$, and as can be seen from Fig 2, both sets of results coincide. It is immediately clear that the results obtained using the present method (BUCCAL) lie below those of Ref (2). The initial buckling load obtained by BUCCAL is, however, coincident with the value given by Timoshenko (12), whereas the result from Ref (2) is slightly higher. In this particular example the only major difference between the BUCCAL analysis and that of Ref (2) is the number of significant figures used in the constants of the Euler beam functions. It is clear from the results presented here that the present method is to be preferred.

The second and third examples concern the postbuckling of a fully clamped, antisymmetric laminate. Each layer of the laminate has the following 'high modulus' properties,

$$E_L/E_T = 30 \qquad G_{LT}/E_T = 0.5 \qquad \nu_{LT} = 0.25$$

where E_L is the modulus in the longitudinal, fibre, direction; E_T is the transverse modulus; G_{LT} is the in-plane shear modulus, and ν_{LT} is the Poisson's ratio. In the first of the two examples, the laminate has a cross-plied lay-up, $0^\circ/90^\circ/0^\circ/90^\circ$, and in the second example an angle-plied lay-up, $45^\circ/-45^\circ/45^\circ/-45^\circ$. The results are presented in Figs 3 and 4 respectively. In both cases results are given for $r = 2$ and $r = 3$, and similar results have been taken from Ref (2). It is immediately evident that the

conclusions drawn from the analysis of the isotropic plate also apply here. It is also worth noting that the two sets of results for each example produced by BUCCAL lie very close to each other.

Attention is now turned to the analysis of a curved, composite plate. Each layer of the plate is assumed to have properties consistent with those of a high strength carbon fibre reinforced plastic, that is,

$$E_L = 142\,000\ \text{N mm}^{-2}$$
$$E_T = 8500\ \text{N mm}^{-2}$$
$$\nu_{LT} = 0.317$$
$$G_{LT} = 5860\ \text{N mm}^{-2}$$

the lay-up is $(0^\circ/90^\circ/\pm45^\circ)_{2s}$.

The plate has the following overall dimensions:

$$a = b = 100\,\text{mm}, \quad r = 250\,\text{mm}, \quad h = 2\,\text{mm}$$

The postbuckled response of this plate when subjected to uniform end loading, is given in Fig 5 for three different boundary conditions, and both perfect and imperfect plates. The actual imperfection tends to vary from one set of boundary conditions to another, due to the nature of the function used to represent the initial imperfections. In all cases $\bar{W}_{11}$ in Eqn (17) is put equal to $0.1\,h$. The postbuckling of curved plates is characterised by the fall-off in stiffness and load carrying capacity that occurs after buckling and this is clearly illustrated in Fig 5. It is also clear that the boundary conditions have a significant effect on the initial buckling loads of the plate, and that after buckling the plates with greater edge restraint start to recover stiffness at an earlier and faster rate than the plates with less edge restraint. Initial imperfection can be seen only to effect the early postbuckled history of the plate.

The postbuckling of the plate described above, under both uniaxial and shear loading is now considered. In this case it is assumed that the ratio between the intensity of the shear loading and the direct loading is two to one. The results for both perfect and imperfect plates with SSSS, CSCS and CCCC boundary conditions are shown in Fig 6. The results are of a similar form to the uniaxial example considered above, except that the presence of the shear load has in general a destabilising effect.

It would have been appropriate to present some experimental results at this stage. However, there does not appear to be any available data for the uniform end loading boundary conditions modelled here. Most experimental tests are conducted with a uniform applied end displacement, and so direct comparison between such results and those presented in this paper would not be valid. As is usual in these cases the actual, practical, loading mechanism lies between the two idealised uniform end load, and uniform end displacement situations. The present numerical model would therefore have a rôle in the design process. Experimental verification of the present work would be highly desirable, but would not be an easy task to undertake.

5   CONCLUSIONS

An analytical technique which predicts the postbuckling response of curved, composite plates with opposite edges either clamped or simply supported, subjected to uniform direct or shear loading and with or without initial imperfections has been described in this paper.

The analysis utilises Euler beam functions in the stress and spatial fields. It has been shown that the constants in these beam functions must be specified to at least fifteen significant figures in order to obtain consistent results.

A series of results has been presented to demonstrate the numerical performance of BUCCAL and the effects of boundary conditions on the postbuckled response of a particular curved plate.

ACKNOWLEDGEMENT

This work has been carried out with the support of the Procurement Executive of the Ministry of Defence, UK.

REFERENCES

(1) PRABHAKARA, M.K. and CHIA, C.Y. Postbuckling behaviour of rectangular orthotropic plates. J. Mech. Eng. Sci., 1973, 15, 25.

(2) CHIA, C.Y. and PRABHAKARA, M.K. Postbuckling behaviour of unsymmetrically layered anisotropic rectangular plates. J. App. Mech., 1974, 41, 155.

(3) TURVEY, G.J. and WITTRICK, W.H. The large deflection and postbuckling behaviour of some laminated plates. Aero Quart., 1973, 24, 77.

(4) HARRIS, G.Z. The buckling and postbuckling behaviour of composite plates under biaxial loading. Int J. Mech. Sci., 1975, 17, 187.

(5) PRABHAKARA, M.K. Postbuckling behaviour of simply supported cross-ply rectangular plates. Aero Quart., 1976, 27, 309.

(6) BANKS, W.M. The postbuckling behaviour of composite panels. Proc. Int. Conf. on Composite Materials, Geneva, 1975.

(7) CHIA, C.Y. Nonlinear Analysis of Plates. McGraw-Hill, 1980.

(8) ZHANG, Y. and MATTHEWS, F.L. Postbuckling behaviour of curved panels of generally layered composite materials. Composite Structures, 1983, 1, 115.

(9) ZHANG, Y. and MATTHEWS, F.L. Postbuckling of cylindrically curved panels of generally layered composite materials with small initial imperfections. Proc. 2nd Int. Conf on Composite Structures, Glasgow. Elsevier Applied Science, 1983.

(10) ZHANG, Y. and MATTHEWS, F.L. Postbuckling behaviour of anisotropic laminated plates under pure shear, and shear combined with compressive loading. AIAA, 1984, 22, 281.

(11) ZHANG, Y. and MATTHEWS, F.L. Initial buckling of curved panels of generally layered composite materials. Composite Structures, 1983, 1, 3.

(12) TIMOSHENKO, S. Theory of Elastic Stability. Mcgraw-Hill, 1936.

NOTATION

a	plate length
b	plate arc length
h	plate thickness
u, v, w^0	displacements in x, y and z directions respectively
$\bar{w}^0$	initial imperfection
w	non-dimensional displacement in z direction
$\bar{w}$	non-dimensional initial imperfection
E_L	extensional modulus in fibre direction
E_T	extensional modulus perpendicular to fibre direction
F	non-dimensional Airy stress function
G_{LT}	in-plane shear modulus
K_R	non-dimensional radius of the plate mid-surface
N_x	stress resultant in x direction
N_y	stress resultant in y direction
N_{xy}	shear stress resultant
$\bar{N}_x$	non-dimensional edge compression
$\bar{N}_{xy}$	non-dimensional edge shear
M_x	moment in x direction
M_y	moment in y direction
P_x	applied compressive edge stress
P_{xy}	applied shear edge stress
$\underline{A}$	matrix of membrane stiffnesses
$\underline{B}$	matrix of membrane-bending stiffnesses
$\underline{D}$	matrix of bending stiffnesses
$\underline{K}$	matrix of curvature terms
$\underline{N}$	matrix of stress resultants
$\underline{M}$	matrix of moment terms
β	aspect ratio
γ_{xy}	shear strain
ε_x	longitudinal strain
ε_y	transverse strain
η	non-dimensional plate arc length
ν_{LT}	Poisson's ratio
ξ	non-dimensional plate width
ϕ	Airy stress function

APPENDIX 1

The governing equations, including the effects of boundary moments, are obtained from the virtual displacement and virtual force principles,

$$\int_0^a \int_0^b \left\{ \frac{\partial^2 M_x}{\partial x^2} + \frac{\partial^2 M_y}{\partial y^2} + 2\frac{\partial^2 M_x}{\partial x \partial y} + \frac{N_y}{R} + N_x \frac{\partial^2 w^0}{\partial x^2} \right.$$
$$\left. + N_y \frac{\partial^2 w^0}{\partial y^2} + 2N_{xy} \frac{\partial^2 w^0}{\partial x \partial y} \right\} \delta w^0 \, dx \, dy$$
$$+ \int_0^b \left[M_x \, \delta \left(\frac{\partial w^0}{\partial x} \right) \right] \Big|_{x=0}^{x=a} dy + \int_0^a \left[M_y \, \delta \left(\frac{\partial w^0}{\partial y} \right) \right] \Big|_{y=0}^{y=b} dx = 0$$

$$\int_0^a \int_0^b \left\{ \frac{\partial^2 \varepsilon_x}{\partial x^2} + \frac{\partial^2 \varepsilon_y}{\partial y} + \frac{\partial^2 \gamma_{xy}}{\partial x \partial y} - \left(\frac{\partial^2 w^0}{\partial x \partial y} \right)^2 + \frac{\partial^2 w^0}{\partial x^2} \cdot \frac{\partial^2 w^0}{\partial y^2} \right.$$
$$+ \frac{1}{R} \frac{\partial^2 w^0}{\partial x^2} + \frac{\partial^2 w^0}{\partial x^2} \cdot \frac{\partial^2 \bar{w}^0}{\partial y^2} + \frac{\partial^2 w^0}{\partial y^2} \cdot \frac{\partial^2 \bar{w}^0}{\partial x^2}$$
$$\left. - 2 \frac{\partial^2 w^0}{\partial x \partial y} \cdot \frac{\partial^2 \bar{w}^0}{\partial x \partial y} \right\} \delta \phi \, dx \, dy = 0$$

where δw^0 is the virtual displacement
$\delta \phi$ is the virtual force

Using Eqn (3) the above two governing equations can be expressed wholly in terms of w^0, $\bar{w}^0$ and ϕ. Non-dimensionalising these equations, and substituting for w, $\bar{w}$ and F from Eqns (11) to (13) leads to the matrix equations (16) and (17).

The above two equations will be recognised as the variational formulation of the equilibrium and compatibility equations of classical plate theory.

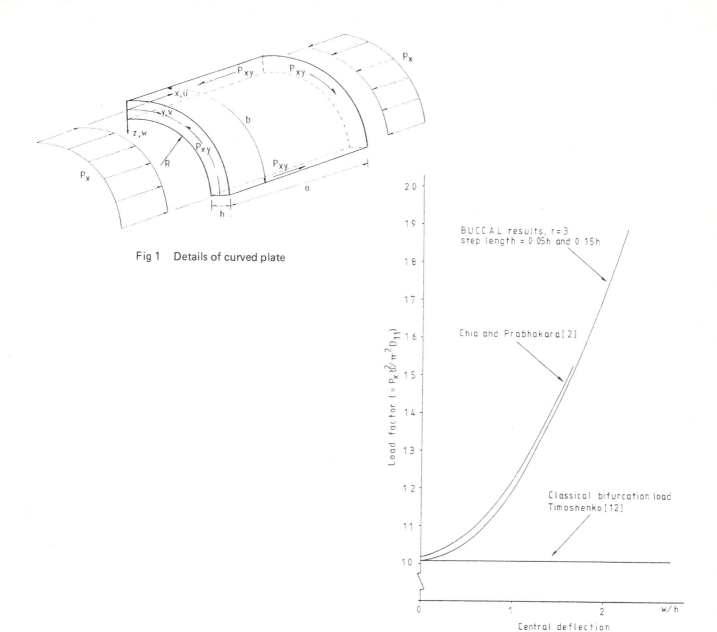

Fig 1   Details of curved plate

Fig 2   Load—deflection curve for square CCCC isotropic
flat plate under direct end load

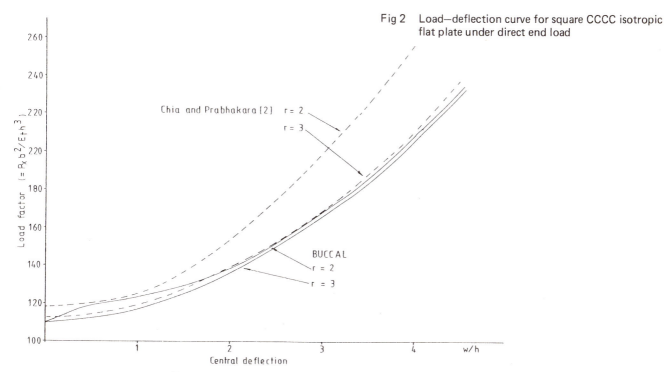

Fig 3   Load—deflection curve for square CCCC
0°/90°/0°/90° composite flat plate under direct
end load

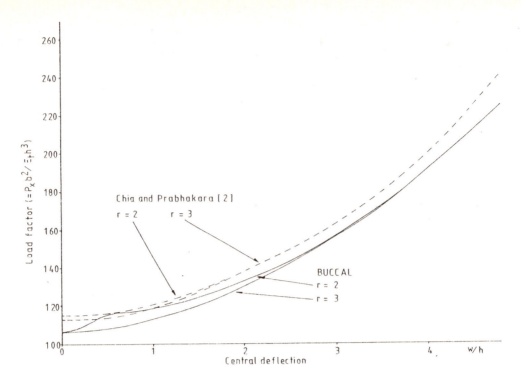

Fig 4   Load—deflection curve for square CCCC
+45°/—45°/+45°/—45° composite flat plate under
direct end load

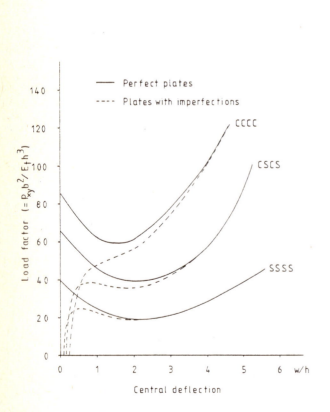

Fig 5   Load—deflection curves for SSSS, CSCS, CCCC
curved composite plate with and without geometric
imperfections under direct end load

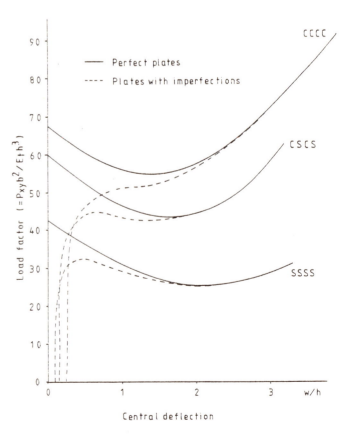

Fig 6   Load—deflection curves for SSSS, CSCS, CCCC
curved composite plate with and without geometric
imperfections under combined shear and direct
loading

C36/86

The characterization of carbon fibre/epoxy pultruded profiles using methods based on TMA and DSC

P R SNOWDON, BSc, J RANDALL, A LAWLER and J A QUINN
Bridon Composites Limited, Runcorn, Cheshire

SYNOPSIS   TMA and DSC are powerful tools for the characterisation of polymeric materials. The use of methods based on TMA and DSC is becoming more widespread with the advent of microprocessor controlled instruments. Some methods and their applications to pultruded profile development and production are described.

1 INTRODUCTION
The development and manufacture of pultruded profiles based on high performance materials requires full evaluation of the resultant products. The properties of interest are as follows:
1) Mechanical Properties
    a) STRENGTH AND STIFFNESS PARAMETERS
    b) FATIGUE BEHAVIOUR
    c) ENVIROMENTAL EFFECTS ON STRENGTH & STIFFNESS

2) Thermal Properties
    a) GLASS TRANSITION POINT (Tg)
    b) COEFFICIENT OF THERMAL EXPANSION
    c) DETERMINATION OF CURE SCHEDULES
    d) THERMAL DEGRADATION

The mechanical properties of profiles can be evaluated using standard test equipment and methods. The characterisation of thermal properties of equal importance to the designer, end-user and manufacturer, involves equipment and methods that are far from being considered standard in the composites industry.

Bridon Composites, engaged in the business of producing profiles for aerospace and related industries, have developed methods for the evaluation of thermal properties using DSC (Differential Scanning Calorimetry) and TMA (Thermo-Mechanical Analysis) and the aim of this paper is to illustrate some of the methods used with this equipment.

2 THE EQUIPMENT

This was supplied by Mettler and comprises of a control and evaluation module which can be used to drive either a DSC or a TMA furnace. The resultant data can be reprocessed by the control module with both the new and reprocessed data being displayed on a printer.

2.1 THE CONTROL UNIT (TC 10 TA PROCESSOR)

This unit controls the measuring cells' temperature rise, power input/output, applied load etc. It records the information from the cells, processes the information from the cells in a manner selected by the operator from the options available and relays the information to the printer. The parameters for each test are selected by the operator and can be stored as a 'method' in the control unit for future use.

2.2 DSC CELL

This comprises a furnace and two sensors.The furnace temperature is controlled by the processor and the differencein power demand between a reference sample and the test sample required to maintain them at the same temperature, is measured as a function of furnace temperature. The maximum furnace temperature is 600 C.

2.3 TMA CELL

This consists of a furnace which heats a sample, any changes in its height being monitored by a pre-loaded probe. The load on the probe can be oscillated at 1/12Hz and the specimen can be either on a flat anvil or used in a 3-point bend configuration.

The maximum furnace temperature is 1000 deg C.

The maximum load is 0.5N (can be raised), while for dimensions the resolution is 4 A

3 APPLICATIONS

3.1 DSC

3.1.1 EVALUATION OF RESIN CURE PARAMETERS

There are many methods currently in use for the evaluation of the cure characteristics of a thermoset resin both in hot and cold cure systems. One group of methods relies on monitoring the exotherm generated by the resin during the reaction, as a means of following the cure. Other methods monitor the increase in viscosity, the decrease in volume or the flow distance of a resin system as indicators of the progress of its cure.
DSC methods rely upon monitoring the exotherm and are capable of generating much more information related to the progress of cure than can be obtained from the more common 'thermocouple in the resin' methods. Cure information can be generated from an isothermal test, from a controlled temperature ramp, or from a pre-programmed combination of the two.

As with 'thermocouple' methods the activation (on-set) temperature, the time to peak exotherm and the height of the exotherm peak can be obtained. The resolution of the equipment is such that this information can be obtained even from systems with low exotherms, eg epoxy resins.

In addition, using the processor to evaluate the cure profile plot(energy output v temperature or time) a number of useful parameters can be obtained. The integration of the "energy output v time plot" (using the processor) in combination with the sample weight gives a quantifiable indication of resin reactivity. Kinetic analysis of this plot, again using the processor, allows the generation of "degree of conversion against time" plots at pre-selected temperatures and the determination of various reaction constants.

These additional data processing features give information pertinent to the selection of optimum pultrusion conditions (line speed, die temperatures, initiator levels and type etc) and can be invaluable as aids to resin quality control.

The data from a typical temperature ramp run are shown in fig 1, 2 and 3.

3.1.2 RESIN TRANSITIONS

As the equipment has the ability to measure endothermic changes in addition to exothermic changes, the glass transition point (Tg) of cured resins and the crystalline melting point (Tm) of thermoplastics can be determined. Tg is an important parameter for epoxy resins used in aerospace applications as it is an indication of upper service temperature. The precise Tg of an epoxy resin is dependent upon a number of parameters. One important variable that effects Tg is the post-cure schedule that the resin experiences, higher Tg's result from longer post-cure times or higher post-cure temperatures. The equipment allows the Tg to be determined, so various post-cure schedules can be evaluated for their effect.

In a similar way Tm of thermoplastics can be determined for quality control or analytical purposes. Figure 4 shows a typical Tm Trace.

3.1.3 PYROLYSIS

Materials can be pyrolysed in the furnace (maximum temperature 600 C) and the resultant exo and endotherms monitored. This gives a reproducible 'fingerprint' of a resin and can be an aid to determining the composition of a pultruded profile. The shape of the fingerprint is obviously determined by the various chemical and physical changes occuring in the furnace (chain scission, cross-linking, evaporation etc) and at this stage it is not claimed that the processes involved can be identified. Nevertheless the technique can be a useful qualitative analytical tool. Figure 5 shows a pryolysis trace of a glass/polyester composite.

3.2 TMA

The position of the loading probe can be determined to 4A on the most sensitive range of the instrument, so very small variations in the dimensions of a sample with temperature can be monitored. This sensitivity allows the instrument to be used to determine a number of material parameters.

3.2.1 COEFFICIENT OF THERMAL EXPANSION (ALPHA)

The variation of coefficient of thermal expansion with temperature can be measured (in any direction of a composite) simply by placing a flat specimen beneath the probe and ramping the temperature. A programme in the processor calculates either the point Alpha or the cumulative Alpha from the probe displacement and specimen thickness. This method can only be used with fully cured systems or at temperatures below post-cure temperatures (since Alpha changes sharply above Tg) otherwise a composite curve combining the effect of thermal expansion and cure shrinkage would result. The generation of such data is invaluable to the Design Engineer when designing structures, as in aerospace, comprised of isotropic and anisotropic materials. The equivalent data for metals are readily available but the data for composites in their many possible forms and over a range of temperatures are not available and still have to be generated for individual systems. Figure 6 shows a table of Alpha v Temperature for a uni-directional carbon/epoxy profile.

3.2.2 COMPOSITE Tg

As mentioned earlier, the Tg of a composite system is an important parameter in aerospace-type systems. Modified dilatometry methods similar to those described in the thermal coefficient section can be used to note any sudden change in Alpha at a specific service temperature. Tg of a composite can also be measured using DSC but often it is the case that the fibre volume fraction is so high that the Tg of the resin is difficult to discern. Empirical work shows that using TMA, the effect of Tg on alpha of a uni-directional carbon fibre composite is most noticable when the specimen length change is measured in the fibre direction. Figure 7 shows the change in Alpha at the matrix Tg.

3.2.3 DETERMINATION OF POST-CURE SCHEDULES OF COMPOSITES

A form of dilatometry can be very useful when determining post-cure schedules for epoxy-based profiles. A method has been developed in which a dimension of a section of profile is monitored during isothermal conditions. The temperature of the isotherm is selected as being a likely post-cure temperature. If the material is not fully post-cured then gradual shrinkage occurs with time, this shrinkage eventually ceasing to be significant. To achieve optimum properties it is suggested that the profile should be post-cured by ramping up to 140 C amd maintaining at that temperature until the shrinkage is effectively complete. As can be seen from Figure 8, the shrinkage has stopped after 50 minutes, ie, the resin reaction is complete.

If a piece of ex-die profile is post-cured in an oven at 140 C for 50 minutes and then isothermed in the TMA cell under the same conditions, very little shrinkage is observed (Figure 9). This behavious is evident when a uni-directional profile is isothermed in any configuration, suggesting that the shrinkage is a cure-related phenomenon rather than a relaxation process.

4 CASE HISTORY

4.1 OPTIMISATION OF POST CURE

A pultruded unidirectional carbon fibre/epoxy resin composite was tested for transverse flexural strength in the ex-die condition. Samples of the profile were then post-cured for different times at 160 C and each sample was tested for transverse flexural strength. An ex-die sample was isothermed in the TMA cell and the shrinkage due to further cure monitored. It is suggested that, from the isotherm on the ex-die material, the optimum post-cure time at a particular temperature is that taken for the sample to attain maximum shrinkage, ie no further resin cure. It can be seen (Figure 10) that samples post-cured under the regime suggested by the isotherming of the ex-die material have achieved optimum transverse flexural strength and those post-cured to a lesser degree have not achieved optimum properties. Samples post-cured for longer than the suggested time do not exhibit any significant change in strength from the samples post-cured for the optimum time. It can be concluded that a reliable post-cure schedule has been derived from a TMA isotherm on ex-die material.

4.2) VERIFICATION OF CALCULATED COEFFICIENT OF EXPANSION

In an application for a carbon fibre pultruded profile, a prime requisite was that it should have a coefficient of thermal expansion in the range of $12-14 \times 10-6$ mm/mm/ C. Using the available resin and fibre data a construction was determined giving the desired Alpha. As a prelude to a plant trial, a laminate was made to verify the theoretical calculations. A sample of the laminate was isothermed in the TMA cell at the recommended resin post-cure temperature and a post-cure schedule determined. Another sample from the laminate was post-cured according to this schedule and then introduced into the TMA cell for a dilatometry run. This run showed that the Alpha of the laminate agreed with the theoretical calculations from 40 C to 110 C. Above 110 C alpha changed dramatically, falling outside acceptable limits. A Tg determination of the resin, after post-cure, using the DSC cell showed the resin to have a Tg of 100 C. When the exercise was repeated with a resin having a determined Tg of 150 C the laminate exhibited the desired range of alpha up to 150 C, a temperature range acceptable for the application. This laboratory exercise is an illustration of how design work can be verified, avoiding the expense of a plant trial, with a combination of techniques based on TMA and DSC.

5 CONCLUSIONS

DSC and TMA are sophistocated tools which can, with method development, generate an enormous amount of useful data both from resins and composites. The instruments available today are very user-friendly but care is necessary when evaluating the data by 'built-in' manufacturers methods. A user should design a series of methods best suited to deriving the specific information required. A case can be made for Standards Institutions compiling standard test methods for adoption by the industry, and particularly by the raw material suppliers. The techniques and methodology of TMA and DSC can produce very pertinent data concerning the behaviour of resins and composites, including pultrusions, in the areas of both quality control and design. That part of the reinforced plastics industry concerned with high performance materials could benefit from a more general acceptance of both techniques.

6 FURTHER WORK

In the short term the equipment is being upgraded to generate information over a wider temperature range and to improve data handling. This should result in Bridon Composites developing a greater understanding of the thermal behaviour of the Company's materials. We are involved in a collaboration with workers at the Polymer Engineering Department, UMIST (J M Methven, H S Snaith) by which it is hoped to model the process. The aim of the project is theoretically to model the process and then to test the model experimentally. TMA and DSC are being used, in both phases of the project, to generate resin parameters and to evaluate resultant profiles. It is felt that it is in areas such as this, where strict control and high resolution are required, that TMA and DSC are techniques that will contribute to the understanding and prediction of behaviour of composite materials.

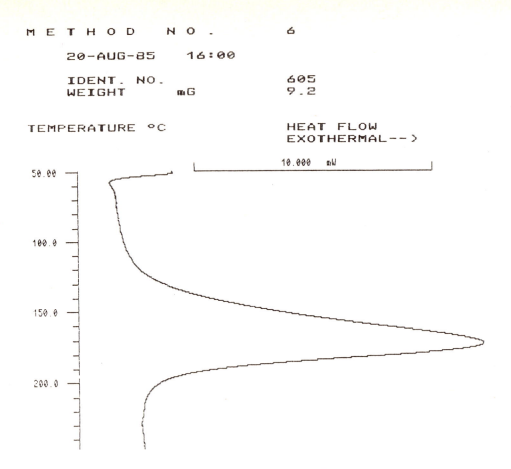

M E T H O D   N O .   6

20-AUG-85   16:00

IDENT. NO.   605
WEIGHT   mG   9.2

TEMPERATURE °C   HEAT FLOW
                          EXOTHERMAL-->

                          10.000   mW

50.00

100.0

150.0

200.0

Fig 1   On-line DSC plot of energy change during a temperature ramp for an
         epoxy resin

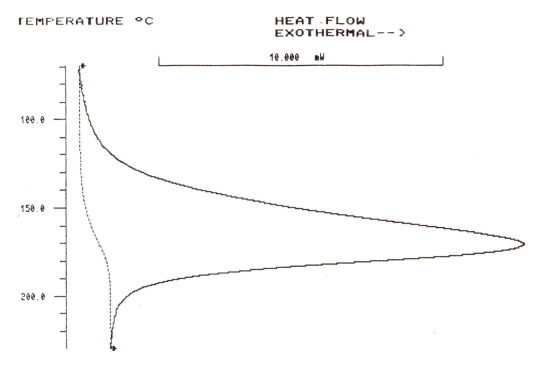

TEMPERATURE °C   HEAT FLOW
                          EXOTHERMAL-->

                          10.000   mW

100.0

150.0

200.0

Fig 2   Rescaled version of Fig 1, prior to peak integration and kinetic analysis

TIME MIN.   ALPHA

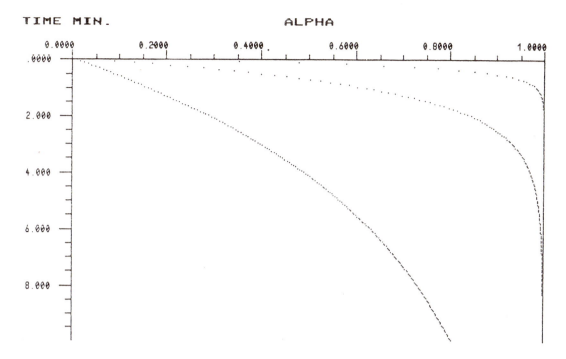

Fig 3   Degree of conversion versus time plots at three different temperatures
         (150°C, 175°C and 200°) — derived from the information of Figs 1 and 2

TEMPERATURE °C   HEAT FLOW
                                  EXOTHERMAL-->

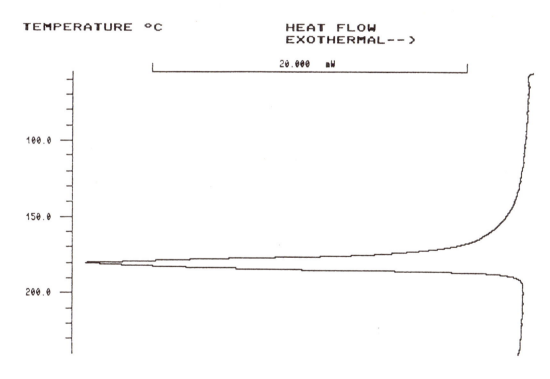

Fig 4   DSC plot illustrating well-defined T_m (material probably acetal)

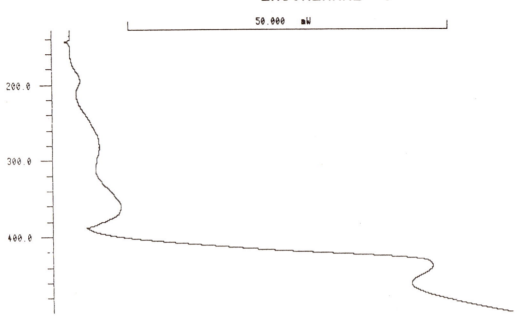

TEMPERATURE °C

HEAT FLOW
EXOTHERMAL-->

50.000 mW

200.0

300.0

400.0

Fig 5   DSC pyrolysis run on glass/polyester composite

TEMP. °C	EXPANSION COEFF. 10E-6 / K		ΔL μm	ΔL/L0 10E-3
45.0 -	1.8		0.00	.00000
55.0 -	4.8 -	5.75 -	0.08	-.05755
65.0 -	5.1 -	5.90 -	0.16	-.11807
75.0 -	4.5 -	5.44 -	0.22	-.16318
85.0 -	2.9 -	4.95 -	0.27	-.19801
95.0 -	4.4 -	4.67 -	0.32	-.23362
105.0 -	2.2 -	4.43 -	0.36	-.26558
115.0 -	4.3 -	4.20 -	0.40	-.29389
125.0 -	3.2 -	4.20 -	0.46	-.33622
135.0 -	4.8 -	4.12 -	0.50	-.37123
145.0 -	5.2 -	4.26 -	0.58	-.42614
155.0 -	4.9 -	4.34 -	0.65	-.47741
165.0 -	7.4 -	4.43 -	0.72	-.53166
175.0 -	12.5 -	4.81 -	0.85	-.62495
185.0 -	23.8 -	5.75 -	1.09	-.80440
195.0 -	18.4 -	6.97 -	1.42	-1.0450
205.0 -	10.5 -	7.43 -	1.62	-1.1892
215.0 -	8.2 -	7.51 -	1.73	-1.2760
225.0 -	9.3 -	7.53 -	1.84	-1.3562

Fig 6   TMA derived expansion data for a unidirectional carbon/epoxy composite,
measured in fibre direction

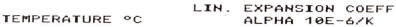

TEMPERATURE °C

LIN. EXPANSION COEFF
ALPHA 10E-6/K

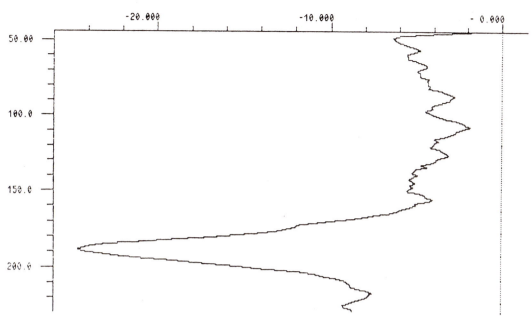

Fig 7   Plot of expansion coefficient versus temperature for a unidirectional
carbon/epoxy composite, measured in fibre direction

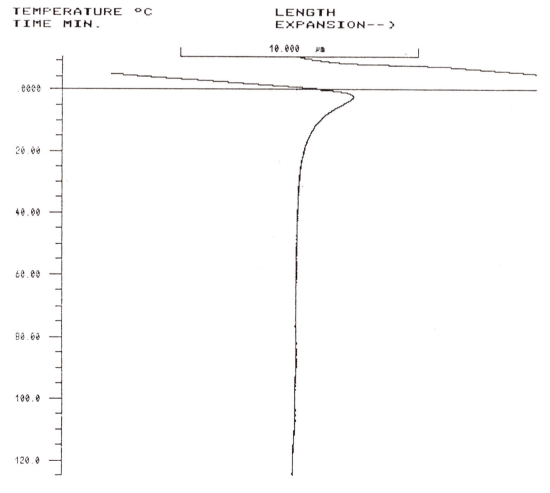

TEMPERATURE °C
TIME MIN.

LENGTH
EXPANSION-->

Fig 8   Effect of 140°C isotherm on carbon/epoxy specimen thickness before
post-cure

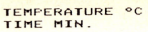

TEMPERATURE °C
TIME MIN.

LENGTH
EXPANSION-->

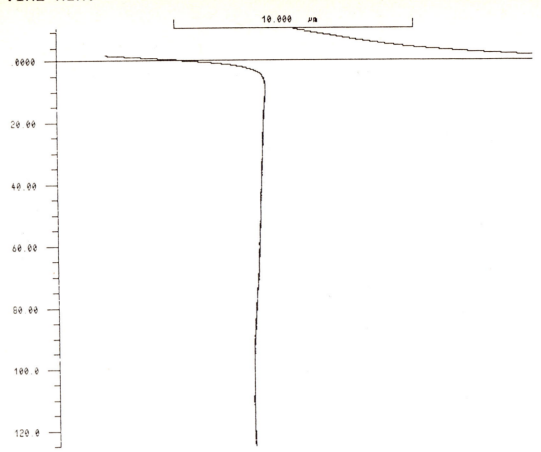

Fig 9   Effect of 140°C isotherm on carbon/epoxy specimen thickness after
        post-cure

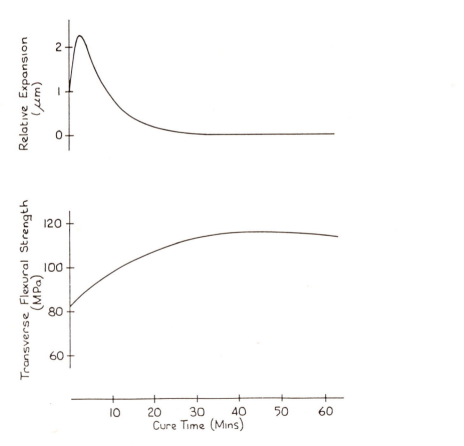

Fig 10 Effect of 160°C isotherm on (a) the thickness and (b) the transverse
        flexural strength of a unidirectional carbon/epoxy composite

C40/86

Compact disc fracture mechanics testing of injection moulded short fibre reinforced thermoplastics

K E PUTTICK, PhD, BSc, MInstP and **R D WHITEHEAD**, MSc
Physics Department, University of Surrey, Guildford
D R MOORE and **D C LEACH**
Petrochemical and Polymer Division, ICI plc, Middlesborough, Cleveland

SYNOPSIS   The compact disc fracture mechanics test is briefly described and its use in the evaluation of stress intensity factors and strain energy release rates of short fibre reinforced materials outlined. Measurements of K_{1c} and G_{1c} for PEEK based moulded discs are presented and discussed: it is shown that the validity of the results, and the mode of crack propagation, depend strongly on the orientation of the milled starter notch.

1   INTRODUCTION

This study formed part of a collaborative project between the Physics Department of the University of Surrey and ICI PLC on fracture testing of injection mouldings. In other work to be published elsewhere we found that the widely used three-line bend fracture mechanics test was inadequate for the evaluation of tough advanced performance materials such as PES and PEEK for two reasons: the results depended on specimen thickness (an effect predicted by fracture mechanics theory), and the crack front shape and surface topography were far from the requirements of plane strain deformation. Accordingly we developed for this purpose the ASTM compact disc test, using single side gated injection moulded circular plaques in which were milled standard chevron notches which enabled the crack to nucleate at a single point and subsequently to develop into a nearly straight fronted flat crack. This specimen was pin loaded in the way shown in Fig 1 and cracks could be propagated in a stable manner by suitable programming of the applied force so as to yield values of K_{1c} and G_{1c} at various points along the crack path. Fig 2 represents a typical force-displacement record, showing how G_{1c} may be derived independently of any detailed analysis of the deformation of the specimen. This paper describes an extension of the work to short fibre filled PEEK mouldings.

2   EXPERIMENTAL METHODS

2.1   Specimen discs

The original mouldings were discs with a diameter of 113 mm: the gate mark on the side was used as a reference point for all orientation tests, and notches were milled at 0°, 45° and 90° to the diameter from the gate. Three grades of PEEK were supplied by ICI in this disc form: glass-fibre filled 4520 GL (20% glass), 4530 GL (30% glass) and carbon-fibre 4530 CA (30% fibre). It was noted that some of the specimens had "cold slugs" in the form of jetting lines which it was anticipated would lead to scatter in the results from different parts of individual discs.

2.2   Experimental method

Each disc was prepared by drilling out the pulling pin holes, milling the chevron notch using a

specially designed jig and fixture, and pre-straining the specimen by a careful routine to introduce a sharp crack. This test piece was then slowly strained to grow the crack in a controlled way, its position along the diameter being measured, recorded and timed so as to calculate the crack speed.

From the force-displacement record the breaking force for a given a/W (where a is crack length and W the distance from the notch tip to the end of the corresponding diameter) is found. The work done to open an incremental area of crack is determined by measuring the corresponding sector under the force-displacement curve (see Fig 2). K_{1c} is calculated using Newman's (1981) formula for this configuration:

$$K_{1c} = \frac{P}{\sqrt{W}} \cdot H.G. \qquad (1)$$

where

$$H \equiv \frac{2+R}{(1-R)^{3/2}}$$

$$G \equiv 0.76 + 4.8R - 11.58R^2 + 11.43R^3 - 4.08R^4$$

$R \equiv a/W, \qquad a \equiv$ length of crack from notch

$P \equiv f/B$

$B \equiv$ thickness of disc

G_{1c} is calculated by the Gurney-Hunt (1967) sector method

$$G_{1c} = \frac{\Delta W}{\Delta A} \qquad (2)$$

$\Delta W \equiv$ work to form incremental area of crack ΔA.

3   RESULTS

3.1   Stress intensity factors

The K_{1c} values are presented in Fig 3. Two preliminary points are noteworthy:

(i)   A number of specimens broke in an unstable manner during the crack introduction stage in which the load was controlled manually. These values are designated 'hand'. (The remainder are classified according to pulling speed and notch orientation relative to the gate).

(ii)   The mode of crack propagation is strongly dependent on orientation. Whereas the cracks initiated from notches at 0° to the gate diameter propagated along the diametal direction in the regular way observed in previous work on PES, cracks started from other orientations deviated markedly from this line. This behaviour was particularly striking for 45° notches: the crack tended towards the gate diameter, and in those specimens with cold slugs the latter had a marked local effect. The 45° and 90° notch cracks were very jagged, with the microscopic directions following on average the local fibre orientation.

These comments of course represent a strong qualification of the K_{1c} figures for the 45° and 90° orientations. Whereas the 0° results are based on cracks broadly obeying the postulates of fracture mechanics testing, the requirements of straight cracks and plane fracture surfaces are not met in the remaining experiments. Within the 0° group, however, the limited comparisons possible suggest that the values are reproducible within normal experimental error, and that there is little difference between the stress intensity factors derived from stable and unstable cracking. Perhaps more surprisingly, there seems to be no significant difference between the values for 20% and 30% glass fibre concentration, or between the glass fibre filled material and the single result for 30% carbon fibre.

A different type of stress intensity factor derivation is indicated by Figs 5 and 6, demostrating the extent to which specimens with different R values obey Newman's formula. In each of these figures the force required to propagate a crack is plotted against the quantity $\sqrt{WB/HG}$, which according to (1) should be a straight line through the origin with slope K_{1c}. Fig 5, giving the results for the 0° notches, shows that for each test piece there is a convincingly linear relationship, with an intercept not significantly different from the origin, and the slopes for the three specimens agree within experimental error. In Fig 6, on the other hand, the correlations are much less good, with considerably larger dispersions both of regression coefficients and intercepts. These wide confidence limits place in doubt the reality of the apparent small increase of K_{1c} with angle.

3.2 Strain energy release rates

The values for G_{1c} shown in Fig 4 are naturally subject to the same qualifications as the K_{1c} results. The irregular crack surfaces on the 45° and 90° test piece must of course possess a surface area than the nominal area calculated on the premise of planarity: consequently the strain energy release rates for those discs must be lower than the measured values. The same factor must also be responsible in part at least for the wide dispersion of results. At the same time, however, it is worth noting that the strain energy release rate determined in this way is probably of greater fundamental significance than the stress intensity factors calculated from Newman's formula: the conditions for applying the latter are not fulfilled, while the use of the expression $\Delta W/\Delta A$ is limited only by lack of knowledge of the true value of ΔA. We may therefore conclude that (i) the 45° and 90° values are upper limits to the true values (ii) better results could be obtained if a method of

obtaining the correct surface area could be found.

With these caveats, we direct attention to two features of the figure. First, the 0° values for 20% and 30% glass fibre concentrations are reasonably consistent within each group, though the 30% distribution is skewed by a few anomalously high values. Second, there appears to be no significant difference between rates for 20% and 30% fibre concentrations.

3.3 Effect of crack speed

K_{1c} and G_{1c} levels were correlated with the rate of crack growth $\dot{a}$ and cross-head speed. No significant dependence was found.

4. DISCUSSION

The compact disc trials show that the test may be useful in three ways. In the first place, it is a valuable qualitative indicator of the overall influence of fibre orientation in injection mouldings, and of the extent to which fracture mechanics is applicable to a given composite. Second, for those crack orientations for which fracture mechanics is appropriate, it may furnish quantitative values of critical stress intensity factor, and is capable of determining point-to-point variations in this quantity (an important attribute in short fibre reinforced material). Third, it allows measurements of G_{1c} independent of the usual postulates of fracture mechanics, and with a little development may provide quantitative information on crack orientations well away from local mean fibre directions.

Taking these in order, we observe that the deviation of crack direction from the normal crack path along a diameter (dictated by the geometrical constraints of the specimen) must be related to the local fracture properties, no doubt because of fibre orientation and perhaps density. In other words, cracks probably form predominantly by parting of interfaces, so that the fracture properties must be traceable to the growth of cracks along fibres created by constraints between the fibres and a plastically deforming matrix. This model indicates that the values of critical stress intensity factor in those situations to which fracture mechanics applies should also emerge from such a plastic-elastic model at the microscopic level. The scatter in the K_{1c} values of the 0° orientation group may be regarded as inherent in the material rather than the test, and reflect genuine inhomogeneities in fibre distribution. The broad agreement of the results of the 30% and 20% concentrations of glass fibre, on the other hand, is remarkable and raises the possibility that fracture properties may saturate at some critical concentration as yet undetermined around the 20% mark.

Finally, as already observed, the G_{1c} values derived from (2) give values independent of any detailed model of crack growth, but require accurate values of fracture surface area increment. It therefore seems to us worthwhile directing attention to the problem of correcting the apparent value by a numerical value to be determined experimentally. We therefore make the following recommendations for the use of the compact disc in evaluating the fracture properties of injection mouldings of short fibre reinforced composites:

(i) Exploratory work using notches in different orientations to establish the extent
of fibreorientation effects. This should
include observation of crack paths.

(ii) For orientations which lead to straight
crack paths, detailed measurements of K_{1c}
by the break-force method and G_{1c} by the
sector method.

We recommend also that methods of measuring true
surface areas be examined in order to use the
sector method to calculate G_{1c} for cracks which
deviate from the theoretical crack path.

ACKNOWLEDGEMENTS

We acknowledge gratefully the help received by
discussions with colleagues at the University
of Surrey and ICI PLC, in particular Dr. J.G.
Rider and Dr. S. Turner.

REFERENCES

(1) GURNEY, C., and HUNT, J. 1967: Proc. Roy.
Soc., A299, 508.

(2) NEWMAN, J.C. Jr., 1981: Int. J. Fracture,
17, 567.

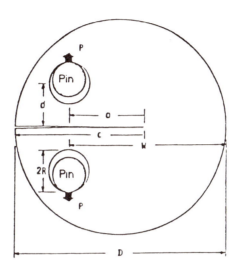

Fig 1   Configuration of compact disc test

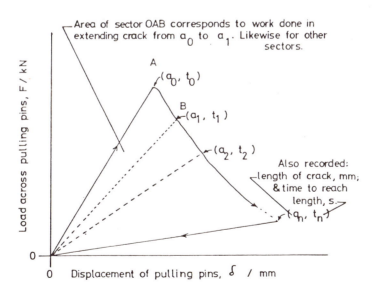

Fig 2   Idealized force—displacement record

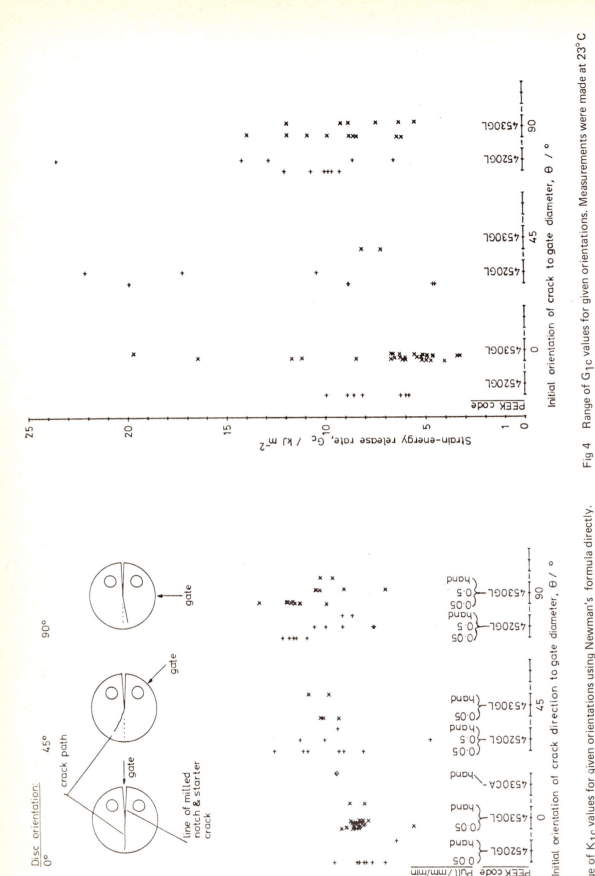

Fig 3   Range of K_{1c} values for given orientations using Newman's formula directly. All measurements were made at 23°C and 50 per cent rh (relative humidity)

Fig 4   Range of G_{1c} values for given orientations. Measurements were made at 23°C and 50 per cent rh

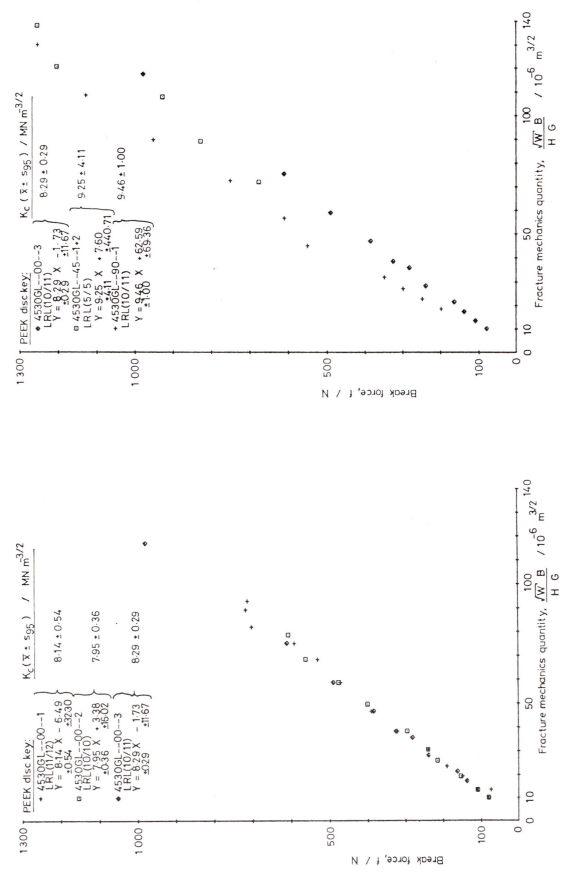

Fig 5 Comparison of three identical discs of PEEK 4530GL with 0° crack orientation using plot of break force versus fracture mechanics quantity. Measurements were made for a pulling rate of 0.05 mm/min at 23°C and 50 per cent rh

Fig 6 Effect of varying angle of initial-crack orientation for PEEK 4530GL using plots of break force versus fracture mechanics quantity. Measurements were made for a pulling rate of 0.05 mm/min at 23°C and 50 per cent rh

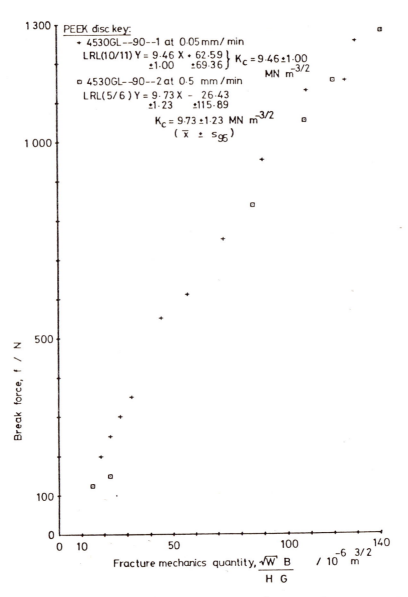

Fig 7   Effect of varying pulling speed for PEEK 4530GL of 90° orientation using plots of break force versus fracture mechanics quantity. Measurements were made at 23°C and 50 per cent rh

C39/86

Computer modelling of hybird reinforced thermoset composites

D S CORDOVA, H H ROWAN, BSc, DIC and L C LIN
Allied-signal Corporation, Petersburg, USA

I   INTRODUCTION

New advances in the use of organic reinforcing fibers, developed under the trademark COMPET$^{(R)}$, thermoset composites have introduced a new factor in the formulation of these materials. Previous papers by the authors reviewed the technical aspects (1,2,3) and the general utilization of these organic fibers in composite production systems (4,5). The findings of these publications show that the fundamental physical properties of Compet/glass fiber reinforced composites did not follow the traditional behaviour of all glass fiber reinforced composites. For instance, it was suggested by this earlier work that impact improvement may be gained not only by replacing an all glass fiber loading by an equivalent Compet/glass fiber loading but can be achieved with lower total hybrid fiber loading as depicted in Figure 1.

To define and address this possibility, a specific program was undertaken where the fundamentals of BMC injection molding compounds were studied by reformulating and analyzing the effects of all the major components. In consideration of the relatively larger number of components of a typical BMC formulation in addition to the injection molding conditions, the approach selected was to use a statistically designed set of trials aimed at definition of a mathematical model.

II   STATISTICAL EXPERIMENTAL MODEL

To define the number of batches that needed to be considered for a meaningful predictive model, each variable had to be considered with respect to its role in the compound formulation. The variables studied were standard formulation variables such as:

>Resin Loading
>Filler Loading
>Fiber Loading

as well as the fiber hybridization ratio:

>Compet/glass Fiber
and
>Compet Fiber Length

This experimental design study defined the specific batches to be compounded including the definition of both the lower and upper bounds of the three key variables. Levels of composite variables were included beyond the normal standard limits for glass fiber reinforced composites, so that potential new limits for hybrid fiber

reinforced composites could be investigated. The limits chosen were as follows:

>15-35% Weight Resin Loading
>160-490 PPH Resin Filler Loading
>10-30% Weight Fiber Loading

A trial in which each batch is tested once rarely gives adequate precision; it must usually be replicated in order to reduce the experimental error. To keep the error variance as small as possible, the experiments were arranged in blocks, one to two replicates per block, ie. if the block is large enough , then the comparisons of the batches could be made within blocks. The minimum number of batches required in order to have a representative group of results for a quarter replica were 55. These 55 batches were formulated, compounded, injection and compression molded, followed by appropriate physical property testing.

III   EXPERIMENTAL METHODS

For each specific formulation, a fifty pound batch of compound was prepared. The general compound formulation utilized typical standard components such as an isophthalic polyester resin, calcium carbonate, and half inch long BMC type glass fiber. Allied's Compet 1W71 polyester fiber was used as the organic fiber reinforcement.

The compounding utilized a sigma blade mixer and standard mixing procedures previously described in other Compet publications[3] were followed. The subsequent compound molding utilized a 300 ton C.A. Lawton injection molding press (plunger type). Test specimens were submitted to physical testing following ASTM procedures. These were:

>Impact
>>Notched
>>Unnotched
>>R-notched
>>Instrumented
>Tensile Strength
>Flexural Strength
>Flexural Modulus
>Profilometer
>SEM Analysis
>Specific Gravity
>Fiber, Resin, Filler Loading Analysis

An average of ten specimens were tested per procedure.

IV COMPUTER MODELING

The assessment of the large amount of results generated by the statistically designed experiments was carried out by computer analysis. The basic mathematical model utilized to correlate these data has the general form of:

$$y = y_o + \sum_{i=1}^{\eta} \sum_{j=1}^{\eta} a_{ij} x_i x_j + \sum_{i-1}^{\eta} b_i x_i$$

Where y is the composite physical property to be studied

$x_i x_j$ are variables

η is the number of parameters studied

y_o is a constant

a_{ij} is a constant

b_i is a constant

These constants were obtained from the basic data via linear regression analysis. Correlation factors for this model are in the 90% range.

From this basic mathematical model, a computer program named "MAPSI" was developed specifically for the Compet/glass fiber data base. The "MAPSI" program basically provides a listing of predicted composite physical properties as a function of compound formulations.

The standard form of data interpretation is to utilize a two dimensional graph such as the one shown in Figure 2. This allows one to study the response of a physical property to the change in one component, in this case, the fiber loading effect on impact strength. While the information generated through this type of graph is useful and confirms traditional trends, additional important information is being hidden. With the help of the "MAPSI" program, important information can be obtained through computer generated "contour plots" which reveal additional trends not shown in the previous figures. An example of this type of computer generated information is shown in Figure 3 which depicts in a tabular form the impact strength values as a function of both resin and fiber reinforcement loadings. These "contour plots" allow one to interpret the data by formulation regions and to add a cost factor for each data point considered.

For example, by negotiating any vertical column, these figures indicate the physical property trends as a function of fiber loading at constant resin loading. Figure 3-A represents a typical study of all glass fiber formulations. The contours represent constant impact values at different compound formulations. The impact data shown here represent trends widely described before in the literature; that is, the more fiber the better the impact. This diagram also helps to understand why the industry has concentrated their efforts in the regions around 25% resin and 25% glass fiber.

The case of Figure 3B represents a typical study of a 50/50 Compet/glass fiber reinforcement. Note that at the regions where all glass fiber

compounds are formulated (approximately 25% resin - approximately 25% fiber) COMPET offers already a substantial improvement in impact. The circular contour trends depicted here indicate also the great potential for modifying a formulation to achieve the same impact at lower cost. For instance, Point B shows an impact value of 4.27 ft lb/in utilizing 31% resin with 27% fiber reinforcement; on the other hand, Point A would deliver the same impact but utilize 19% resin with 13% fiber loading. Clearly, utilizing more of an inexpensive filler results in a more cost effective formulation. Furthermore, hybrids systems indicate that compounding could be possible in regions where filler loadings are too high for all glass fiber based formulation. For example, attempts at compounding formulations using all-glass reinforcement in the neighborhood of 19% resin loading proved extremely difficult. Under these conditions, there did not seem to be adequate wetout and excessive heat was generated, raising the temperature of the compound to where the paste almost set up in the mixer. On the other hand, the 50/50 Compet/glass fiber mix at the same resin loading did not show such behavior. This is believed to be due to the approximately 45% better resin wetout characteristics of Compet fiber versus typical fiberglass.

V COMPOSITE DESIGN

The development of the computer "contour plots" utilizing the "MAPSI" program resulted in two other programs called "Compet I" and "Search 5". "Compet I" predicts composite physical properties for a given compound formulation and fiber geometrical characteristics.

The "Search 5" program is the inverse of "Compet I". It graphicaly maps out one or more compound formulations for a particular set of prerequired composite physical properties. A typical output of this program is shown in Figure 4.

To illustrate the capability of these programs, a standard automotive formulation was modified by reducing the resin and fiber loadings, and increasing the filler in an equivalent number. Table 1 presents the predictive and experimental values for test specimens which show a good agreement, even though the prediction program is based upon a general purpose formulation. Also note that two process systems are compared here, BMC vs. TMC. Table 1 also shows the impact values for this formulation used to make an injection molded automotive GOP. The reported values are in both the vertical and horizontal direction and they show a dramatic improvement over the control. A significant feature here is that almost identical impact values are found in both directions (vertical and horizontal); behaviour not found before with the injection molded glass fiber compound. Moreover, the formulation was shown to have similar or better cosmetic appearance than the control.

Another industrial grade example is shown in Figure 5. The customer's standard electrical formulation was modified utilizing the previous described approach. The principle design objective was impact strength increase at lower compound cost. Figure 5-A and B shows an additional feature of this formulation modification. When an all glass (non-modified formulation) electrical box (Figure 5-A) is impacted at 30 ft/lb by a drop weight cylinder, it shatters. The COMPET/glass reinforced electrical box (Figure 5-B) impacted at the same level shows a much different failure mode, the cylinder literally bounces upon impact and the part cracks instead of shattering.

VI   CONCLUSIONS

The formulation modification approach described in this paper has been found to have excellent agreement in several industrial scale trials completed to date where the potential for more cost effective formulations has been demonstrated. Additional work is in progress to calibrate the model for different resins, filler types, and other additives. This further work is not confined only to formulation variables, but also to process. Models for BMC using compression molding and for SMC, TMC, and fabric laminates are planned. All of the described computer programs were developed by Allied Corporation as an effort to expand the capabilities of thermoset reinforced plastics.

REFERENCES

1. D.S. Cordova, D.R. Coffin, J.A. Young, and H.H. Rowan, Proceedings of the 39th SPI Annual Conference, Reinforced Plastics, Section 3-B (1984).

2. D.S. Cordova, D.R. Coffin, J.A. Young, and H.H. Rowan, Conference Proceedings of the 42 nd Annual Technical Conference - SPE, PP 663 (1984).

3. D.S. Cordova, H.H. Rowan, and J.A. Young, Proceedings of the 40th SPI Annual Conference Reinforced Plastics, Section 20-B (1985).

4. C.W. Cordova, H.H. Rowan, and J.A. Young, Proceedings of the 41st SPI Annual Conference, Reinforced Plastics (1985).

5. Hillermeier, K; Blumer, H., Research Report T83-155, IKV Aachen, 1983.

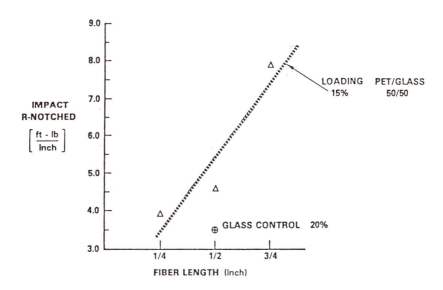

Fig 1   The effect of fibre length and loading on injection moulded composite impact properties

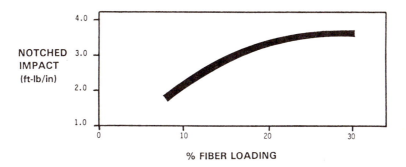

Fig 2   Compet/glass fibre reinforcement — injection moulding

RESIN (% WGT)

FIBER (% WGT)	15.00	19.00	23.00	27.00	31.00	35.00
7.00	1.64	2.03	2.39	2.73	3.05	3.35
9.00	1.75	2.09	2.42	2.73	3.02	3.30
11.00	1.89	2.19	2.47	2.75	3.02	3.27
13.00	2.07	2.32(A)	2.56	2.80	3.04	3.27
15.00	2.29	2.48	2.68	2.88	3.08	3.28
17.00	2.54	2.67	2.82	2.98	3.15	3.32
19.00	2.32	2.90	3.00	3.12	3.25	3.39
21.00	3.14	3.15	3.20	3.28	3.37	3.48
23.00	3.49	3.44	3.43	3.46	3.51	3.59
25.00	3.88	3.76	3.70	3.67	3.68	3.72
27.00	4.30	4.11	3.99	3.91	3.88(B)	3.87
29.00	4.75	4.49	4.30	4.18	4.10	4.05

(a)

RESIN (% WGT)

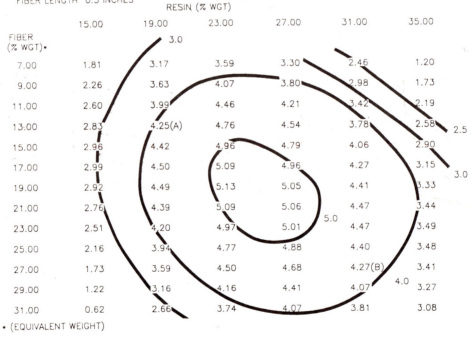

FIBER (% WGT)•	15.00	19.00	23.00	27.00	31.00	35.00
7.00	1.81	3.17	3.59	3.30	2.46	1.20
9.00	2.26	3.63	4.07	3.80	2.98	1.73
11.00	2.60	3.99	4.46	4.21	3.42	2.19
13.00	2.83	4.25(A)	4.76	4.54	3.78	2.58
15.00	2.96	4.42	4.96	4.79	4.06	2.90
17.00	2.99	4.50	5.09	4.96	4.27	3.15
19.00	2.92	4.49	5.13	5.05	4.41	3.33
21.00	2.76	4.39	5.09	5.06	4.47	3.44
23.00	2.51	4.20	4.97	5.01	4.47	3.49
25.00	2.16	3.94	4.77	4.88	4.40	3.48
27.00	1.73	3.59	4.50	4.68	4.27(B)	3.41
29.00	1.22	3.16	4.16	4.41	4.07	3.27
31.00	0.62	2.66	3.74	4.07	3.81	3.08

• (EQUIVALENT WEIGHT)

(b)

Fig 3   Computer generated contour plots
(a)   Glass formulations
(b)   50/50 Compet/glass formulations

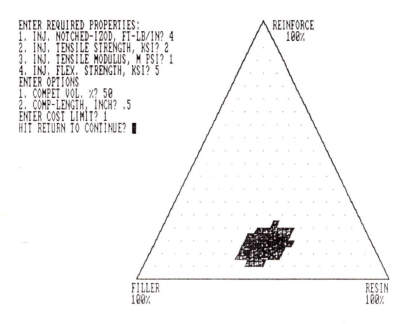

```
ENTER REQUIRED PROPERTIES:
1. INJ. NOTCHED-IZOD, FT-LB/IN? 4
2. INJ. TENSILE STRENGTH, KSI? 2
3. INJ. TENSILE MODULUS, M PSI? 1
4. INJ. FLEX. STRENGTH, KSI? 5
ENTER OPTIONS
1. COMPET VOL. %? 50
2. COMP-LENGTH, INCH? .5
ENTER COST LIMIT? 1
HIT RETURN TO CONTINUE? ▮
```

REINFORCE
100%

FILLER
100%

RESIN
100%

Fig 4   Computer output for 'Search 5' Program

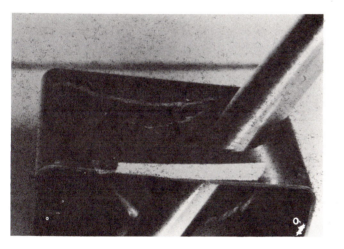

Fig 5a Glass reinforced formulation

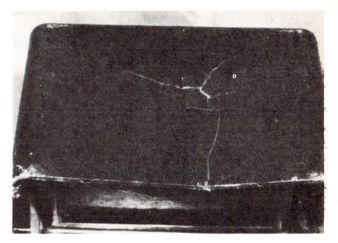

Fig 5b Compet/glass fibre reinforced formulation

Table 1   Automotive TMC-injection moulding

I.D.	FIBER LOADING %	FILLER LOADING %	RATIO COMPET/ GLASS	FIBER LENGTH (INCH)	IMPACT NOTCHED (FT LB/IN) PRED.	EXP.	GOP IMPACT R-NOTCHED* (FT LB/IN) HORIZONTAL	VERTICAL
CONTROL	20	50.7	0/100	0.50	3.2	2.3	1.99	0.63
FORMULATION B	16	57.73	50/50	0.50	4.9	4.4	3.96	3.01

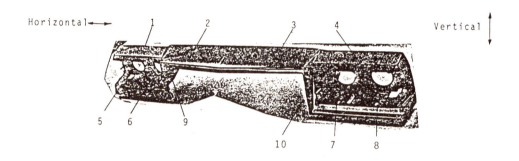

*NOTE: FROM THE 10 POSITIONS DEPICTED ABOVE TEST
         SPECIMENS WERE CUT IN BOTH DIRECTIONS

C19/86

Thick section fibre reinforced thermoplastics injection moulding

M J BEVIS, BSc, PhD, CEng, FInstP, FIM, FPRI, **P S ALLAN**, BSc, PhD, MInstP and **P C EMEANUWA**, BSc, MSc
Department of Materials Technology, Brunel University, Uxbridge, Middlesex

SYNOPSIS New processes have been developed for the production of thick sectioned fibre-reinforced thermoplastic injection mouldings. The processes provide for ways of influencing the development of micromorphology, residual stresses and the elimination of micro-porosity in the form of voids and cracks in thick-sectioned mouldings. The processes are based on the use of separately controlled oscillating melt pressures through single of multi-gates during the solidification of polymer melts. The simple application of the new processes results in the elimination of micro-voids and micro-cracks from complex thick-section mouldings, modulus enhancement and substantial improvements in internal weld line strengths through preferred alignment of fibres, and control over residual stresses. The modification of the properties of internal weld lines in thick-sectioned glass fibre reinforced polypropylene injection moulding is used to illustrate the 'live' feed moulding processes.

1 INTRODUCTION

Two types of internal defect can result when thick sectioned components of fibre reinforced thermoplastics are produced by conventional injection moulding. The first defect is that of a voided zone at the centre section of the moulding. This is caused by the gate or narrow sections separating thick sections freezing off prior to the solidification of the thick cores of the moulding. Tensile stresses at the centre of the moulding result from the shrinkage of the molten core material and cause voids to form in the remaining molten material. This is more likely to happen in fibre reinforced materials than their unreinforced resins, because the greater stiffness of the fibre reinforced material resists the deformation which causes surface sinking in the unreinforced materials. The second type of defect occurs when the gate or narrow sections in a moulding freeze off at approximately the same time as the centre of the moulding. The subsequent differential cooling of the centre and the outer surfaces of the moulding to room temperature induces stresses in the solid material which are in excess of its fracture stress. This results in the formation of fine cracks in the body of the moulding which tend to follow the line of fibre orientation and are more likely to occur in the strong brittle fibre reinforced thermoplastics.

The two types of defect referred to above result in a substantial reduction in the mechanical properties of fibre reinforced thermoplastics.

The mechanical properties of fibre reinforced thermoplastics may also be reduced by the presence of internal weld lines which form where two melt fronts meet in a mould cavity. The preferred fibre orientation tends to be in the plane of the internal weld and therefore normal to the primary injection moulding flow direction and, results in a substantial reduction in the tensile strength of the internal weld region.

All three types of defect referred to above may be eliminated from thick sectioned injection moulded components by use of profiled oscillating packing melt pressures rather than use of the traditional profiled packing pressure. The oscillating packing pressure and 'live' feed mould processes are summarised in section 2 and 3 respectively, and the results of the application of a double live feed system to a range of thick sectioned glass fibre reinforced polypropylene bars is summarised in section 3.

2 OSCILLATING PACKING PRESSURES

The main object of using an oscillating packing pressure in injection moulding is to ensure that thick sections in moulded components have sufficient time to freeze off in the mould cavity before the gate, sprue, runner or any narrow section separating the thick section of the mould cavity from the injection moulder nozzle has frozen off. This is achieved by the periodic compression and expansion of the melt in the thick sections which causes enhanced shearing of the melt in the narrower sections. The result of this action is that the internal heat generated by the melt shearing in the narrow sections delays the freezing-off time for these sections. In addition, a significant amount of control over the freeze-off time for the thin sections can also be obtained by changing the oscillating pressure profile.

The mould packing technique which is preferred at Brunel (1) uses an auxiliary mould packing device (MPD); coupled with the optional facility of flow moulding for the initial filling of the mould cavity. The MPD is situated between the end of the barrel of the injection moulder and the fixed platen as shown in Fig.1. The piston of the MPD is driven by an auxiliary power pack which is controlled by a microprocessor. The auxiliary MPD is caused to operate when initial mould cavity filling is complete and replaces the conventional packing stage of the injection moulding cycle. Typical cavity pressure profile obtained with (OPP) and without (SPP) the use of the MPD are shown in Fig.2.

When used in conjunction with flow moulding the MPD overcomes many of the deficiencies associated with low pressure moulding and, therefore provides for the production of relatively large volume mouldings on small injection moulding machines.

It has been found that (1) with the use of appropriate profiling of the amplitude and frequency of the oscillating packing pressure, it is possible to eliminate microcracks and microvoids. Mouldings up to 110mm thick have been successfully produced from 20% glass fibre reinforced PEEK by this technique, and a variety of other thick sectioned mouldings have been produced from materials such as carbon fibre reinforced PEEK and nylon and glass fibre reinforced PES, nylon and PEEK.

Our results have also shown that oscillating packing pressures can influence the residual stresses in fibre reinforced thermoplastics, although the relaxation of these stresses is more complicated than in the unreinforced thermoplastics (2-4).

3 MULTIPLE 'LIVE FEED' INJECTION MOULDING

The efficiency of using an oscillating pressure for mould packing depends on the movement of molten material within the narrow sections of the moulding. This, in turn, depends on the compressibility of the molten material in the thick section of the mould which lies further from the gating point than the narrow section. The amplitude of oscillations in cavity pressure required for the packing of fibre reinforced resins was found to be substantially greater than for unreinforced thermoplastics. As a consequence of these and other observations the authors (5) have developed the new concept of utilising a plurality of 'live' feed points to the mould cavity as a way of controlling the microstructure in moulded components. The basic principle of the live feed process is illustrated by the schematic diagram in figure 3 of a two live feed device. Two packing chambers 1 and 2 are fed via the two channels 3 and 4 by the single feed 5 from the injection moulder barrel. Additional independent pressure can be supplied to the two feeds at the twin nozzles 6 and 7 by the reciprocating pistons 8 and 9. The pistons movements are controlled by a microprocessor via an auxiliary hydraulic power pack which contains two pumps and two electrically controlled pressure valves.

Shearing of molten material between the two feeds in the mould cavity can be effected in a controlled manner while the component is solidifying. This results in the following important attributes for the live feed process:

use of lower melt pressures than a single feed device for oscillating packing and the elimination of microvoids and microcracks,

the development of preferred orientation by design in moulded components,

the modification of the micromorphology and mechanical properties of internal weld lines in moulded components.

The results of the application of the two live feed device for the elimination of microvoids and the enhancement of the mechanical properties of internal weld lines in thick sectioned glass reinforced polypropylene injection mouldings are described below.

The use of this technique requires the provision of additional sprues and runners to the mould cavities. No other special design requirements for the mould cavity should be necessary. The technique may lead to a reduction in scrap levels for some mould geometries. In cases where additional runner systems increase the scrap, these may be justified by an increase in physical properties.

4 APPLICATION OF THE TWO LIVE FEED DEVICE

Double end-gated mould cavities were used with the two live feed device to produce mouldings with cross-sections of 3 x 20mm, 6 x 20mm and 20 x 20mm and a length of 160mm. These mould cavities are represented by item 10 in figure 3. Three classes of mouldings in glass fibre reinforced polypropylene were produced using optimum processing conditions.

A. Single end-gating with asymmetric activation of live feeds.

B. Double end-gating without activation of live feeds.

C. Double end-gating with asymmetric activation of the live feeds.

The full range of mouldings in terms of the cross-sections and processing conditions were prepared from 30% glass fibre reinforced polypropylene. The 20% glass fibre reinforced polypropylene mouldings were produced using only the 3 x 20mm and 6 x 20mm cross-section cavities.

The room temperature tensile properties of the as-moulded 3 x 20mm and 6 x 20mm cross-section polypropylene mouldings were gained using a 5cm per minute cross-head speed. Summaries of the results are presented in Table 1 and 2 respectively, where each of the quoted values for tensile strength is the mean of ten tests for every glass fibre concentration and process condition.

Table 1

	Tensile Strength (MPa) 3mm Test Specimen	
	20% glass fibre	30% glass fibre
single gate	52.2	67.3
double gate (static) (internal weld)	26.3	26.6
double gate (asymmetric oscillations) (modified internal weld)	48.2	56.5

Table 2

| | tensile strength (MPa) 6mm test specimen | |
	20% glass fibre	30% glass fibre
single gate	56.1	62.8
double gate (static) (internal weld)	30.5	25.5
double gate (asymmetric oscillations) (modified internal weld)	57.2	62.2

The 20 x 20mm cross-section mouldings were sectioned and then polished into six tensile test bars each approximately 3mm in thickness. These zones were labelled from 1 to 6 through the thickness of the moulding where zones 1 and 6 represent opposite skins of a moulded block. Three moulded 20 x 20mm cross-section blocks were produced from 30% glass fibre reinforced polypropylene for each of the three processing conditions referred to above providing three tensile test bars for every position 1-6 and set of processing conditions.

The processing conditions were selected to give minimum microporosity in the case of the static double end-gated bar, and to give optimum packing and not necessarily high levels of anisotropy in the live feed single and double end gated mouldings.

The results of tensile tests are summarised in Table 3-5.

Table 3 Tensile test results - double end-gated moulding with static packing pressure

Zone	Weld strength (MPa)	Weld strain to failure	Remarks
1	25.6	0.04	Microvoids
2	29.9	0.04	
3	14.9	0.03	
4	-	-	Large voids
5	10.7	0.03	Small voids
6	8.4	0.02	Microvoids

Table 4 Tensile test results - double end-gated mouldings with asymmetric activation of live feeds.

Zone	Weld strength (MPa)	Weld strain to failure
1	49.4	0.06
2	78.4	0.08
3	69.2	0.06
4	59.3	0.06
5	71.9	0.06
6	52.2	0.08

Table 5 Tensile test results - single gate with asymmetric activation of live feeds

Zone	Weld strength (MPa)	weld strain to failure
1	62.0	0.08
2	73.2	0.06
3	57.7	0.06
4	59.6	0.09
5	65.4	0.07
6	52.5	0.08

The results presented in Tables 1-5 show that the tensile strength of internal weld lines can be substantially improved by the application of a plurality of live feeds. The weld line strengths in mouldings produced under static conditions are reduced to the weld line strengths of unreinforced polypropylene (25 MPa) and even less in the case of the very thick 20 x 20mm cross-section mouldings.

The action of the asymmetrically activated live feeds causes the weld strength of the 6mm thick fibre reinforced mouldings to increase to that of the strength of the single gate mouldings without internal weld lines. The activation of the live feeds eliminates macro and micro-porosity in the 20 x 20mm cross-section double end-gated mouldings and raised the internal weld strength to that of the single end-gated mouldings produced with the live feed process.

A substantial increase in internal weld line strength from less than 50% to greater than 85% of the strength of weld line free specimens was recorded for the 3mm thick mouldings. This result demonstrates the power of the multiple live feed process for enhancing the mechanical properties of injection mouldings in the 3mm to 20mm thickness range. Mechanical property improvements are attributed to the suppression of microporosity and the preferred alignment of glass fibres to be normal rather than parallel to internal weld planes. The live feed process provides a way of retaining the optimum mechanical properties of fibre reinforced thermoplastics throughout the volume of thick-sectioned parts containing internal weld lines.

REFERENCES

(1) ALLAN, P.S. and BEVIS, M.J. Producing void-free thick-section thermoplastic and fibre reinforced thermoplastic mouldings
Plas Rubb Int, 1984, 9, 32-36

(2)(3) ALLAN, P.S. and BEVIS, M.J. The production of void-free thick-section mouldings
I Shotweight and dimensional reproducibility
Plas Rubb Proc & Appl, 1983, 3, 85-91
II Preferred orientation and residual stress measurements
Ibid, 1983, 3, 331-336

(4) ALLAN, P.S. and MORTAZAVI, M.J. The effect of oscillating packing pressures on the residual stresses in thick-section polyethylene injection mouldings
Plas Rubb Proc & Appl, 1985, 5, 71-78

(5) ALLAN, P.S. and BEVIS, M.J.(in preparation)

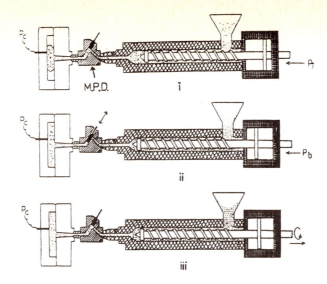

Fig 1   Schematic representation of an injection moulding cycle using the mould packing device (MPD):
(i) mould filled by injection pressure p_i (ii) mould packed by MPD with a back-up pressure of p_b from the moulder
(iii) moulder resets for next cycle

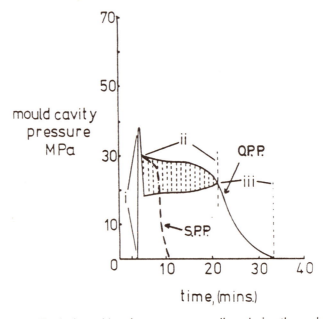

Fig 2   Typical mould cavity pressure recordings during the cycle illustrated in Fig 1. The pressure responses during the three phases shown in Fig 1 are indicated for an oscillating packing pressure (OPP) used to pack a 35 mm thick polyethylene bar. A static packing pressure (SPP) profile is indicated by a broken line

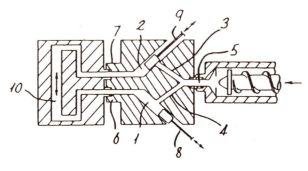

Fig 3   Schematic diagram of the two 'live' feed mould packing device which replaces the mould packing device (MPD) in Fig 1

C32/86

Behaviour of long fibre reinforced polypropylene and Nylon 66 in injection moulding

A C GIBSON and A N McCLELLAND
Departmant of Metallurgy and Materials Science, University of Liverpool

SYNOPSIS Reinforced thermoplastic moulding materials are now available with a fibre length of 10 mm. The flow behaviour of 50%(wt) glass reinforced polypropylene and nylon 66 is described, along with the morphology of the mouldings. The surprising ease of processability of long fibre materials and the lack of fibre degradation is accounted for by the persistence of fibre bundles or domains of parallel fibres during flow. There is evidence that long fibre materials form dead zones when flowing in a convergence.

1 INTRODUCTION

The most widely used route at present for compounding fibre reinforced thermoplastics involves blending in a melt extruder. Fibre breakage during wet-out results in an aspect ratio which is far short of that required for maximum reinforcing efficiency. Although degradation can be minimised by careful design of processing machinery, there is still a severe limitation on the aspect ratio attainable by this route. It can be argued that wet-out takes place when the fibres are in an approximately random orientation. This is the least efficient means of packing fibres. The maximum fibre volume fraction, V_{MAX}, that can be incorporated decreases substantially as the aspect ratio, L/d, of the fibres increases (1), following a relationship of the form (2):

$$V_{MAX} = k \frac{d}{L} \qquad (1)$$

k has a value in the range 4-6.

As a result of this constraint the fibre length attainable in conventional compounding is limited to values around 0.1-0.3 mm. There has been interest therefore in alternative processes. In one of these, a continuous tow of fibres is passed through a crosshead die and impregnated with thermoplastic melt. The process resembles both pultrusion and wire-coating. After cooling, the resulting strand is then chopped to produce injection mouldable pellets. Although processes of this type have been described in the literature (3,4) and have also been used commercially, there were often problems with achieving complete fibre wet-out. Moreover, it was by no means obvious that, even if the wet-out problem could be solved, the material would be injection mouldable without substantial degradation.

The work reported here, which was performed in collaboration with ICI (Petrochemical and Plastics Division) involved moulding materials containing 50% by weight of 10 mm long glass strands. These were produced by a proprietary strand impregnation process using matrices of polypropylene and nylon 66. The latter material, Verton, (5), is now a commercial product.

The term 'long fibre' (LF) will be used to denote materials containing discontinuous fibres substantially longer than those encountered in conventional melt-compounded product. The latter materials will be referred to as 'short fibre '(SF). The materials examined in this study are listed in table 1.

It was surprising to find that the LF materials could be moulded with relative ease on a conventional injection moulding machine without undue fibre degradation, provided some care was taken not to work the compounds too heavily. Figure 1 shows a number average fibre length distribution obtained from mouldings of LFPP. The weight average fibre length was 6.8 mm, indicating good fibre length retention. The processing behaviour of both LFPP and LFN were fairly similar to that of their short fibre counterparts. This, combined with the low level of fibre degradation, was fairly remarkable, given that the combination of aspect ratio and volume fraction violates the limit of equation 1 by a considerable margin.

The key to fibre length retention appears to be that, if fibres are parallel to one another during wet-out, they tend to remain locally parallel during processing, even through the flow régime of an injection moulder screw. The continued existence of regions of parallel fibres during flow explains the ability of the material to avoid the limitation of random packing.

This paper reports an investigation of the flow behaviour of LF, SF and unreinforced materials using an instrumented injection moulding machine as a rheometer. Details of the instrumented nozzle head of the machine are shown in figure 2. The fibre orientation patterns found in nozzle flow and in the final product will also be discussed.

2 THEORY OF FLOW BEHAVIOUR

The pressure drop when a polymer melt flows through a nozzle into a capillary can be separated into components due to the capillary pressure drop,

P_C, and the entrance pressure drop, P_E. These two components can be evaluated by performing flow experiments with capillaries of different length. P_C is related to the apparent shear viscosity of the melt, η_A, by

$$P_C = \frac{8Q l \eta_A}{\pi r_1^4} \qquad (2)$$

where Q is the flow rate, l and r_1 are the capillary length and radius.

The capillaries used in the present work all had an entry semi-angle, θ, of 45° and a bore diameter of 4 mm.

The apparent wall shear rate in the capillary, $\dot{\gamma}_A$, is given by

$$\dot{\gamma}_A = \frac{4Q}{\pi r_1^3} \qquad (3)$$

Entry flow turns out to be very important in the present case. Theory to describe this is less well-established. However an analysis has been proposed by Cogswell (6) and this has been shown to apply for another type of reinforced material, DMC (7,8), at die angles up to 65°. The value of P_E given by Cogswell's analysis is

$$P_E = \dot{\gamma}_A \left[\frac{\lambda_A}{3} \tan\theta + \frac{2\eta_A}{3 \tan \theta} \right] \left[1 - r_1^3/r_o^3 \right] \quad (4)$$

λ_A is the apparent extensional viscosity and r_o is the nozzle entrance radius.

Equation (4) can be simplified somewhat. Since r_1^3/r_o^3 is usually much smaller than one, the last term can be taken as unity. Also, under most conditions, except at low die angles, the extensional viscosity term is much larger than the shear viscosity term, so equation 4 can be reduced to

$$P_E = \frac{\dot{\gamma}_A \lambda_A}{3} \tan \theta \qquad (5)$$

When the ratio of λ_A/η_A is large, flowing melts tend to converge at a narrower angle than the die angle, giving a dead zone of stagnant material. At present there is no suitable theory that can be used to predict such behaviour. It should be borne in mind, therefore, that the effective convergence angle may be less than θ, and in some circumstances values of λ_A calculated using equation (5) may be underestimates.

3 RESULTS AND DISCUSSION

3.1 Polypropylene Compounds

Flow measurements were performed on LFPP, SFPP and PP at $\dot{\gamma}_A$ values in the region $10^3 - 10^4$ s^{-1}. Results are shown as plots of pressure against capillary length in figure 3. The large intercept values (P_E) show that, in each case, the resistance to flow at the die entry is large. The similarity between SFPP and UFPP at high shear rates is in agreement with previous work (9).

It is quite surprising that LFPP, where the fibre aspect ratio is at least ten times higher shows broadly similar behaviour to SFPP.

One pronounced difference between the materials is in die swell behaviour. SFPP shows very little die swell, compared with pp. By contrast, the LF material bursts forth from the capillary to form a fluffy fibre foam, suggesting that fibre bundles have been deformed elastically during nozzle flow and spring back on emerging.

Using equations (2) and (5), shear and extensional viscosities were calculated and are shown in figure (2), plotted against $\dot{\gamma}_A$. The most striking features are the high λ_A/η_A ratios (which are of the order of 10^2 or higher) and again the apparent similarity in behaviour of the materials.

There are two possible origins for the high λ_A/η_A values, both of which are associated with the effects of the extensional flow field at the nozzle entry. The first is molecular orientation of the resin. This effect would increase with increasing strainrate and also with increasing molecular weight. It has already been demonstrated that unfilled melts can show a high λ_A/η_A ratio at convergences (6).

The second effect is fibre alignment. This has been shown to produce a substantial increase in λ_A in liquids where there are no molecular alignment effects (10). The extensional viscosity of a suspension of aligned rods is given by (11, 12):

$$\lambda_A = \eta_{RES} \left[3 + \frac{4}{3} V_f \frac{(L/d)^2}{\ln(C/Vf)} \right] \qquad (6)$$

where η_{RES} is the shear viscosity of the unfilled resin and the value of C is in the region 0.907-3.628.

In the absence of fibres the λ_A/η_A ratio can be seen to revert to 3, the value for a simple fluid. When fibres are present, however, the right hand term usually dominates and a very strong dependence on aspect ratio is predicted.

Equation 6 can be modified to include effects of molecular alignment:

$$\lambda_A = \lambda_{RES} + \frac{4}{3} \eta_{RES} V_f \frac{(L/d)^2}{\ln(C/Vf)} \qquad (7)$$

At the moment it is not possible to assess conclusively the relative magnitudes of the two effects mentioned above, as the molecular weight of the polymer differed between the moulding materials, being highest for the homopolymer and lowest for the LF material. However, if the two effects were of similar size in the SF material, this would help to explain the similarity between this and the unfilled polymer.

The similarity between the LF and SF materials is remarkable, given a probable factor of ∿30 difference in the aspect ratios and the squared aspect ratio term in equation 7. A likely explanation is a fundamental difference in the mechanism of flow between the two materials. Flow in LF materials may not involve much relative movement between adjacent fibres, but movement of bundles or domains of parallel fibres. The effective aspect ratio of these regions would be much lower than that of the individual fibres, probably around 30-100.

The morphology of moulded LF components is different to that found with SF compounds. In both cases there is an outer layer where the orientation is parallel to mould flow and a central core, usually oriented transverse to this. With LF mouldings the central core region is much thicker than with short fibres and is quite inhomogeneous, showing evidence of swirling clumps and bundles of fibres with local orientation. There is a strong possibility that at least some of this bundle-like structure was present upstream of the nozzle. Figure 5 shows a section from the sprue of a moulding. This has a highly oriented outer layer with a core, again consisting of swirling bundles which must have passed through the nozzle of the machine. The origin of this type of structure will be discussed further in the following section on LF nylon.

Despite the macroscopic inhomogeneities mentioned the LF polypropylene showed substantial mechanical property improvements over its SF counterpart.

3.2 LF Nylon 66

Figure 6 shows λ_A and η_A values for the LF nylon material obtained from similar flow measurements to those described above. The results are qualitatively similar to those for the LF poly-propylene material.

Figure 7 shows a section through a seed tray moulding in LF nylon. The section is perpendicular to the mould flow direction. Again this section shows a thicker core than would be present with an SF compound. The core again shows evidence of a bundle-like texture, although the degree of inhomogeneity is less than in the case of the LFPP. There is also evidence of a resin-rich layer between the core and the wall layer, and a thin surface layer.

In order to gain some information on the flow pattern in the nozzle region, 1% of a compound containing LF carbon was added to the moulding material as a tracer and a slug of material was allowed to solidify in the barrel after flow had occurred. A polished section taken from this slug is shown in figure 8a. Well upstream of the nozzle the lines of fibre orientation have an approximately parabolic contour, with transverse orientation at the centre of the section and a movement towards axial orientation near the wall. This orientation pattern is due to the combined effects of the screw rotation which produced the moulding charge and the forward flow which took place prior to solidification.

On approaching the nozzle convergence, material near to the axis accelerates and the orientation profile becomes sharper. On reaching the convergence the transverse orientation of this material is destroyed as it is broken up and accelerated into the nozzle. Assuming a similar pattern to exist with the LFPP material, this is probably the origin of the disrupted clumps of material observed previously at the centre of the sprue. Only material which had previously been aligned transverse to the axis could flow through the nozzle without substantial axial orientation occurring.

There is evidence of dead zone formation

near to the die exit. To confirm this, some injection experiments were performed with a change from unpigmented LF nylon to the material containing the tracer. After running the screw until empty on the first compound, the second was added and flow shots were performed until the tracer was observed in the extrudate. A slug was then solidified. The polished section in figure 8b shows clear evidence of the dead zone, which extends upstream for some distance.

Of course the existence of dead zones casts some doubts on the calculation of λ_A using equation 5. The actual convergence angle within the nozzle was found from figure 8b to be 19°. Repeating the calculation with equation 5 and this value gave the higher λ_A curve shown in figure 6, indicating an even higher value of λ_A/η_A than had previously been expected.

Work is continuing on the characterisation of LF moulding material, particularly in relation to flow behaviour and flow orientation.

4 CONCLUSIONS

(i)   LF moulding materials show surprisingly good processability and fibre length retention.

(ii) Flow involves movement of local regions of orientation which have a lower aspect ratio than individual fibres.

(iii) There is a tendency for dead zones to form in regions of convergence.

5 ACKNOWLEDGEMENT

The Composites Processing programme at Liverpool University is supported by S.E.R.C. A.N. McClelland acknowledges the support of D.E.N.I. and ICI.

6 REFERENCES.

1. MILEWSKI, J.V., "Handbook of filler and reinforcements for plastics", Van Nostrand Reinhold, 1978.

2. EVANS, K.E. and GIBSON, A.G., Composites Sci. and Technol., 25, 1, 1986.

3. BADER, M.G. and BOWYER, W.H., Composites, 5, 150, 1973.

4. FOLKES, M.J. and KELLS, D., "Fibre Reinforced Composites '84", 3-5 April 1984, University of Liverpool, Plastics and Rubber Institute, 1984.

5. Product data on 'Verton long fibre reinforcement', ICI Petrochemicals and Plastics Division, Wilton, Cleveland, U.K.

6. COGSWELL, F.N., Polymer Engineering Science, 12, 64, 1972.

7. GIBSON, A.G. and WILLIAMSON, G.A., Polymer Engineering Science, 25, 968, 1985.

8. GIBSON, A.G. and WILLIAMSON, G.A., Polymer Engineering Science, 25, 980, 1985.

9. CROWSON, R.J., SCOTT, A.J. and SAUNDERS, D.W.,
   Polymer Engineering Science, 21, 748, 1981.
   CROWSON, R.J., FOLKES, M.J. and BRIGHT, P.F.,
   Polymer Engineering Science, 20, 925, 1980.

10. MEWIS, J. and METZNER, A.G., J. of Fluid
    Mechanics, 62, 593, 1974.

11. BATCHELOR, G.K., J. of Fluid Mechanics, 46,
    813, 1971.

12. LAMB, D.W., Ph.D. Thesis, University of
    Liverpool, 1985.

TABLE 1 MATERIALS

CODE	MATRIX	GLASS CONTENT (W/W)	FIBRE LENGTH
LFPP	Polypropylene	50%	10 mm
SFPP	Polypropylene	50%	∼ 0.1–0.3 mm
PP	Polypropylene	O	–
LFN	Nylon 66	50%	10 mm

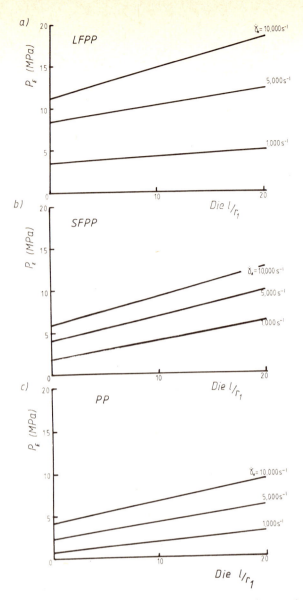

Fig 3 Die pressure plots for LFPP, short fibre polypropylene (SFPP) and polypropylene (PP) at 240°C. Die bore diameter, 4 mm. Die entry semi-angle, 45°

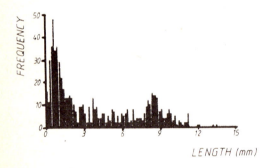

Fig 1 Fibre length distribution for long fibre polypropylene (LFPP) after moulding

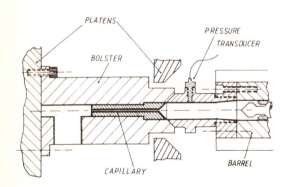

Fig 2 Nozzle and capillary configuration for rheometry measurements

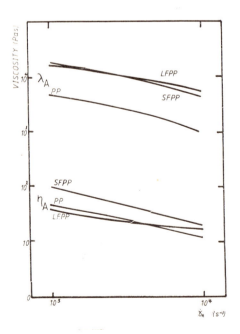

Fig 4 Shear and extensional viscosity data for LFPP, SFPP and PP at 240°C

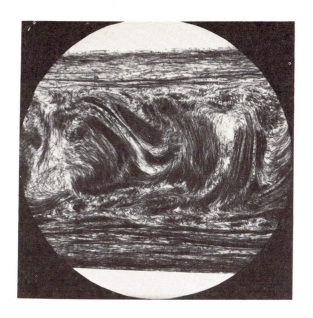

Fig 5   Microradiograph of section through sprue of LFPP moulding. Flow direction ⟶ diameter, 5mm

Fig 7   Microradiograph of section through a LFN seed tray moulding of 4.11 mm thickness. Section cut parallel to mould flow direction

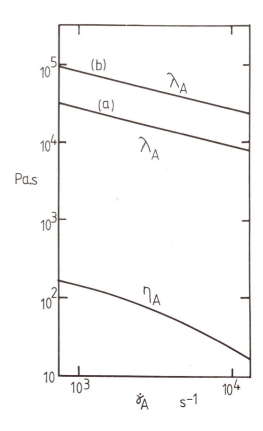

Fig 6   Shear and extensional viscosity data for long fibre nylon (LFN) at 285°C λ_A curves calculated assuming (a) 45° convergence angle, (b) 19° convergence angle with dead zone

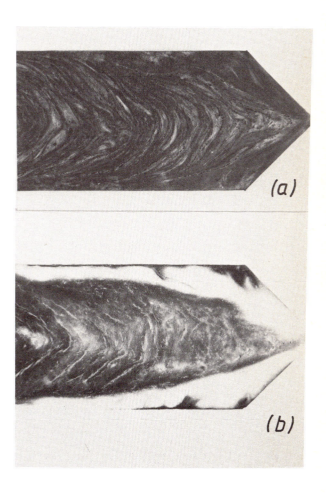

Fig 8a   Polished section through material solidified in barrel after flow. LFN with 1 per cent tracer of LF carbon nylon

Fig 8b   As for (a), with material change, showing dead zones

C22/86

Section pultrusions of continuous fibre reinforced thermoplastics

G COWEN, U MEASURIA, BSc and R M TURNER, BSc, MSc, PhD
Imperial Chemical Industries (ICI) plc, Wilton, Middlesborough, Cleveland

SYNPOSIS   The feasibility of section pultrusion with thermoplastics is demonstrated.   Examples of tape, rod, slab, tube, angle and channel profiles have been produced using a range of reinforcing fibres and thermoplastic resins. The combination of carbon, glass and Kevlar 49 fibres with matrices such as nylon 6-6, polypropylene, polyethylene terephthalate (PET) and PEEK illustrates the versatility of the process.   The efficiency in wetting individual fibres with the viscous polymer melts is shown and this reflects in excellent mechanical properties. Effects of process variables such as line speed on impregnation are exemplified.   The advantages of these thermoplastic pultrusions both as products in their own right and as inserts in mouldings is explored.

1   INTRODUCTION

The pultrusion of glass fibre reinforced thermosets has been commercialised for many years.   It relies on penetrating densely packed fibre arrays with low viscosity monomer systems which are subsequently crosslinked in heated shaping dies. This reaction limits the line speed of the process to 1 m/min, although more recently the advent of methacrylate resins has increased productivity by a factor of three (1).

Thermoset pultrusions are undoubtedly structurally efficient in their utilisation of the fibre reinforcement but they have the disadvantage of a brittle and intractable matrix.   The combination of continuous reinforcing fibres with a tough linear chain thermoplastic to give sections which can be further manipulated, if required, by thermoforming is therefore attractive.

A review (2) of recent progress in thermoplastic impregnation would suggest that solution processing, solution presizing followed by melt coating, in situ polymerisation of monomer or prepolymer, cross-head extrusion coating, powder impregnation and fibre hybridisation could all lead to a viable thermoplastic pultrusion technology.

Thermoplastic preimpregnated tapes based on high loadings of continuous collimated carbon fibres fully wetted with a high performance polyether etherketone (PEEK) resin were introduced by ICI in 1982 as 'Aromatic Polymer Composites' (APC) (3). More recently a new range of long fibre injection moulding compounds 'VERTON' based on pultruded glass nylon 6-6 rod chopped to 10 mm lengths has been launched (4).

Modification of the current APC and 'VERTON' impregnation technology together with the design of water cooled forming and polishing dies has allowed us to successfully produce a variety of thermoplastic section pultrusions. The efficiency of the process in wetting individual fibres is crucially important when pultruded stock is to be used without further thermal processing.

2   EXPERIMENTAL

2.1   Materials

A range of reinforcements were available for impregnation and are listed in Table 1.

Polymers incorporated included nylon 6-6, polypropylene, polyethylene terephthalate (PET) and polyether ether ketone (PEEK) all obtained from ICI PLC.

2.2   Equipment and Process

The number of rovings and the size of the heated exit die from the impregnation unit are used to control the level of reinforcement in the section.

The molten pultrudate is shaped and subseqently rapidly cooled to form stable continuous sections.   The position of the forming dies is changed for different line speeds and also for different polymers. Pressure is essential to consolidate the section and hence retain the void free impregnation and good fibre wetting.

A series of water cooled forming dies were constructed to produce a variety of sections and examples are shown in Figure 1. A Betol tube exit die was modified to enable hollow sections to be formed.

Table 2 summarises the range of shapes and sizes of products made in this work. Some of the shapes are pictured in Figure 2. In the case of glass fibres up to 63 rovings of 4800 and 2400 tex were required to form the largest section.

In the course of the work line speeds were varied from 0.25 to 3 m/min and reinforcement levels from 40 to 70% by weight. It was found that line speeds had to be reduced with increase in section to guarantee good wetting and impregnation.

3 RESULTS

3.1 Microscopy

Pultruded sections were embedded in epoxy resin and polished by hand on wet grinding papers 220, 320, 500 and 1000. A final polish was given with 0.3 μm alumina paste on a Selvyt cloth before studying by reflected light microscopy.

Magnified sections of the pultrudate are shown in Figures 3 and 4. In all cases they indicate complete void free impregnation and excellent fibre wetting at high levels of reinforcement. The fibre distribution is good.

3.2 Effect of Reinforcement Fibre

Tensile properties of uniaxial beams (10 x 1.5 mm) with the various fibre types in nylon 6-6 polymer are shown in Table 3.

In the case of the carbon and glass fibres examination of the failed samples showed good wetting. For Kevlar 49 there was evidence of many loose fibres and fibre pull-out indicative of poorer wetting or a poorer interface although calculations indicated some 95% efficiency in modulus transfer. With these high efficiencies of fibre property utilisation the advantages of continuous fibres over short fibre reinforcement is obvious. Comparative flexural properties are presented in Table 4.

3.3 Effect of the Matrix

For the case of the OCF 429 XY glass flexure results with the various matrices are listed in Table 5. The flexural strength for polypropylene is low because of the onset of premature compressive failure as the matrix is unable to support the stiff fibres. Otherwise for the engineering polymers the fibre properties dominate.

This is also seen for carbon fibres with nylon 6-6 and PEEK in Table 6. For both polymers short beam shear testing on slabstock, 10 mm x 3 mm, gave a compressive yield failure rather than interlaminar cracking even with span to depth ratios lower than

(see Table 7). This indicates that the materials are clearly well consolidated, bonded and tough.

3.4 Effect of Glass Type and Finish

Using nylon 6-6 polymer a series of composites of 62% weight fraction were pultruded at a 0.5 m/min line speed with various glass fibre reinforcements.

Flexural properties are given in Table 8. Vetrotex RO99 is a direct roving, ie the filaments are all drawn together from the same bushing and are therefore equi-tensioned, and is preferred over the assembled variety from both the processing and the property points of view. It is interesting that although the direct rovings used here had a finish not designed to be specifically compatible with polyamides the physical characteristics of equi-tension and good 'wet-out' performance produced better mechanical performance than for example the polyamide compatible P388 assembled roving.

3.5 Effect of Line Speed

Figure 5 illustrates the variation in mechanical properties of sections pultruded at different line speeds using Vetrotex RO99 glass with nylon 6-6.

Flexural strength is found to fall off with speed while initial modulus is much less affected. Strength reduction is attributed to increases in void content and this, in part, is influenced by the efficiency of consolidation and cooling.

3.6 Properties at Elevated Temperature

A Du Pont 981 DMA apparatus was used to measure stiffness from RT to 300°C on uniaxial pultrusions of carbon, glass and Kevlar 49 in both nylon 6-6 and polypropylene. Samples were heated at 5°C/min. Modulus retention with temperature is plotted in Figures 6 and 7 by normalising the data to room temperature stiffness. There is a dramatic improvement with continuous fibre reinforcement especially in the case of carbon fibre reinforced nylon which offsets the fall in stiffness of the matrix associated with the T_g. The carbon-nylon composite retains some 70% of its modulus at 200°C and only falls off above 240°C.

3.7 In Mould Reinforcement

To demonstrate this approach pultrusions were inserted on opposite faces of simple moulded bars or sheets. These have proved compatible in that flexural modulus has improved and strength has been sustained. Data obtained on rectangular test bars of 145 x 12.5 x 6.4 mm reinforced with 10 x 1 mm section pultrusions on both faces are collated in Table 9. Process conditions were similar for both 'with' and 'without' reinforcement so that polymer properties should be constant.

The use of the inserts is also found to toughen the mouldings as demonstrated in Figure 8 for both filled and unreinforced polypropylene co-polymer. This shows the force/deflection output from an instrumented falling weight impact test (5). The energy absorbed in these tests is given in Table 10.

4   DISCUSSION

Pultrusion is basically a simple process that gives true continuous production of a reinforced plastic component which is extremely stiff in the longitudinal direction. Thermoplastic pultrusions offer a number of benefits over their thermoset counterparts mainly through their melt processing convenience and their toughness. Stock sections may be post formed off-line to suit applications and overwrapped if required at selected areas. The pultrudate may be used as a reinforcing member or insert in compression or injection moulding. The use of focussed high intensity heat sources allows selected tape sections to be further used in filament winding and tape laying.

The process for thermoplastics also broadens the range of available matrix materials from the traditional polyesters and epoxies to ones which may satisfy extremes in corrosion resistance, fire or high temperature requirements.

This work has concentrated solely on uniaxial fibre rovings and in the future combination with woven fabric or random mat to give some transverse reinforcement needs to be tackled. ICI has developed separately a woven form of APC-2 (6). Other essential work is required on the engineering side to improve consolidation, cooling and surface finish by better die design, and also to scale up the existing equipment.

5   CONCLUSIONS

The pultrusion of sections with thermoplastic matrices is now feasible based on advances in impregnation technology. When fully developed the process will offer a class of materials with a number of benefits over their thermoset counterparts such as greater versatility in processing, inherent matrix toughness and resistance to hostile environments.

REFERENCES

1   HOWARD, R.D. and SAYERS, D.R.   The development of new methacrylate resins for use in pultrusion. 40th Annual Conference, Reinforced Plastics/Composites Institute, SPI, Jan 28 - Feb 1, 1985 paper 2-A.

2   COGSWELL, F.N.   The mechanical properties of reinforced thermoplastics, edited by Clegg D.W. and Collyer A.A. Applied Science Publishers, 1984, Chapter on Continous fibre reinforced thermoplastics.

3   BELBIN, G.R. BREWSTER, I. COGSWELL, F.N. HEZZELL, D.J. and SWERDLOW, M.S.   Carbon fibre reinforced polyether etherketone. SAMPE Conference, Stresa, 1982.

4   CUFF, G.   Injection moulding of long fibre reinforced thermoplastics.   14th Reinforced Plastics Congress, Brighton, BPF, 1984, paper 19.

5   HOOLEY, C.J. and TURNER, S.   Mechanical testing of plastics.   Automotive Engineer 1979, 4, 48.

6   MEASURIA, U. and COGSWELL, F.N. aromatic polymer composites : broadening the range. 30th National SAMPE Symposium, Anaheim, 1985.

Table 1   Range of reinforcements

MANUFACTURER	CODE	TYPE	DIAMETER μm	FINISH
Carbon				
Courtaulds	XAS-N	12K	7	Epoxy
Hercules	AS-4	12K	7	Unsized
Kevlar 49				
Du Pont	T-968	5K 7890 decitex	11.9	None
'E'Glass				
Owens Corning	429 XY	Assembled 2400 tex	10	Silane thermoplastic compatible
Fibreglass	FGRE 5/2	Assembled 2400 tex	14	Polypropylene finish
Vetrotex	P388	Assembled 2400 tex	10	Polyamide finish
	RO99/P103	Direct 2400 tex 4800 tex	17	Polyester, vinyl ester, epoxy compatible. Good 'wet-out'

Table 2   Pultruded sections

RODS diameter mm	EXIT DIE mm	SPEED m/min	NO OF STRANDS	WT/METRE (g/m)
1.5	1.5	1.82	1 (2400 tex)	3.3
2.0	2.0	1.82	3 (2400 tex)	9.3
4.5	4.5	.46	12 (2400 tex)	24
6	6.0	.46	20 (2400 tex)	60
15	15	.3	42 (4800 tex)	311
SLABS mm				
6 x 0.3	1.5	1.82	1 (2400 tex)	3
12 x 0.45	2.0	1.82	3 (2400 tex)	9
10 x 1.5	4.5	.46	6 (2400 tex)	24
10 x 3	6	.46	20 (2400 tex)	60
47 x 4	15	.3	42 (4800 tex)	310
100 x 2.5	Slot	.3	42 (4800 tex) 21 (2400 tex)	436
TUBE				
OD – 6 mm ID – 5 mm		.3	15 (2400 tex)	50
OD – 9 mm ID – 5 mm		.3	12 (4800 tex) 3 (2400 tex)	90
ANGLE	Slot	.46	30 (2400 tex)	90
2.5 mm	7 x 4.5			

11 mm

CHANNEL	Slot			
	7 x 4.5	.3	20 (2400 tex)	61

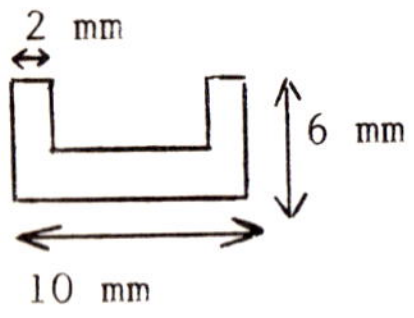

2 mm

6 mm

10 mm

Table 3   Effect of reinforcement fibre in nylon 6-6

FIBRE	WF %	UNIAXIAL MODULUS GN/m^2	% THEORETICAL MODULUS	TENSILE STRENGTH MN/m^2
Carbon Hercules AS-4	56	92.1	89	1180
Kevlar 49	40	43	95	951
'E' Glass Vetrotex RO99	68	32.7	91	874

Table 4   Flexural properties of nylon 6-6 with various reinforcements

SAMPLE	WF	FLEXURAL MODULUS GN/m^2	FLEXURAL STRENGTH MN/m^2
Unfilled	0*	2.8	105
Short Glass	33*	8.7	250
Short Glass	50*	12.0	320
Continuous Glass	68	31	664
Short Carbon	20+	16.8	290
Short Carbon	30+	20.0	355
Short Carbon	40+	23.5	420
Continuous Carbon	56	102	965

*ICI and LNP+ Data Sheets

Table 5   Effect of the matrix with glass fibres

MATRIX	WF %	FLEXURAL MODULUS GN/m^2	FLEXURAL STRENGTH MN/m^2
Polypropylene	65	20.1	173
PET	58	20.5	508
Nylon	63	20.9	459
PEEK	58	25.7	523

Table 6   Effect of the matrix with carbon fibres

FIBRE	MATRIX	FLEXURAL MODULUS GN/m^2	FLEXURAL STRENGTH MN/m^2
Hercules AS-4	Nylon 6-6	102.4	965
AS-4	PEEK	105.1	930

Table 7   Short beam shear strengths

SPAN : DEPTH RATIO	SHORT BEAM SHEAR STRENGTH MN/m^2	
	carbon / nylon 6-6	carbon / PEEK
5:1	79	55
4:1	90	66
3:1	96	71

Table 8   Effect of glass type and finish in nylon 6-6

	FLEXURAL MODULUS GN/m^2	% THEORETICAL MODULUS	FLEXURAL STRENGTH MN/m^2
OCF 429 XY	31.0	97	664
Fibreglass FGRE 5/2	28.9	90	836
Vetrotex P388	27.0	84	820
Vetrotex RO99/P103	31.2	97	885

Table 9   In mould reinforcement

SAMPLES	% GLASS w/w	FLEXURAL MODULUS GN/m^2
Polypropylene (PP) copolymer	0	0.9
+ pultrusion	49.4	11.6
30% short fibre reinforced PP	30	5.1
+ pultrusion	58.7	15.7
PET	0	2.4
+ pultrusion	29	13.4
30% short fibre reinforced PET	30	7.1
+ pultrusion	44	16.9
Nylon 6-6	0	2.8
+ pultrusion	20.2	7.8
30% short fibre reinforced nylon 6-6	30	7.2
+ pultrusion	40.6	9.8

Table 10   Impact energy absorption

| | IMPACT ENERGY (J) | |
	PEAK	TOTAL
Polypropylene copolymer	2.5	5.3
+ pultrusion	1.2	11.8
30% short fibre polypropylene	2.7	3.0
+ pultrusion	1.4	14.7

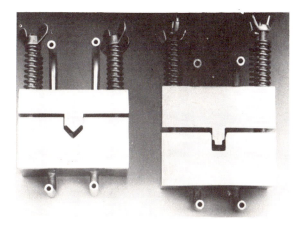

Fig 1   Water cooled forming dies

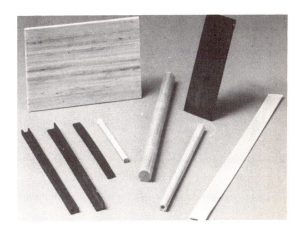

Fig 2   Pultruded sections

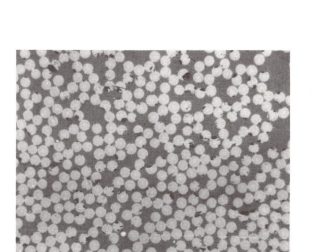

Fig 3   Microstructure of carbon fibre/PEEK pultrusion

Fig 4   Microstructure of glass fibre/polypropylene pultrusion

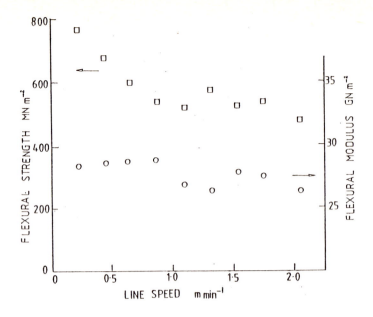

Fig 5   Effect of line speed on mechanical properties

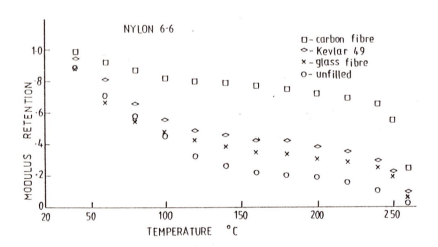

Fig 6   Modulus retention with temperature for nylon 6—6 pultrusions

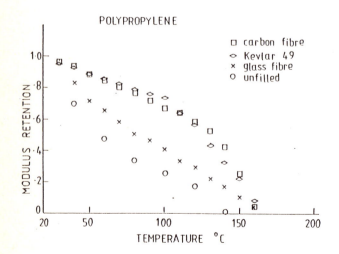

Fig 7   Modulus retention with temperature for polypropylene
pultrusions

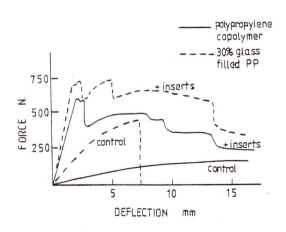

Fig 8   Impact results on in-mould reinforced polypropylene

C26/86

Flexible manufacturing of composite aerospace structures

D G JOHNSON, BSc
Department of Electronic Engineering, University of Hull
J J HILL, BSc, PhD, CEng, MIEE
Department of Engineering, Bristol Polytechnic

SYNOPSIS   This paper examines the application of an industrial robot to the assembly of satellite antenna dishes from carbon-fibre. A sensory gripper is described which uses vision sensing and force sensing to destack and then accurately align the pre-cut composite pieces onto the mould tool.

1. INTRODUCTION

Flexible automation should make the manufacture of composite aerospace structures more cost-effective by reducing some of the high labour costs traditionally associated with this technology. In collaboration with British Aerospace, this project seeks to develop a flexible manufacturing environment for the assembly of satellite antenna dishes in small batch sizes. This particular assembly encompasses many of the problems associated with automation in the industry.

British Aerospace manufacture a large range of aerospace structures , an increasing number of which involve carbon-fibre composites. Supplied in either rolls or sheets, the composite must first be cut to the desired shape and subsequently laid-up on a mould tool in a specified pattern. Backing paper protects the composite prior to use and this must be removed to allow the material to be fixed in place.

An analysis of assembly problems involving carbon-fibre revealed a number of tasks which could potentially be automated. These are :

1. Cutting carbon-fibre from sheets or rolls into specific shapes using data from a CADAM source.

2. Laying-up the carbon-fibre onto the mould tool.

3. Drilling the completed structure.

4. Acoustic testing of the completed structure.

In a practical system, in order to economically justify the robot, it would probably be necessary to combine two or more of the above tasks in a work cell. Each task presents problems from the point of view of automation. Cutting the carbon-fibre could be accomplished by a water-jet or reciprocating knife, but mounting the cutting tool on the robot would not be a trivial task. The lay-up of the composite must be done to a fairly high precision, to ensure accurate butt-joining of the composite. The flexible nature of the material may make it difficult to predict the exact position of the composite in relation to the robot gripper.

Some solutions to the drilling problem are already operating in industry, for example (1). Usually a template is used to guide the robot-held drill using a compliant mounting on the robot wrist. The problem here is the expense of the template, which although small for large batch sizes, still remains a significant manufacturing cost for small batch production. A system which obviates the need for such a template would offer significant cost savings.

Acoustic testing of the completed structures involves scanning a probe across the surface. Although special-purpose equipment is available to do this, a robot-based system would offer a more cost-effective solution for some applications.

The particular problem addressed by this paper is the lay-up. Labour costs associated with lay-up are a significant portion of the total product cost. This is especially true in aircraft manufacture where the number of layers of composite can be 50 or more.

2 THE ASSEMBLY PROBLEM

The lay-up problem is itself significant and accounts for the majority of the assembly time. Pre-cut pieces of carbon-fibre composite must be laid precisely on the mould tool, ensuring accurate butt-joining of adjacent profiles. The robot gripper must be designed to hold the profile and to apply it to the mould tool ensuring satisfactory joining to the previous piece. The curved nature of the surface presents problems in the design of the gripper. One solution is to have a gripper capable of moulding itself to the shape of the structure, but this must be rejected on the grounds of complexity. Alternatively, the gripper must be able to follow the contours of the surface, applying the profile during the course of its motion. This paper will concentrate on the lay-up of the antenna dish, but the gripper described is intended as general purpose and references to the aforementioned assembly serve only to illustrate the functional role of the gripper. The antenna dish assembly involves laying down four layers of composite, each layer consisting of a number of pie-segments of about 500mm in length. A central crown section is first laid using reference points on the side of the mould tool as a guide. The pie profiles of the first layer are then fitted ensuring that adjacent profiles constituting a layer butt against each other to an accuracy of less than 1mm down the length of the joint. Upon completion of the first layer, subsequent layers are applied in a similar manner to complete the lay-up.

One major problem in automating the lay-up is the wide variety of shapes and sizes of profiles typically encountered. To some extent this problem could be alleviated through a 'design for assembly' approach (2), but the problems of handling large and small pieces are likely to remain. Solutions to the problem of handing different sizes of profiles have included adjustable span grippers (3) or very large grippers with a number of individually controlled air chambers. This particular problem is not considered in this paper because a single size of profile is used (with the exception of the central crown piece), allowing attention to be directed at the more fundamental problem of sensory assembly. The gripper to be described in the next section has been designed to handle the 500mm pie segments, and manipulate them accurately. It has been installed on a Puma 560 industrial robot.

3. GRIPPER CONSTRUCTION

The profiles are supported by means of six suction cups on the underside of the gripper. These cups are connected through rubber tubing directly to a vacuum pump. Sufficient vacuum is generated so that with five cups open to the atmosphere the single remaining cup can still hold the profile. Furthermore, the profiles can slide across the cups without too much trouble.

Because the material is flexible and needs to be positioned accurately, it was found necessary to include sensing on the gripper. Small errors in the profiles position can be accommodated by feeding error information forward to adjust locations in the work-cell. Two different sensors are used, namely vision sensing and force sensing. Visual sensing is used to monitor profile position on the gripper and to provide a quantitative quality check on the final butt-joint. This is achieved by two 256-element CCD (Charge Coupled Device) linear array cameras mounted on opposite ends of the gripper. The use of gripper-mounted linear-array cameras for robot vision has been previously reported (4). In this application, the marginal amount of extra information obtainable from area array cameras does not justify the considerable extra expense. A single line of pixels provides all the necessary information to determine the edge position of the profile. A mirror is used in conjunction with each camera to enable two slits, each of width 25mm to be viewed. The available resolution is therefore 10 pixels/mm which is considered adequate for this application. Each slit requires its own source of illumination for the camera, and this is provided by high-power light emitting diodes (LEDs), two for each slit. The peak sensitivity of the CCD devices corresponds with the peak emitted wavelength of the LED. The LEDs are only turned on during the exposure period, allowing high intensities to be used and minimising power consumption when the vision system is not required. When both LEDs are illuminated, the observed intensity distribution across the scene is quite uniform and the varying sensitivity of the CCD sensing elements provides the major cause of non-uniformity. Each LED can, however, be controlled individually to achieve side illuminated images. This is important in the derivation of surface profile information to detect overlap, see (5).

Prior to use, the vision system runs through a calibration phase in which the optimum exposure time of the CCD sensor is found. The average pixel intensity of a captured image is determined and used to increase or decrease the exposure period, producing an average intensity lying within a specified band. A maximum of eight images are captured to achieve this and the whole process takes less than half a second. The optimization of the exposure time is necessary to accommodate different coloured backing paper.

The video date from the CCD sensors is digitized into a 6 bit word by the vision system and stored in memory for subsequent image processing. The high contrast images are easily processed to determine edge positions and gap sizes.

The second sensing mechanism is force sensing which is used to allow the robot to follow the curved surface of the tool. The robot-mounting pod is attached to the active surface of the gripper through a pivot, allowing the gripper to rotate about the pod. A potentiometer is mounted on the shaft and can be used to provide the necessary information for surface following. Changes in the potentiometer reading can be used to measure changes in surface height over known linear motions of the gripper. This allows the gripper to follow the surface contours of the mould-tool with no prior knowledge of the surface. For regular surface shapes, for example the parabolic section of the satellite antenna dish, a world model can be rapidly derived .

Overall control of the system is performed by an IBM PC/XT microcomputer, which communicates information between the robot and the sensors using a master-slave protocol. The overall control program is written in the C programming language and resides on the IBM. Information to the Puma 560 robot is sent down the robots serial communication line, which is normally reserved for the terminal.

4. THE GRIPPER IN USE

The pre-cut composite profiles are stacked and are collected by the gripper from the dispenser. Small errors in the position of the profile on the gripper are detected using the vision sensors and corrections are fed forward to adjust the lay-up coordinates. Once on the gripper, the lower backing paper is removed. This is currently done manually, although an automated method is under development. The profile is then offered to the mould-tool and sufficient pressure applied to ensure a bond between one end of the profile and the mould-tool. One problem here is the inconsistent tac (stickiness) of the composite which can present difficulties. The gripper is moved along the tool surface using information from the shaft resolver to move the gripper up or down and hence following the surface. Because the profile is now fastened at one end, it slides across the surface of the gripper during the lay-up. The rubber suction cups maintain sufficient vacuum to hold the profile but still permit this sliding to occur. The gripper is shown doing the lay-up in figure 1 .

5. EXPERIMENTAL RESULTS

Early results performed without sensory feedback were unsatisfactory, with gap sizes and overlaps of up to 6mm at the perimeter of the dish. However, with sensory feedback the requirement to achieve less than 1mm gap size or overlap has been met. A number of problems have become apparent which may necessitate modifications and enhancements to the gripper. One of these problems is undesired slip of the composite across the mould tool during the lay-up. This is caused by poor initial adhesion of the profile to the tool and is especially prevalent whenever the tac of the composite is low. In manual assembly this would be compensated for by applying a greater force or by applying the same force over a greater area. With a robot, the curved surface makes it potentially difficult to apply an even contact force over a large area. A flexible gripper may be one solution.

6. CONCLUSIONS

The objectives of the current research programme are to investigate the problems of handling and lay-up of composite carbon fibre structures through a particular case study. A prototype gripper has been described which allows profiles to be handled and accurately

positioned on a mould tool. The vision and force sensing on the gripper allow errors in the position of the profile on the gripper and the position of the tool to be tolerated. Work on this problem will be continued and extended to other related assemblies.

ACKNOWLEDGEMENTS

The partial support of the Science and Engineering Research Council in providing support for this project is gratefully acknowledged. The help of British Aerospace in providing raw materials, basic equipment and financial support is also gratefully appreciated. The authors acknowledge helpful discussions with Mr.J.Dickinson and Mr.M.Higgs of British Aerospace and Professor Alan Pugh of the Department of Electronic Engineering, University of Hull. The help of John Hodgson in constructing the gripper is also appreciated.

REFERENCES

1.   Wright J., "Compliance enhances positioning accuracy", Eureka, May 1984.

2.   Boothroyd,G. and Dewhurst,P., "Design for assembly handbook", Department of Mechanical Engineering , University of Massachusetts, Amherst, Massachusetts.

3.   Bubeck,K.B., "Advanced composite material Handling application", 13th International Symposium on Industrial Robots, Chicago, April 17-21 1983.

4.   Johnson,D.G. and Hill,J.J., "A modular linear array camera for robot vision", Digital systems for industrial automation, Vol.2.3, June 1984.

5.   Johnson,D.G. and Hill,J.J., "A sensory gripper for composite handling", 4th International conference on robot vision and sensory control, London, 1984.

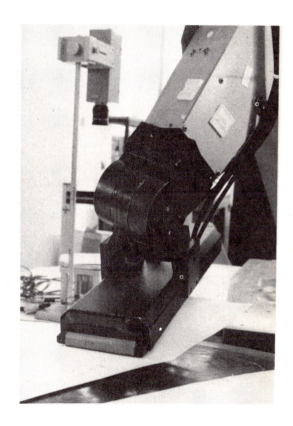

Fig 1   Gripper laying down a piece of composite onto mould tool

C29/86

Computerized numerical control filament winding for complex shapes

M J OWEN, BSc, MS, PhD, CEng, FIMechE, FPRI, **V MIDDLETON**, BSc, PhD, CEng, MIMechE,
D G ELLIMAN, BSc, PhD, MBCS, **H D REES**, MPhil, CEng, MIMechE, MIProdE,
K L EDWARDS, BEng, AMIMechE, **K W YOUNG**, BSc and **N WEATHERBY**, BTech, MSc
Department of Mechanical Engineering, The University of Nottingham

SYNOPSIS   Filament winding is an old established mechanised process for the production of fibre
reinforced plastics (FRP).   In the simplest concept a payout eye is mechanically traversed relative
to a mandrel whilst laying a band of parallel fibres in a helical pattern on the surface of a
rotating mandrel.   Products from pipe to large rocket motor cases have been made by this method for
many years.   With the advent of CNC it has become possible to devise and control filament winding
machines which can produce more complex movements of the payout eye relative to the mandrel and hence
more complex shapes than had previously been thought possible.   The paper describes via examples the
practical advantages and limitations of filament winding for complex shapes.   The examples highlight
the need to reduce the labour intensive activity of part-program development.   Part-program
development can be expedited by CAD.   The filament winding machine has been interfaced to a larger
computer where it is already possible to model axisymmetric surfaces and to develop and display
geodesic winding patterns.   The geodesic winding patterns can be used to generate corresponding part-
programs and transmitted to the filament winding machine controller.   The authors are about to
interface the system to a larger computer capable of handling a more sophisticated surface modeller.
In the foreseeable future it will be possible to link into finite element stress analysis programs.
Meanwhile, design data are being developed for filament wound materials.

1   INTRODUCTION

Filament-winding   is   an   old   established
mechanised process for the production of fibre-
reinforced plastics (FRPs).   In its simplest
form, the filament winding machine is similar to
a lathe, the payout-eye replacing the cutting
tool and a cylindrical mandrel the workpiece.
As the mandrel rotates, resin impregnated filaments are pulled onto its surface via the
payout-eye which is mechanically traversed
parallel to the mandrel axis producing a helical
pattern.   On completion of the winding, the
resin is cured and the mandrel removed to leave
a FRP shell.   To wind filaments on a mandrel of
more complex shape, a machine is required which
is capable of producing more complex movements
of the payout-eye relative to the mandrel.
Recently   multi-axis   computerised   numerical
control (CNC) winding machines have become
available.

The filament winding programme at the
University of Nottingham (1) makes use of a
prototype CNC 5-axis machine by Pultrex Ltd.
(Fig 1).   The overall programme is split into
four SERC funded programmes and two industrially
funded programmes.   Commissioning the machine,
development   of   winding   and   programming
techniques for simple and complex shapes,
identification of mandrel systems for complex
shapes and methods of incorporating inserts into
filament structures formed the core of the
programme.   In parallel with this work has taken
place on the development of a CAD/CAM system for
axi-symmetric shapes, the definition of material
properties,   and   non-geodesic   winding.

Industrially funded work on an automotive component and an aerospace component keep the above
programmes directed towards practical applications.   The primary objective of the overall
programme is to provide a facility for designing
and manufacturing complex shaped filament wound
components.   This paper describes the methods
available for developing software for winding
complex shapes.   The main advantages and limitations of the methods are highlighted.   The main
limitations of the prototype winding machine and
some practical winding considerations are also
described.

2   THE WINDING MACHINE

The winding machine is a Pultrex MODWIND 1S-5NC
with a single spindle and 5 independently
numerically controlled axes of movement (mandrel
rotation, three mutually perpendicular carriage
movements and a rotary movement of the payout
arm).   The machine can accept mandrels up to 3
metres long by 1 metre diameter.   The maximum
speed of a linear axis is 40 m/min and the
maximum speed of a rotary axis is 100 r/min.
The machine is controlled by a Bosch micro 8 CNC
machine tool controller.   The controller uses
standard CNC commands to position the axes as
well as a customer parametric cycle (CPC)
facility enabling arithmetical and logical
operations to be carried out.   A teach-in
facility is also provided.   The majority of the
CNC commands are redundant because the payout-
eye is not coincident with the surface of the
mandrel, but moves relative to it in laying
fibres on a mandrel.

3   THE DEVELOPMENT OF SOFTWARE FOR WINDING

Extensive software development was undertaken based on winding techniques, theoretical analysis and part-programs for a variety of simple and complex shapes. Four separate routes were identified for developing part programs for the Pultrex winding machine (see Fig 2). The advantages and disadvantages of these are as follows.

Method 1   Theoretical (CPC)

- uses a small amount of memory (only the part-program and the current data being operated on need to be stored)

- generalised programs are possible

- high accuracy and repeatability

- long program development period

- mathematical analysis can become very complicated

- limited by the computing power of the micro 8

- high initial program development cost, but cost per component is reduced for a large number of components

- high level of skill required

Method 2   Graphical

- relatively short program development period

- involves storage of large amounts of co-ordinate data

- generalised programs are not possible

- accuracy limited to accuracy of drawing

- high costs if a large number of different shaped components are involved

- high level of skill required.

Method 3   Experimental (teach-in)

- relatively short program development period

- low accuracy

- suitable for very complicated shapes, especially when non-geodesic paths are involved

- high cost if a large number of different shaped components are involved

- generalised programs are not possible

- involves storage of large amounts of co-ordinate data

- winding machine is required for the program development period

- low level of skill is required.

Method 4   CAD/CAM

- very long system development period

- developed system able to generate co-ordinate data rapidly

- able to generate part-programs without involving winding machine, i.e. does not interrupt winding process

- requires another computer

- high accuracy and repeatability

- high initial development cost

- able to optimise mandrel shape and winding pattern quickly and without any need to manufacture hardware

- system designed to be used by operators with minimal computational experience.

This paper concentrates on the CPC method with an introduction to the CAD/CAM approach. Other papers will be devoted to the other work.

4   CNC PROGRAMMING

4.1   CPC method

The micro 8 controller has only a limited memory capacity (see Section 5) which severely limits the size of part-programs. Part-programs comprising of long lists of axis co-ordinate data derived from graphical and experimental methods rapidly exceed the memory capacity. However part-programs based on the CPC facility (2) generate axis co-ordinate data during the winding operation. Memory is only required for the program and the current co-ordinate data.

It was decided to build-up a library of structured part programs for a series of simple and complex shapes. This has enabled compound shapes to be wound by combining both the shapes and their part-programs in a modular manner. Also, using parametric input to describe the characteristic features of a winding, a variety of different sized parts can be wound from just a few programs. The shapes chosen for the library were circular cylinders, prisms, hemispheres, cones, branched cylinders and flat plates. These were then combined to produce shapes such as spheres, dome-ended cylinders, cone-ended cylinders and tee-pieces.

The first step in winding a filament pattern on a particular shape is to derive a mathematical relationship for the payout-eye path. From the mathematical relationship an algorithm can be devised and a CPC part-program created. Stable (slip-free) filament paths are usually based upon geodesics (3). A geodesic curve on a developable surface for example becomes a straight line on the development. This provides the basis for mathematically deriving the filament path on developable surfaces such as cylinders and cones. Fig 3 shows the geodesic filament paths for a cylinder and a cone. A sphere though does not have a developable surface, but the geodesic paths are great circles.

A general procedure was identified for mathematically deriving the payout-eye path for any shape:

(a) Derive the equations for the filament path based on geodesic lines or other criteria.

(b) Derive the equations of the tangents to the filament path.

(c) Obtain the equations of the payout-eye path from the intersection of the tangents with an arbitrary and imaginary surface surrounding the mandrel (4).

In (c), the surrounding surface may be chosen to be any shape, although regular shapes simplify the mathematical analysis for determining the payout-eye path. For a surface which is not a scaled-up version of the mandrel shape, the payout-eye clearance will be variable. It is therefore important to have a minimum clearance for the enveloping surface to prevent interaction of the payout-eye and mandrel. Fig 4 shows an example of the payout-eye path for winding the hemispherical dome of a dome-ended cylinder. For axi-symmetric shapes, the application of the payout-eye path equations to a part-program can be simplified because the mandrel can be rotated about its longitudinal axis and the payout eye need only be traversed parallel to the mandrel axis (X movement).

For branched shapes, rotation of the mandrel merely complicates the payout-eye motion. It is therefore convenient to fix the mandrel in a convenient orientation and wrap the branch by payout-eye movement (see Fig 5) using the X, Y and Z axes. Clearly, as the eye moves around the branched surface, there is a tendency to twist the filaments. To overcome this problem, another degree of freedom is available, a rotation of the payout-eye (B axis). Associated with the rotation of the payout-eye is a need to rotate the whole filament package to prevent it from twisting. The solution to the problem was to mount the package on the carriage of the winding machine. The package was mounted on a frame, such that the frame could rotate simultaneously with the payout arm. This modification has been achieved for pre-preg materials, but for wet-winding a specially sealed resin tank would be necessary.

The link between theory and the creation of a part-program is a suitable algorithm. The algorithm is particularly important for using CPC to enable the most efficient use of its limited capabilities. The CPC program, like most high level computer languages, can be structured into subroutines based upon the algorithm adopted. Basic winding patterns are held in the subroutines, while a main program controls the scheduling and quantity of winding patterns (subroutine calls). For a different combination of winding patterns it is only necessary to use a different main program.

For filament winding, the majority of general CNC commands ('G' functions) of a machine tool controller are irrelevant. There is no need for drilling, milling and tool offset commands. Fortunately, the CPC facility enables mathematical statements and branching to be carried out. This overcomes the deficiency caused by the loss of the 'G' functions.

CPC programs are activated by special cycle call statements. In the cycle call, numerical values can be assigned to several parameters. This has provided a convenient means of describing winding patterns and operating winding programs. The parameter addresses are used to represent the parameters which uniquely define the physical characteristics of a mandrel and winding pattern, while the corresponding numerical values of the parameter addresses allow a particular mandrel and winding pattern to be specified. This overcomes the usual problem with conventional CNC programs which are specific and constructed from finite numerical values (co-ordinate data). The CPC programs introduce flexibility and generality and with parameter input allow mandrels and winding patterns to be specified easily and quickly. The main (scheduling) program for a series of CPC programs simply has to contain a list of CPC call statements. Structuring programs in a modular manner is also useful. This allows both programs and mandrel shapes to be combined to allow the winding of compound shapes.

4.2 Teach-in method

For very complicated shapes, in the absence of a CAD/CAM system, the teach-in system is invaluable. The teach-in facility allows the payout-eye to be moved manually over the mandrel surface in order to lay a filament over an experimentally determined filament path and digitise co-ordinate values simultaneously. The experimental filament path is obtained by stretching string over the mandrel surface. The method has the drawback that a large amount of co-ordinate data storage is required. The motion of the axes during a teach-in operation is obtained via a joystick remote control but is nevertheless rather slow and cumbersome. The teach-in method would be improved if the axes could be moved by hand.

It is important during a teach-in operation to digitise points at smaller intervals on the mandrel surface where the radius of curvature is small and only extend the interval where the radius of curvature is large. This prevents the payout-eye short-cutting and interacting with the mandrel as the machine employs linear interpolation between digitised points.

4.3 CAD/CAM of filament winding

A CAD/CAM system for filament winding offers the advantage of being able to achieve an optimum design without any need to manufacture hardware or occupy a winding machine. This not only saves time and reduces costs but also allows the winding machine to be used without interruption. The proposed CAD/CAM system is shown in Fig 6. The CAD system developed enables:

(a) The mandrel surface to be defined.

(b) Allows filament paths to be laid upon the mandrel's surface.

(c) Defines the filament paths, directions, and laminate thickness.

(d) Computes the payout-eye paths required to wind the filament paths.

At present the system does not include a strength or deflection calculation capability. A data base of material properties is in preparation. The CAM system developed converts the payout-eye paths into a part program for a CNC winding machine via a post processor.

The system has been developed for axisymmetric shapes only. The mandrel is designed as a series of digitised points forming a two-dimensional curve which is then rotated about an axis to form the mandrel surface. To enable filaments to be laid on the mandrel, it is converted into a series of triangular surface patches (5). These are stored as a database containing surface normals and node co-ordinates. The size of the triangular patches can be varied to model areas of complex curvature. The smallest practicable number of patches is desirable to reduce the number of calculations and memory usage. The filament path is calculated as a series of steps in which the filaments cross one boundary and one patch. The filament path thus developed is stored as a sequence of points and tangent direction cosines. This is converted into a payout-eye path by fixing the position of the payout-eye on another surface of revolution surrounding the mandrel and finding where the filament tangent intersects this surface. The payout-eye path is also stored as a series of co-ordinates.

The machine part program is then developed by resolving the payout-eye motion into radial and axial components (X and Z machine axes in Fig 1). By rotating the mandrel by the correct amount between successive positions, the part program is generated.

5 WINDING MACHINE LIMITATIONS

The main limitations have been identified as:

(a) The operating speed of the winding machine is limited because the micro 8 is a point-to-point control system and lacks a look-ahead capability, i.e. the controller does not look beyond the next point programmed. This results in high accelerations and decelerations between movements. The problem is compounded by the mass of the carriage and a lack of rigidity in the carriage supports.

(b) The small memory capacity of the micro 8 (40K bytes) limits the size of programs. Large programs, usually created via teach-in, have to be divided and transferred from an external memory device during a winding operation. Unfortunately, a direct numerical control (DNC) interface supplied for automatic transfer of programs between the micro 8 and external devices uses non-standard protocol and requires both hardware and software development. Fortunately, a link via the teletype (TTY) port has been established for program transfer, but manual operation is only possible. This causes interruptions in the winding process.

(c) The machine has a bulky Y axis and B axis assembly and limited Y axis travel. This seriously restricts the mobility of the payout-eye for winding branched shapes.

Subsequent machines manufactured by Pultrex Ltd feature a continuous path controller with a look-ahead capability, a larger memory capacity of 128K bytes and a DNC interface which uses standard protocol (a Fanuc System 6M). The machines also feature stiffer carriage supports.

However, like the micro 8 or any other machine tool controller, there is no software available for filament-winding. It must be developed by the user or machine builder.

6 SOME PRACTICAL WINDING CONSIDERATIONS

(a) Once the payout-eye path has been established for winding a basic pattern on a shape it is necessary to advance the pattern to enable it to be repeated. The aim being to produce full coverage of the shape. Two methods have been identified for regular shapes:

(i) Indexing - the mandrel is rotated with the payout-eye stationary at the end of a repeating pattern, such that the next pattern will be laid alongside the previous one.

(ii) Progression - the mandrel is progressively advanced during the pattern such that when the pattern is finished it is in the correct position to wind the next one.

For branched shapes pattern indexing and progression are not possible. Coverage can only be obtained by devising a set of unique patterns. Full coverage may not be possible and the proportion of each pattern will depend on the applied loads.

(b) Non uniform cross-section mandrels, such as a cone, lead to changes in winding angle when a geodesic path is followed. The winding angle increases from a reference angle at the base to 90° at a point depending on the height and angle of the cone before returning to the base again at a negative reference angle. Associated with the increase in winding angle is an increase in thickness of the wound composite skin. This presents a major problem with winding multiple layers in that subsequent layers are wound upon a surface profile quite different to the original surface profile of the mandrel.

(c) Tapering or converging surfaces e.g. cones, spheres, etc., provide a returnable geodesic path. For constant sections, a non-geodesic path is required to reverse the helix e.g. helix reversal at the end of an open-ended cylinder. Instantaneous payout-eye reversal limits the winding angle to high values only. Gradual payout-eye reversal considerably extends the winding angle range, but increases the wound cylinder length and build-up at the ends. The built-up regions are normally removed after winding.

(d) Helically wound filaments under tension on a cylinder are self compacting. The lateral face of a prism however does not provide a compacting force normal to the surface. Therefore, after winding it is necessary to apply an external force normal to the flat prism faces to induce consolidation. To impart consolidation during the winding operation it is necessary to provide a slightly convex profile to the prism's faces. Also, to minimise damage to the filaments it is necessary to use blend radii on the edges of the prism mandrel. Concave surface profiles are also inadvisable because filaments tend to span the concave portions and loose contact with the surface.

(e) In order to make the best use of the filaments strength it is necessary to arrange the filaments along the principal stress directions. In general, this requires non-geodesic paths. To wind non-geodesic filament paths, several techniques have been identified, but not fully explored in this programme.

   (i) Tacky pre-preg materials

   (ii) Rapid UV light curing resins

   (iii) Pegs on the mandrel's surface

   (iv) Grooves in the mandrel's surface

   (v) Welding thermoplastic materials (6).

7 CONCLUSIONS

The use of a standard machine tool CNC system for controlling a filament winding machine renders the majority of the controller software redundant. The development of part-programs via mathematical, graphical and experimental methods are labour intensive, occupy the winding machine and require the use of mandrels. A CAD/CAM system however, allows winding patterns to be developed quickly, does not involve the winding machine, and does not require the use of mandrels. Only after a suitable mandrel shape and winding pattern have been found is the program transferred to the winding machine. It is envisaged that part-program development via mathematical, graphical and experimental routes will find more use in relatively simple shapes or in short term projects while CAD/CAM will find more use in winding very complicated shapes or in long term projects.

In order to provide a capability for winding more complex shapes the winding programme at the University of Nottingham has been extended with the addition of a further three programmes.

An anthropomorphic robot is to be interfaced to the machine to assist in the placement of fibres and inserts. The CAD/CAM work is being extended to complex non-axisymmetric shapes. A purpose made multi-axis controller for filament winding has been designed and is under development.

ACKNOWLEDGEMENTS

The authors are grateful for the financial support provided by the Polymer Engineering Directorate of SERC, Ford Motor Complany Ltd and British Aerospace Plc.

REFERENCES

(1) OWEN, M. J., MIDDLETON, V., ELLIMAN, D. G., REES, H. D., EDWARDS, K. L., YOUNG, K. W. and WEATHERBY, N. Filament winding for complex shapes in FRP. SERC Polymer Engineering Directorate Review Conf., Loughborough University, 1985.

(2) EDWARDS, K. L. Advanced CNC filament winding of complex FRP shapes. PhD Thesis, Mechanical Engineering Department, Nottingham University, 1985.

(3) LYUSTERNIK, L. A. Shortest paths - variational problems. Popular Lectures in Mathematics, 13, Pergammon Press Ltd., 1964.

(4) HILLE, E. A. Filament winding technique - methods for producing geometric complicated structural parts. Dissertation Thesis, Technical University of Aachen, W. Germany, 1981.

(5) BRAUN, W. Problems of determining and preparing control data for non-rotationally symmetrical winding bodies explained using the geodetically wound elbow. Proc. 3rd Tech. Conf. SAMPE, European Chapter, Engineering with Composites, v2, Paper 17, 1983.

(6) BOWEN, D. H. Filament winding in the 1980's. Fibre Reinforced Composites - '84 Conf., University of Liverpool, 1984.

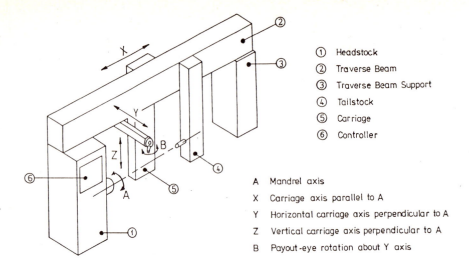

① Headstock
② Traverse Beam
③ Traverse Beam Support
④ Tailstock
⑤ Carriage
⑥ Controller

A   Mandrel axis
X   Carriage axis parallel to A
Y   Horizontal carriage axis perpendicular to A
Z   Vertical carriage axis perpendicular to A
B   Payout-eye rotation about Y axis

Fig 1   Schematic of the filament winding machine

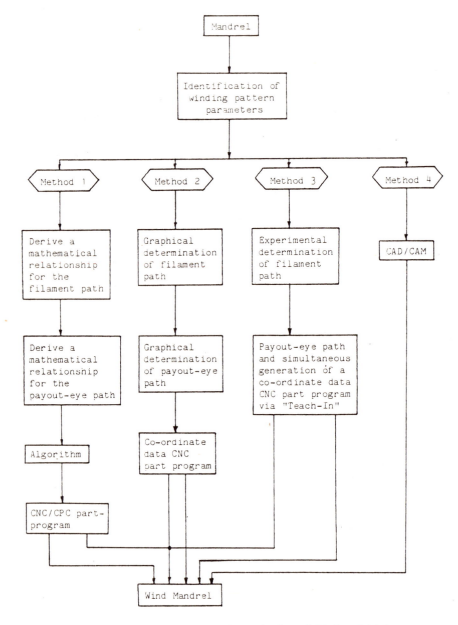

Fig 2   Block diagram showing the methods available for obtaining
computerized numerical control (CNC) part-programs for the
Pultrex winding machine

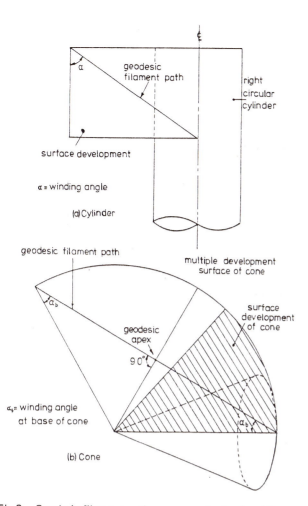

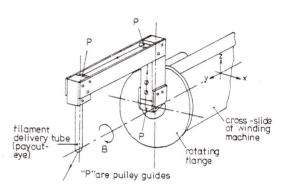

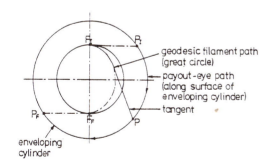

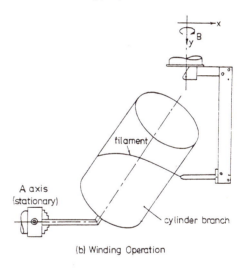

Fig 3   Geodesic filament paths on example developable surfaces

(a) Cylinder

α = winding angle

geodesic filament path

right circular cylinder

surface development

geodesic filament path

multiple development surface of cone

geodesic apex

90°

surface development of cone

α_b = winding angle at base of cone

(b) Cone

filament delivery tube (payout-eye)

cross-slide of winding machine

rotating flange

"P"are pulley guides

(a) Payout Arm

A axis (stationary)

filament

cylinder branch

(b) Winding Operation

Fig 5   The winding of a branch

geodesic filament path (great circle)

payout-eye path (along surface of enveloping cylinder)

tangent

enveloping cylinder

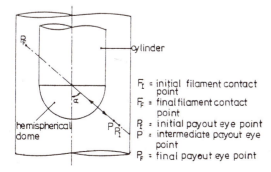

cylinder

F_I = initial filament contact point

F_F = final filament contact point

P_I = initial payout eye point

P = intermediate payout eye point

P_F = final payout eye point

hemispherical dome

Fig 4   Intersection of tangents with an enveloping cylinder to form the payout-eye path for winding the dome end of a hemispherical dome-ended cylinder

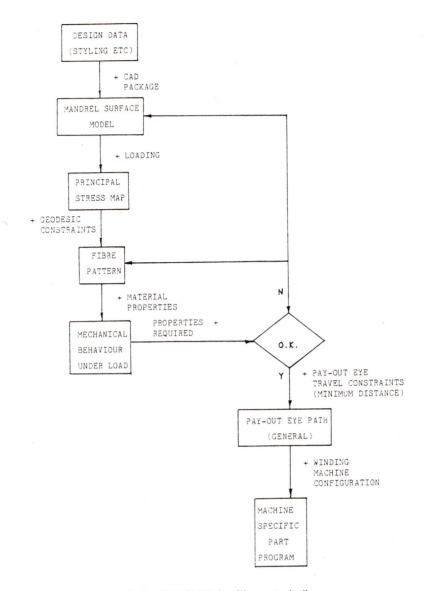

Fig 6   CAD/CAM for filament winding

C27/86

Mechanized manufacture of composite main rotor blade spars

D HOLT, PhD, CEng, FIEE, FIProdE, FIIM, FBIM
Westland plc, Yeovil, Somerset

SYNOPSIS This paper describes the latest developments in a long term programme directed at the integrated design and manufacture of composite main rotor blades at Westland Helicopters.

The availability of high performance composite materials enabled the design of rotor blades with considerably enhanced performance, relative to the standard metal blades. The use of 'pre-preg' materials called for entirely new methods of manufacture to be devised, and mechanisation of the process was essential if performance and quality standards were to be guaranteed. A close liaison between design and manufacturing engineers has resulted in a series of mechanisation projects culminating in an f.m.s cell for the manufacture of main rotor blade spars.

1 INTRODUCTION

In the mid-1970s a major UK Demonstrator Programme was launched with the aim of introducing composites technology into rotor blade design and manufacture. The first component to be developed was a tail rotor blade based on Sea King Helicopter tail rotor blade geometry but using a highly cambered section to substantially improve performance. An identical metal blade was produced for comparison purposes which clearly demonstrated two of the major advantages of composite materials; the economic reproducible production of sections difficult or impossible to manufacture using metals, and the greatly improved fatigue life of composite rotor blades. The necessity of manufacturing rotor blades in metal-tooling was also demonstrated.

The next project was the production of main rotor blades, and again the Sea King design was chosen for the initial programme. Considerable design skills were necessary to deal with the twisted high aspect ratio main rotor blades of the Sea King, and this requirement stimulated the use of very powerful computer-aided, 3-dimensional design techniques. The system employed by Westland is SYSTRID which is a product of the Battelle Institute of Geneva.

The first essential step in the process is to define blade sections at a number of spanwise locations and then to 'patch' these sections radially using a variety of curve types. Grids employing around 2000 data points allow intermediate sections to be checked, drawn by CAD etc as required.

The most pressing problem in blade design is invariably time. The blade must be aerodynamically defined, drawn, stressed (utilising aeroelastic information) and if necessary redrawn, restressed etc. Substantial headway has been made in developing interlinking routines in which data can be freely transmitted from one specialised area to another using common formats and coordinate systems. In particular,

the use of SYSTRID has promoted a natural merging of the lofting and jig-and-tool functions. The direct computer linking of the tooling planner to the NC machine has resulted in a substantial reduction in timescales, a further bonus being the greatly reduced finishing times off the machines. Manufacture of our own tooling has also been beneficial when introducing the minor modifications invariably required as a result of early practical processing trials.

Requirements for main rotor blades include high fatigue strength, dynamic tuning in three directions, complex aerodynamic shape, weight balance and damage tolerance. These parameters are enhanced by the application of composite materials, however additional measures need to be taken to achieve adequate performance in erosion protection and lightning strike protection.

There has been steady development of resins and fibres leading to the high performance pre-pregs and wet-lay-up systems available today. The Royal Aircraft Establishment Farnborough has played a vital role in the development and characterisation of composite systems whilst Westland has taken advantage of the systems on offer both in wet-lay-up (structural components) and pre-preg field.

The pre-pregs used at Westland cure at 120°C and are based on Ciba Geigy Fibredux 913, a high toughness, low bleed resin. The basic fibres in use are XAS Carbon and E glass. The range of materials includes:-

Unidirectional Pre-preg
CIBA-GEIGY FIBREDUX - CARBON 913C XAS 5
                       HYBRID 913C XAS/913GE10
                       GLASS 913G E10

Woven Pre-preg
CIBA-GEIGY FIBREDUX - 913G 7781
BROCHER VICOTEX   - 145/40/759

Mixtures of these materials are needed to achieve the stiffness and strength requirements.

A conventional metal blade comprises some 250 easily handled and assembled components, and therefore blade manufacture has been a labour intensive process. Composite blades on the other hand comprise thousands of components which are of the order of .02mm thick, up to 10 metres long and 30cms wide, are tacky, and easily damaged. Packs of up to 100 plies are required, and it is extremely difficult to achieve the necessary quality by manual techniques. There is a clear need to mechanise the process, but there is little experience in this area and there are few experienced machine tool suppliers. It was necessary for Westland to develop the skills and equipment in-house.

2   MECHANISATION

Mechanisation of main rotor blade manufacture was introduced to the Sea King blade project. The composite blade was designed as a retrofit for the existing metal blade and is aimed ultimately to provide a blade life of 50000 hours. This will have a significant effect on the life cycle costs of the helicopter; for example a 2-4% reduction for civil operations. Composites materials have enabled the designer to tailor the characteristics to control mass distribution and stiffness and ensure dynamic compatibility. The composite blades are interchangeable with metal blades as a set. The Sea King blade (figure 1) uses pre-impregnated unidirectional carbon and glass epoxy composites in the spar, and woven pre-preg glass on an epoxy honeycomb core in the trailing edge. The blade incorporates heater mats and a titanium erosion shield bonded to the spar.

The spar is a D-cross-section component and is assembled from four packs of plies on an inflating mandrell. The assembly is placed in a metal tool for consolidation and curing. The four packs and approximate dimensions are;

        two side walls 10m x 10cms
            back wall 10m x 6cms
            nose moulding 10m x 5cms

The blade is thicker at the root end, and therefore short infills are interleaved with the sidewall plies. During development work it was possible to manufacture 2m test specimens by hand lay-up, but extending this technique to 10m lengths created wrinkling and bubbling problems. The problems were overcome by developing a machine which formed laminates of composite material of constant width, and of full blade length.

Reels of resin-impregnated material from the supplier were slit to the required width and mounted on a tape-laying head which traversed a flat bedway some 11m long. The head was designed to remove the protective front film and backing paper, and to lay the material onto the bedway using a rubber roller as the laying and consolidating medium. The material was sheared at the end of each ply using a rotating knife mounted at the end of the lay-up surface.

Ten plies of material were laid to form a very good quality laminate; this then had to be transferred manually from the bedway into the curved mould tools. Six persons were required for this task; any slight twisting or bending of the laminate caused detrimental permanent wrinkling of the plies.

The side walls were thus built up in the tool by hand from 10 ply-machine laid laminates with hand laid root-end infills.

The back wall and nose mouldings are tape laid and pressed to shape (also de-bulking the laminate) before being transferred to the tool. The inner and outer $\pm45°$ wraps were hand laid to the mandrell and tool respectively before the four packs were assembled and the tool closed. The inflating mandrell provided consolidation pressure during the curing cycle and was withdrawn from the completed spar before fitting the trailing edge assembly and completing the finishing operations.

This experience of mechanisation of spar manufacture was invaluable in encouraging us to look towards a system where there would be no hand laying of materials nor manual transfer of sub-assemblies thus further improving quality and reducing cost.

3   AUTOMATION

The fundamental problem for automation is to lay material directly into the twisting and tapering tool cavities. The breakthrough was the introduction of brush laying. Consequently an extensive programme has been undertaken to develop the necessary material packaging technologies and brush laying techniques to enable the full range of spar wall materials (including unidirectional warp sheet, $\pm45°$ cross-plied and $\pm45°$ woven) to be laid at will.

This breakthrough coincided with the need to manufacture blades for the Lynx-W30 series of helicopters which are a more ambitious application of composite materials. The blade was developed under the British Experimental Rotor Programme - BERP, and involves utilising the properties of composites to give aerodynamic improvements which result in a substantial thrust increase for a given rotor system weight. The blade (figure 2) has advanced aerofoil sections and a novel planform tip shape which would not be possible in metal manufacture at economic cost.

The blade construction (figure 3) is in three major components; the spar which carries the great majority of the loads, the trailing edge which primarily completes the aerofoil shapes, and the erosion shield. The spar is of the modular 'D' type with walls constructed principally of pre-preg glass and carbon unidirectional fibre hybrid sandwich with inner and outer wraps of 45° unidirectional carbon and one outer layer of glass fibre; it contains an acrylic foam core of $4.4lb/ft^3$ density which is used in the moulding process to provide consolidating pressure (typically 100 psi).

Although carbon fibre reinforced plastic is a very efficient structural material it is avoided in thick sections due to explosive failure mechanisms. The use of glass/carbon hybrid has been found to be an attractive compromise having an acceptably high modulus and fibre-dominated failure mechanisms similar to those of glass fibre reinforced plastic.

The automation of the spar manufacture was undertaken in several stages. First an inexpensive risk-reduction investigation was undertaken to demonstrate the feasibility of brush laying. The major problems were shown to be:-

- the achievement of alignment accuracy
- the avoidance of wrinkling

There were three design compromises:-

- the spar inner and outer wraps must be discontinuous
- all spar side walls must be parallel
- infills must be rectangular.

Continued development and close liaison between design and manufacturing engineers together with the construction of a technology proving rig showed that the design compromises were acceptable, and that the technical difficulties could be overcome.

The proving rig accommodates a single reel of slit-to-width material. The pre-preg is unreeled, the facing and backing films are removed, and the material is laid directly in the mould tool at the rate of one layer per pass. The rig drives itself along rails mounted on the edges of the mould tool. Consolidation of the irregular surfaced stack is by brushes which accommodate to the tool shapes. Careful control of tension and material guidance systems is necessary to achieve wrinkle-free stacks.

The third stage of automation resulted from an analysis which showed that the major justification for automated blade manufacture came not only from tape-laying, but also from the elimination of manual intervention in the tape-laying process at material changeovers or reel replacement times. The result was the development of the Westland multi-cassette MkIV tape layer (figure 4). The tape layer was itself designed on the CADAM system and is a computer controlled production standard machine of which two are in full production and more are under construction.

The machine is 16 metres long and will accept a range of mould tools for blade spars. The laying head traverses the length of the rig on X-axis slideways and selects one of eight material cassettes, each cassette coming complete with its own set of brushes. The eight cassettes are parked in magazines, three at one end of the machine and five at the other and permit a selection of material types and widths (26mm to 400mm) to be loaded. Any faults in the material or resin system are 'flagged' on the cassettes by arrangement with the supplier. The information of fault type and position, for each cassette of material is entered into the computer system. A detector

on each magazine recognises flags eleven metres ahead of the fault. Since the maximum length of any ply is less than eleven metres, the fault will not occur in the ply currently being laid. The machine has data on the length of the next ply to be laid and can determine if there is sufficient material before a fault to complete a lay. In cases of insufficient material for a ply, the laying head traverses to a scrap-laying area where it will lay material until the fault is passed.

A computer based weighing station is included in the machine which monitors the mass of material laid in the tool at any time during lay-up and compares with target masses. The targets are always the minimum such that correction material must be added, never removed. This is necessary since Westland blades, in common with those of other manufacturers, are made to a weight tolerance of $\pm2\%$.

The machine has a sophisticated material guidance system with an accuracy of ±0.75mm, and will lay plies of varying length at any point in the mould tool.

4 FLEXIBLE MANUFACTURING CELL

The EH101 helicopter project is an Anglo-Italian collaborative programme designed from inception to produce a long-range, large capacity helicopter to meet the future requirements of naval, military and commercial markets. A comparison with the Sea King illustrates the increased performance of the rotor system.

	All-up-Weight	Rotor Diameter
Sea King	21,400 lbs	62 feet
EH101	31,500 lbs	61 feet

Each helicopter has a five-bladed main rotor system, but that of the EH101 features an advanced aerodynamic shape with advanced planform tip region similar to that of the Lynx. After extensive investigation of alternative designs it was decided that joints in composite materials were unacceptable in this blade and therefore the composite plies must include the necessary root end and tip shaping. This decision ruled out a development of the MkIV tape layer and demanded a rethink of the entire manufacturing concept. The shapes to be laid are illustrated in figure 5, and after considerable debate it was determined that the optimum technique for the long spar components would be to profile the pre-preg during the laying process and to profile and place by robot the tip shapes. The experience of material guidance, tension control, and machine control gained during the development and operation of the MkIV machine gave us confidence of the soundness of these decisions. An advantage of this approach is that only two raw materials - hybrid glass/carbon, and $\pm45°$ carbon pre-preg - in a single width are required.

Several techniques for profiling have been examined including water-jet, laser, and ultrasonic knife. Following extensive tests, the ultrasonic knife has been selected for the cell. Given a suitably flat baseplate and vacuum holddown system it is possible to profile the pre-preg material without cutting into the backing paper at 40m/min.

The cell is illustrated in figure 6, and consists of three main elements:-

- a profiling and tape placement machine for full length plies

- an X-Y profiling machine with an ultrasonic knife head to cut out plies for infill packs

- a transfer system to present and lay-in infill packs to the tool on-demand in a predetermined sequence.

Computer integrated manufacturing (CIM) capabilities today will ensure that the design information from SYSTRID and CADAM linked to the f.m.s cell will eliminate many of the problems of the manufacture of complex assemblies by ensuring that each ply is produced 'just-in-time' for the assembly to proceed uninterrupted. Quality is built-in.

Clearly this f.m.s cell can cope with any blade design which fits within its operating envelop and can readily be extended to accommodate a wider range of materials.

Computer control of the cell is essential; a particularly interesting feature is the material guidance system for the full length plies. All the edges of the ply are cut, and therefore datums are difficult to establish. It is proposed to use artificial vision in the control system to align the material. This technology is familiar to us from work on sheet metal manufacturing automation. The placement machine incorporates:-

- 14 electrical servo-controlled axes
- 6 pneumatic control axes
- 8 video cameras.

It is planned to commission the spar manufacturing f.m.s cell in 1986.

6   ACKNOWLEDGEMENTS

The author is grateful to his many colleagues at Westland Helicopters who have contributed material for this paper.

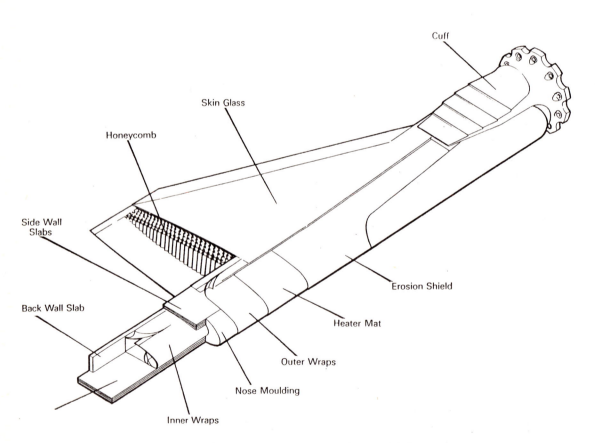

Fig 1   Sea King composite main rotor blade

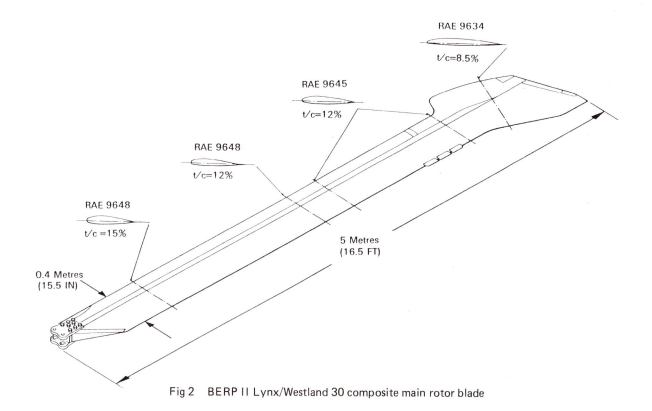

Fig 2   BERP II Lynx/Westland 30 composite main rotor blade

RAE 9634
t/c=8.5%

RAE 9645
t/c=12%

RAE 9648
t/c=12%

RAE 9648
t/c =15%

5 Metres
(16.5 FT)

0.4 Metres
(15.5 IN)

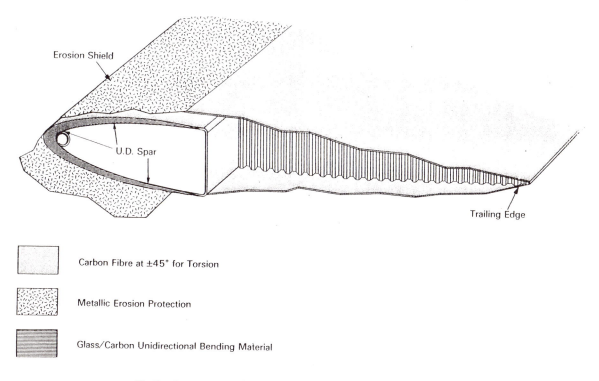

Erosion Shield

U.D. Spar

Trailing Edge

Carbon Fibre at ±45° for Torsion

Metallic Erosion Protection

Glass/Carbon Unidirectional Bending Material

Fig 3   Lynx composite main rotor blade — general section

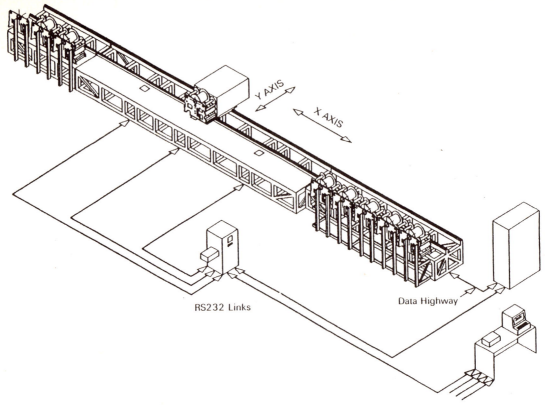

Fig 4   Mark 4 tapelayer

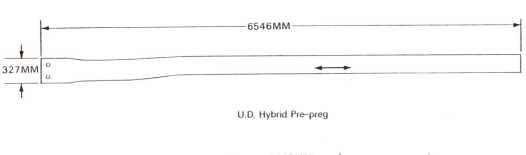

U.D. Hybrid Pre–preg

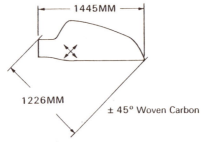

Fig 5   Typical 'pre-preg' shapes for EH101 blade

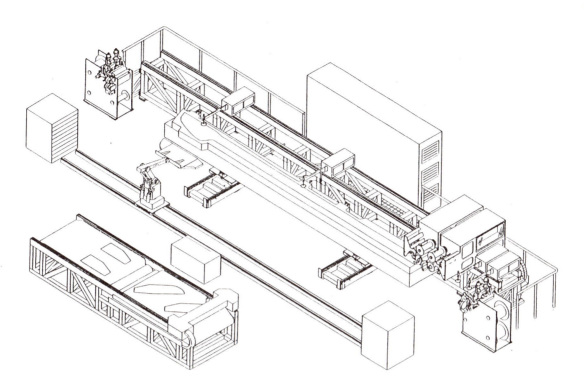

Fig 6   EH101 spar production cell

C41/86

Large scale manufacture of three-dimensional woven preforms

S TEMPLE, BA
Cambridge Consultants Limited, Cambridge

1.   BACKGROUND

Stephen Temple is a consultant engineer with Cambridge Consultants Limited who do contract research and development for clients. For some time now, CCL have been carrying out internally funded seed developments on composite manufacturing techniques. Whilst Stephen has been applying his extensive knowledge of the textile industry to the problem of making suitable fibre preforms, the analytical group within CCL have been working on structural analysis and failure mechanisms in composites.

This paper describes some developments in the manufacture of 3D woven fibre preforms, with the emphasis laid upon the machinery to make such components. However, it cannot be emphasised too highly that a successful development of a particular composite product depends upon cooperation between the structural designer and the manufacturing technology expert.

1.1   3 Dimensional Weaving

The interest in 3D weaving goes back to 1971 when Rolls Royce had difficulties with laminated composites. It seemed obvious that the problem of delamination which they encountered in their fan blades, could be solved by making a preform with fibres running in three principle directions - a 3D woven fibre reinforced composite. However, seeing such a solution is not quite the same as actually designing a piece of machinery to produce that composite. It was some years before any ideas of substance were developed. CCL's next point of contact with composite manufacture was when asked to advise on methods being used to make large 3D woven blocks of material by hand-specifically, to improve the productivity, since it took something like a man year to produce a block whose dimensions were of the order of 1 foot by 1 foot by 2 foot. This introduced the author to the practicalities of producing a 3D block and led him to examine a great quantity of patent literature relating to all sorts of 3D weaving techniques.

At about this time, his father, who is archaeologist, introduced him to a form of weaving which has largely died out but which is still practised in some of the remoter parts of Afghanistan. It is called Tablet Weaving, and uses a mechanism for opening the warp threads ready for the weft to be passed through which is quite different from that of a modern loom. What is interesting about Tablet Weaving is that it intrinsically produces a 3D woven structure, typically consisting of two layers, generally in different colours, with the colours passing from one layer to the other so as to form a pattern on each side of the material. The resulting weaves are singularly attractive but they also implicitly contain a solution to the problem of making a 3 dimensional material on a 2 dimensional loom.

2.   PROS AND CONS OF 3D PREFORMS

Before describing the development of manufacturing techniques for 3D weaves, it is worth looking at the possible advantages and disadvantages of using a 3D preform. It is obvious that the best use of a high strength fibre is to support a stress in one principle direction only, and to lay as much fibre in that particular direction as you can possibly press into a given space. But such a material is going to suffer from a very low strength indeed in any direction other than that in which the fibres lie. To compromise by laying fibres in criss-cross directions results in a dilution of the proportion of fibre in any one of those directions. Figure 1 shows this effect. It is assumed that we have three principle directions, X, Y and Z, where the proportions of yarn in the X and Y directions are the same and different from that in the Z direction. This assumption is made because it is a natural consequence of the type of weaving technology brought to bear on this problem. The extreme right of the graph shows the unidirectional composite with the all the material in the Z direction and the extreme left of the graph shows a bidirectional composite with equal amounts of fibre in the X and Y directions. As these proportions are altered so the nett proportion of yarn in any one direction changes, and somewhere in the middle you get an isotropic material, where all three proportions are the same. Because of simple geometry, it is inevitable that spaces are left between the yarns and these reduce, not only the proportion of fibre in any one direction, but the total volume fraction of fibre in the whole block. There is a minimum volume of fibre which occurs in the isotropic case and is about 30%. It is possible to compare the strength to weight ratio of the proportion of fibre in any principle direction with that of other well known materials. For this purpose, it is assumed that the fibre used is carbon, that each yarn within the 3D structure has the same density as a conventional pultrusion, and that there is no wasted space other than that which is absolutely necessary between the yarns. On this basis are marked the strength to weight ratios of Titanium and Steel for a rough comparison with what you might achieve with the fibre reinforced composite.

2.1 Delamination Resistance

The main aim of a 3D fibre reinforced composite is to prevent the notorious delamination that occurs in more normal methods of manufacture. Delamination in itself can be caused by using a beam of relatively thick proportions wherein stresses are induced transverse to the direction of the beam as well as along the beam, when it is bent, rather like bending a pad of paper and seeing the leaves separate. These secondary stresses are usually quite small by comparison with the principle stress, and it is assumed that the effect of making a 3D woven composite would be to carry these secondary stresses in a very small proportion of fibre placed in the appropriate directions. Such a construction would lie somewhere close to the extremes of the graph, giving a high strength in one or two directions, with good delamination resistance in the others.

2.2 Damage Resistance

A particular manifestation of delamination occurs when damage has been done to the composite before it is subjected to a load. This damage can either be as a result of poor quality control in the manufacture, resulting in air bubbles or other inclusions within the matrix, or it can be as a result of what is often euphemistically referred to as 'run-way stones' – that is to say, stones thrown up by the wheels of an aircraft to hit the underside of the wing. More likely is accidental damage arising from spanners dropped during maintenance. Whatever the source of damage it can result in buckling of the fibres on the compression side of a beam, with consequential delamination occurring at the high stress concentration where the damage ends. It is this mode of failure which causes very sudden collapse of composites when their working stress is exceeded and which can reduce that strength by very large factors indeed. A 3D composite should resist this type of failure by effectively providing crack stoppers at frequent intervals all over the material.

2.3 Work of Fracture

Another major problem with composites is that, on failure, they absorb very little energy unlike metals which deform plastically once pushed beyond their elastic limit and absorb a great deal of energy in the process. It has been found by practical tests that the work of fracture of 3D woven composites is very much higher than that of simpler laminated constructions. This, obviously, is again a function of the crack stoppers provided by all the criss-crossing fibres. Every time a crack starts, it comes to a point where a crossing fibre must be broken before the crack can propagate any further. The behaviour of a 3D woven composite during failure is much more like that of the plastic behaviour of a metal although, of course, there is no chance of recovery after the damage has been done.

2.4 Product Manufacture

One of the major problems with composites is the difficulty with which a specific product can be made. In general, for more conventional materials, the engineer has a very wide range of manufacturing techniques on which to draw, many of those techiques depending to a large extent upon the plasticity of metal, either at normal working temperatures or when heated in the region of its melting point. Since fibre reinforced composites do not possess plasticity we have to develop techniques which produce the appropriate arrangement of fibres in the final shape of the product itself. As the fibres are extremely fine these techniques all tend to be slow and very often labour intensive. A 3D woven preform can be arranged in such a way as to allow it to be shaped after it has been formed and also to include quite complex shaping at its manufacturing stage. On conventional textile machines this results in a much faster conversion of the yarn into material with reduced handling in order to produce a finished product from the basic preform.

2.5 Fibre Orientations

However, a drawback of 3D weaving is that the orientation of the fibres is dictated by the structure of the weaving machine rather than the desires of the structural engineer. For 3D pre-forms to be of major structural benefit it is necessary for the textile machinery designer to work in very close liaison with the composite structure designer. It is often going to be the case that the demands of the structure and the demands of the manufacturing process are incompatible.

2.6 Fibre Damage in Weaving

A disadvantage of conventional textile machinery when applied to fibre reinforced composites is that all the mechanical manipulations which that machinery is going to subject the yarn to are going to result in some degree of damage to the fibre. The fibres have a very low strain to break compared to most conventional textile yarns, and consequently care is needed when leading the yarns around tight corners and through guides in general. However, modern developments in fibres, with higher strain to break relationships, and developments in sizing techniques so as to hold the yarns together during the weaving process, have made it possible to produce quite complicated preforms with only very minor damage to the yarns.

3. WEAVING TECHNIQUES

There is a rich variety of textile processes which can be brought to bear on the problem of making fibre reinforced composites.

3.1 By Hand

The simplest way of making a 3D fibre preform is to put all the fibres in all the positions that you want by hand - one at a time. This is labour intensive but has yielded great success for a number of applications where cost is not a major consideration. In general the hand process is assisted by a piece of machinery that is referred to as a loom but more in the sense of a wiring loom than that of a conventional textile machine. Usually an array of wires is set up, those wires representing, let us say, the Z direction of the yarns in the block. In the X and Y directions, long hollow needles are used to pass between the wires and lay in a single thread at a time. These layers of X and Y threads are compacted from time to time by using a large number of thin strips of metal and beating down on a set of X Y layers so as to pack them as tightly as possible. After a few months, this process will yield a block consisting of X and Y direction yarns with Z direction wires. This is then taken to a separate operation where each Z wire is stripped out, drawing through behind it a yarn. After a few more months a block is produced consisting wholly of fibre, generally with quite a high volume fraction and ready for impregnation.

3.1.1   Semi-automatic Versions

A great deal of ingenuity has gone into trying to automate or partly automate this manual process. In a conventional loom the Z wires or yarns most closely correspond to those of the warp. The X and Y yarns correspond to the weft of a conventional loom, except, of course, that there are two wefts at right angles to each other. In a conventional loom the warp threads are separated by mechanisms called heddles to provide a relatively large gap for a shuttle or other device carrying a yarn to pass across the width of the machine. (Figure 2). Obviously in the 3D case this opening and separation has to occur both in the X direction, (say the width of the machine), and in the Y direction (say the height of the machine): and providing heddle mechanisms which will carry out this two dimensional opening is obviously a taxing mechanical engineering problem. Nevertheless, some attempts have been made at it with varying degrees of success. Generally the reliability of such machines is low by comparison with that of conventional textile machines and the rate of production is also extremely low as well.

3.2   Tablets

Tablet Weaving is an ancient method of making a fabric. It differs from conventional looms by using tablets to produce the hole or shed through which the weft thread is passed. In the normal form, the tablets have four holes, four warp threads go through each tablet, and a set of such tablets then forms the entire warp. To make the weave the tablets are rotated so that openings between the different warp threads are produced sequentially. To make a pattern the tablets are rotated first forwards and then backwards passing any desired warp thread from the top layer of the fabric down to the bottom and back again. In consequence you get a material with a very rich diversity of pattern, but also with substantial thickness. Continuing to rotate the tablets in one direction only, a material is produced in which the warp threads zig zag continuously backwards and forwards from the top surface to the bottom. In neither the warp threads nor the weft threads is there any of the crimp that is normally present in a conventional weave, so the yarns run straight, without any deviation,

to and from the extremes of the fabric in all directions. A diagram of the tablet weave and loom is shown in Figure 3. By using tablets with more holes it is possible to produce quite thick weaves and it's principally this technqiue which has been used at CCL for making samples of 3D woven fabric. Notice the clever trick whereby two directions in the material (the ±45° directions) are generated by one set of yarns only. Effectively three dimensions have been produced from only two starting, and it is this that is the key to high volume production using fairly ordinary machines.

3.2.1   Tablet Weaves

There are a great variety of weaves which you can produce from tablets; for instance the warp threads do not need to reverse immediately when they reach the surface, they can float, as it's known, over several weft yarns, producing effectively a skinned fabric. Another major variation on a tablet weave is to alternate the zig zagging ±45° warp yarns with sets of straight warp yarns in which the opening for the weft thread is produced by a more conventional heddle arrangement. This in turn produces yet another set of structural properties in the fabric.

3.2.2   Conventional Looms

A major disadvantage of tablet weaving is that the groups of warp yarns twist as the tablets are rotated continuously. It is only possible, therefore, to produce a limited length of fabric on a tablet loom. Furthermore, the tablets do damage to the threads because they slide along them as well as lifting them up and down. Both these problems can be solved by using a conventional loom to carry out a tablet weave.

Whilst we have introduced this weave structure via a tablet loom, it is only a piece of topology and that same topology can be produced on any arrangement of loom that you like. It is known that such weaves have been produced in the past on conventional looms where we have to use one heddle, that is to say one warp carrier, for every hole in the tablets. This means that the thickness of the material one can make is limited according to the structure of the loom and, in general, narrow looms allow more heddles per warp station than wide looms, so that if you want a very thick cloth it must be narrow, and if you want a wide one, then it's going to be thinner. Furthermore, of course, both the thickness of the material and the speed with which it's made will depend upon the thickness of the individual yarns used. It is therefore difficult to give any general idea of how fast one might produce these materials on a conventional loom. However, to take a particular example, a material of the order of 10mm thick and 1m wide could be produced on a loom running at about 300 weft insertions per minute. To round figures, such a material would be produced at the rate of 100mm per minute. Whilst conceding that it is not possible to produce a really large block of the sort described as having been made by hand for about a man year of effort, nonetheless, the loom just quoted would produce the equivalent weight of fabric in about 90 minutes.

4. BRAIDING

Woven structure fabrics can also be produced on braiding machines. These look very much like a maypole, with a set of carriers holding bobbins of thread and running in and out of each other whilst going around the circle in just the fashion of a maypole dance. A classic way of weaving fabric is to use a maypole braider and then slit the resulting tube along a helical line - opened out flat this gives a conventional woven structure to the fabric. This technique can be extended to produce the type of weaves described as being made on tablet looms, with certain slight differences.

4.1 Tubes With Axial Yarns

On an automatic braiding machine the bobbins run around a track guided by gear wheels. A fixed yarn may be passed through the axis of each gear wheel so that the braided or woven yarns are wrapped around stationary yarns that run along the axis of the tube. This produces a tube with a substantial wall thickness having a proportion of material running straight along its axis and another proportion of material braided in a criss cross pattern around the circumference. There are actually two sets of braided yarns, one set going in a right hand helix and the other in a left hand helix. Using a small number of carriers produces helices of relatively fine pitch and you get a structure which in cross section is as nearly as possible identical to that of a tablet weave. However, extending the number of carriers (and by the way the rate of production of the machine), develops a larger pitch to the helices and a fabric that has a rather complex structure, with the axial yarns being described as 0° and the woven yarns being ±45° (or any other angle), but also zig zagging through the wall thickness of the material. By altering the proportions of the yarns in the various directions we can achieve a tube with high bending strength along its axis, a good hoop strength if it is to take pressure, or a good torsional strength. Figure 4 shows the structure of this weave.

4.2 Solid Braids

Normal braiding produces a tube but can be adapted to produce a solid section - in particular there are square braids which produce a very stiff beam. To produce the patterns that are described on a square basis it is necessary that the braid itself is actually slightly out of square, with mutually prime numbers of threads along the two axis of a rectangle. However, this can be quite a small effect. As far as is known it has not been done, but it would clearly be possible by extension of this idea to produce shaped sections such as I-beams. Furthermore, a flat or rectangular section produced on a braiding machine can be shaped afterwards to form an open section such as a channel.

4.3 Octohedral

One final variation on the theme of braiding is to go back to square one and leave out the axial yarns, but nonetheless produce a 3D braid of the same structure as described above. This has a curious property because we have now produced an octohedral cell to the weave, that is to say, like a cube but on its corner rather than on one flat face. Trial pieces of such a fabric have been produced and they have the benefit, from the point of product manufacture, that they can be distorted to a very high degree. In textile terminology it has very good 'drape'. Such a weave can be formed over a hemispherical object. Obviously it would not have the stiffness in any one direction that one would expect of a more conventional structure, but nonetheless it represents a material with a very much higher stiffness than one can get, for example, from short staple injection moulding compounds. The ability to deform it in a plastic fashion suggests the opportunity to do deep draws and produce shaped objects like crash helmets in a single shot.

5. VARIATIONS

So far, the materials described have been uniform, and designed for the fastest possible production. However, all the techniques described are capable of producing much greater variety by adding special features to them. It is in this direction that the manufacturing technology of 3D woven textiles would have a substantial impact on the cost of making composite products. In general, of course, it is true to say that if you add variation and complexity to a piece of machinery you are going to slow it down. So far, however, for all the specific cases looked at, the raw production rates of a loom are so high that they are almost embarassing. To produce special products such as springs for cars or tennis racquets or any other highly shaped component would, in most cases, justify the reduced production rate of a loom that could make a more nearly finished preform.

5.1 Shaped Tablet Weaves

As indicated earlier, the basic tablet loom is a rather inflexible machine from the point of view of producing composite structures. Although an elegant means of patterning the surface of the fabric, the consequential alterations in the internal structure are probably unacceptable from the point of view of structural materials. However, when producing a tablet weave on a conventional loom, the opportunities for producing variations are quite substantial. It is possible to make ribs in both warp and weft directions, although the structure of these would be different from each other. In the warp direction it is obviously possible to have regions of warp in which there are a different number of component threads from other parts of the warp. This will produce variations in the thickness across the width of the material and continuously down its length. Associated with each such variation there would have to be a small weft insertion to provide the weft threads for the rib itself. These devices would come from conventional ribbon loom technology and be added to the basic broad loom. In the weft direction there is a more interesting possibility. The tablet weave can be made with any number of weft threads through the depth of the fabric. There is an opportunity, therefore, to produce a material which, let us say, has 18 or so layers through the thickness but which then splits into two layers of 9 threads which later rejoin to a single thickness again, producing a fabric with a 'button hole' through its width. Furthermore, using an adaptation common in the textile world for materials such as velvet, the two halves of the split can be woven to different lengths, so that it is possible to produce an upstanding rib on one side of the fabric only, or a lug which can be cut to form a means of fixing further items to the basic weave itself. As with warp ribs it is possible that these weft ribs can be less than the full width of the fabric, albeit at the expense of requiring separate weft insertions to carry out the special parts of the weave (Figure 5). In general, with all such features, there is good continuity between the upstanding feature and the main body of the fabric because the warp threads run straight through from one part to the other. Consequently all such features built onto a 3D preform are going to resist both shear and peel forces very much better than the equivalent added on items in a laminated form of construction.

5.2   Tubes

Braiding machines are as flexible as weaving machines, requiring less complex mechanisms to produce substantial variations. The two principle variations possible are in the diameter of the tube and in the components around ' circumference of it.

5.2.1   Diameter - Pipes, Pressure Vessels

Varying the diameter of a braided tube is almost trivially simple. The diameter is related to the rate at which the material is drawn off from the braiding machine. Changes in the speed of draw off will alter the diameter over a very large range indeed. Of course, one is not actually varying the total number of threads at any point in the fabric but rather the density of the woven threads as compared to the axial ones. Typically it is possible to produce a variation in diameter greater than 4:1 over very short distances indeed. An obvious possible application for this technique would be in the manufacture of discrete lengths of pipe. Enlargements could be built at intervals onto the braided tube, so that when the pipe is cut, there is a larger diameter socket into which to fit an adjoining length. A rather more extreme change in diameter would make it possible to produce pressure vessels with small diameter fittings at the ends to accept closures and pressure regulators. A particular technique would be to use a thin walled metal liner in order to produce an impervious layer. This liner would be preformed to a sort of sausage shape, and then the braid made around it. This is very similar to the winding technology used for pressure vessels, but with potentially much higher rates of production and a better integrity of the structure of the fibre composite.

Returning for a moment to pipes, an interesting problem is that of making a corner piece for pipework. Recent experiments have shown that material can be drawn off from the braiding machine differentially on one side as compared to the other. As long as this is done fairly close to the point of braiding then the axial threads can still be drawn through the braided ones causing the material to be curved along its axis.

5.2.2   Axial Variations - I-Beams

It was mentioned previously that it was possible to produce solid braids of a rectangular form. It is also possible to design machinery for more complex shapes such as I-beams. Of particular interest here is the possibility of making the web of the I-beam of different character from the flanges. Generally speaking one wants a large proportion of axial material in the flanges to carry the principle stresses of bending, but in the web there are ±45° stresses arising because of the substantial depth of the beam. In this area it would be desirable to reduce the number of axial yarns and to use only the braided yarns to carry the secondary stresses. It may also be desirable for the braided yarns

in this case to be a different material from the axial ones. For example, a combination of glass braided yarns and carbon axial yarns in the flanges would seem to produce something close to an ideal I-beam construction. Any similar variations in content around the circumference are possible and have been demonstrated at CCL.

6.   IMPREGNATION

In general, impregnation of composites is a non-trivial problem and it is recognised that by producing fairly well packed solid preforms it is not made easier. However, modern developments in resin technology are producing results which are certainly consistent with the idea of a fairly solid preform that is moulded into a final product in a matter of minutes rather than a matter of hours.

6.1   Vacuum

All the samples that have been made so far at CCL have been vacuum impregnated using relatively low viscosity epoxy resisn. As far as actually filling the weave is concerned these techniques are entirely effective and quite quick, but of course the resins themselves are somewhat slow, implying rather long cure cycles for production composites. Precuring before final moulding is obviously a possible way of speeding up this process.

6.2   Resin Injection

However, it is thought that the technology showing most promise for the manufacture of finished components is resin injection. The modern very fast resin injection compounds also have a low viscosity, and for high production rates the cost of the moulds implied by resin injection would be acceptable given the fast cycle times that can be produced. Preforms with the appropriate features would be produced as described, drawn or otherwise shaped to fit the mould, and then injected.

This promises a very simple and automatic approach to composite manufacture on a large scale.

6.3 Future

CCL are currently developing some new techniques for impregnation which would further improve the speed of production. These cannot be disclosed at the moment but they are aimed at high volume production with short moulding cycle times.

7. SUMMARY

Fibre reinforced composites lack both the diversity of manufacturing processes and the well rounded properties which metals enjoy.

Only in the aerospace industry has 3D weaving been used to any great effect – and there only for the improved structural properties and in spite of manufacturing difficulty. It is now apparent that such structures can be produced cost effectively on suitably designed and developed textile machines. By making preforms to the final shape of the product in a single automatic operation, it becomes possible to add 3D weaving to the suite of manufacturing options for less esoteric products. In doing so, development effort applied to the manufacturing technology will yield less labour intensive processes with higher product performance than can be achieved by many existing techniques.

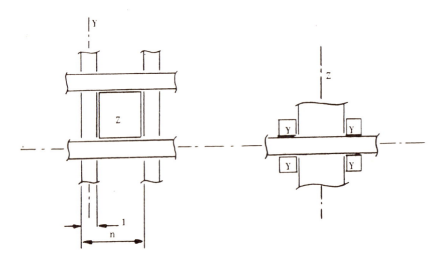

Section Through X-Y Plane   Section Through Z-Y Plane

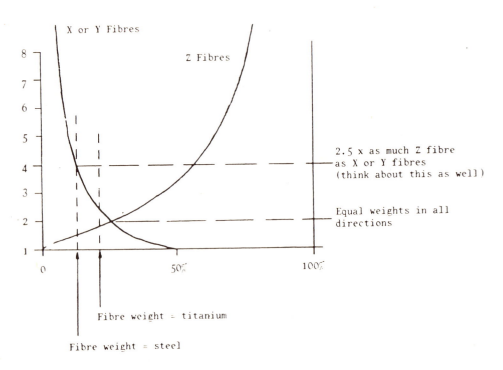

Fig 1   Three-dimensional structures: fibre proportions in each direction

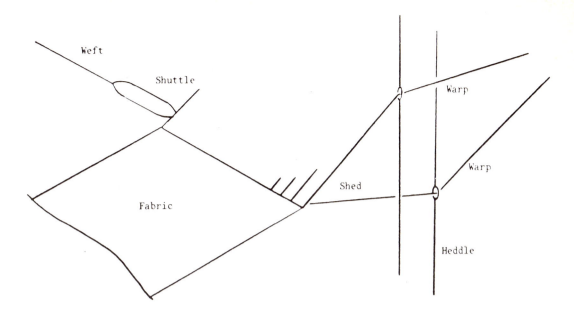

Fig 2   Loom layout and terms

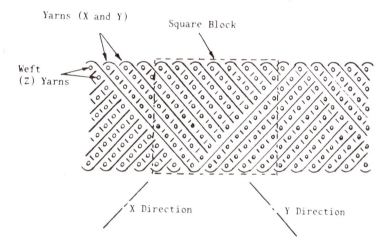

Fig 3a Section through X—Y plane of woven three-dimensional fabric

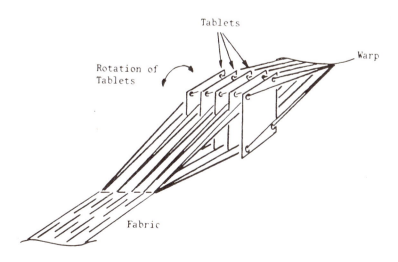

Fig 3b Tablet loom

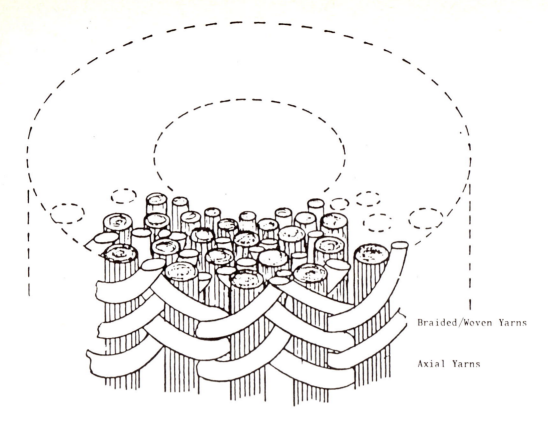

Braided/Woven Yarns

Axial Yarns

Fig 4

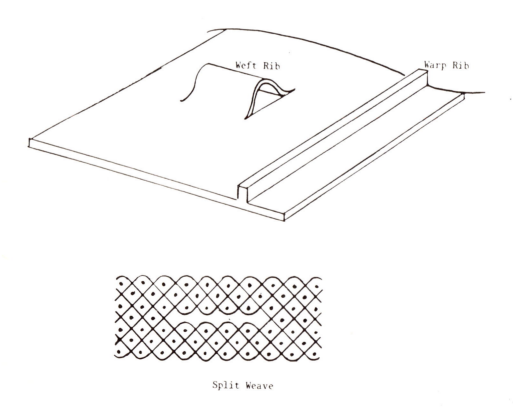

Weft Rib

Warp Rib

Split Weave

Fig 5   Ribbed weaves

C16/86

The novel use of composites in engineering applications

D P BASHFORD, BTech
Fulmer Research Laboratories Limited, Slough, Berkshire

SYNOPSIS Fibre reinforced composites are very adaptable engineering materials whose industrial use
is expanding. In this paper three case studies are presented on the application of composites to
helicopters, minesweepers and cranes. For each study innovative designs and solutions are presented
to highlight aspects of composite design. These can be broadly identified as follows.

a) The need to use efficient geometries to optimise weight saving.

b) The benefits of using high strain to failure matrix materials.

c) Integrated composite designs can reduce the number of individual components.

d) Attention must be made to loading modes other than tensile, i.e., compressive, shear and
    torsion, as being significant in composite design.

1 INTRODUCTION

Composites are attractive materials in their
potential to reduce structural mass either by
partial or total replacement of metallic
components. The range of fibre reinforced
plastics available - based on carbon, Kevlar
and glass fibres - is large to suit the
requirements of differing applications. These
requirements being a combination of mechanics,
manufacturing and cost.

The weight saving characteristics of fibre
reinforced plastics results from the high axial
properties of the fibres and the low density of
combined fibre and resin. The high specific
tensile strength and stiffness of fibres, shown
in Table 1, illustrate the starting level of
performance. But as fibres, and conversely
their composites, are anisotropic materials
these properties are in effect diluted to
ensure suitable properties in more than one
direction or plane; this being necessary for
structural integrity. Table 1 shows the
dilution in tensile properties for simple
multidirectional laminates. However, few
structures rely solely on tensile loading, so
compressive, shear, flexural or torsional
loading modes have also to be considered. To
react these loading modes relies on resin or
matrix dependant properties of the composite.
The strength and modulus of resins are around
two orders of magnitude lower than the tensile
properties of the fibres. The skill in
designing with composites is to locate the
fibres in the optimum orientations and avoid
overstressing the weaker property planes.

The terms strength and stiffness are worth
expanding on in the context of composite
designs. For traditional thermosetting resin -
epoxy and polyester - based composites both
resin and fibre are essentially brittle, low

strain to failure materials. This confers on
the composite an inability to plastically
deform and makes the material sensitive to
notches and stress concentrators. This has
significance in the use of mechanical fastening
as a joining technique, and the need to
avoid/minimise sudden changes in section on
load paths. A composite also relies on the
resin matrix for its integrity. If the
properties of the resin change within the
operating environment and the required life of
the structure, consideration has to be made at
the design stage on the basis of allowable
strengths. The primary depreciation in resin
properties is through combined thermal and
moisture effects. Taking all factors into
account this can give design allowable
strengths as low as 40% of an average measured
strength property.

On the stiffness side a comparison between
metals and composites is an interesting one.
CFRP possesses the highest tensile stiffness to
weight ratio of the composites with KRP second
and GRP a notable third. Metals, i.e. steel
and aluminium, have the same tensile specific
stiffnesses. CFRP is the only composite that
can match the metals on tensile specific
stiffness, as shown in Table 1. How then can
composite structures be made rigid enough and
save weight? The answer lies in two areas

a) The composites possess a higher volume per
    unit weight, therefore greater sectional
    thicknesses, which offers structural
    rigidity in flexure and bending.

b) The ability to position material where it
    is required through the manufacturing
    (moulding) routes available. This enables
    more efficient geometries to be used than
    is possible with metallic constructions
    dictated by cost constraints.

Table 1 Comparison of Mechanical Properties of Some Structural Materials

Material	Young's Modulus (E) GPa	UTS (σ) MPa	Specific Modulus (E/ρ) m x 10^7	Specific Strength (σ/ρ) m x 10^6
E-Glass Fibre	72.3	3170	2.8	1.24
S-Glass Fibre	82.7	4130	3.3	1.65
Kevlar 49 Fibre	137.8	3445	8.1	2.04
HM Carbon Fibre	379	2070	19.8	1.09
XAS Carbon Fibre	235	2900	13.1	1.62
AS Carbon Fibre	200	2340	11.0	1.29
Maraging Steel	193	2070	2.5	0.26
Aluminium 7075	68.9	565	2.5	0.20
Titanium 6A1-4V	103	1070	2.5	0.25
Laminates	$E_{11}=$	$\sigma_{11}=$	$(E/\rho)_{11}=$	$(\sigma/\rho)_{11}=$
E-Glass/Epoxy	$\sim$22	$\sim$951	$\sim$1.07	$\sim$0.46
Kevlar 49/Epoxy	$\sim$42	$\sim$1034	$\sim$2.7	$\sim$0.67
XAS/Epoxy	$\sim$71	$\sim$870	$\sim$4.4	$\sim$0.54

Note: The laminates contain 60% by volume of fibres with 50% 0° plies and the rest $\pm45^{\circ}$ or 90°. The tensile properties are taken in the 0° direction.

This said the lower moduli of elasticity of composites can create problems in resisting compressive loadings leading to crippling or buckling. Cost is an important constraint, and for replacing steel the composites start losing their advantages as the stiffer materials, i.e. CFRP, have the highest cost per unit weight. In this case a compromise may be needed in producing a composite structure of lower overall rigidity than a steel equivalent provided strength is not impaired. For traditional steel constructions an element of over-design can invariably exist which is then exploited in the composites favour.

In this paper three case studies are presented which highlight the novel use of composites, based on benefits gained through anisotropy and low density. In one study the anisotropy creates problems as loading in a resin dominated direction can lead to catastrophic failure. Here, the introduction of tough, high strain resins provides a solution. This is indicative of an overall trend towards polymeric matrices which impart greater damage tolerance and durability to otherwise brittle structural materials.

2 THE DEVELOPMENT OF A COMPOSITE BEARINGLESS ROTOR HUB FOR A HELICOPTER

On a helicopter, rotor hubs connect driveshafts to rotor blades, be this on the main or tail rotor. The function of the hub is to transmit the forces between the blades and the main body of the helicopter as the aircraft manoeuvres. These forces are cyclic and their level and intensity dependent on the weight of the helicopter and its required flight capabilities. The hub must also allow the pitch of each blade to alter on each rotation to create conditions for lift and aircraft motion. In this study the main rotor hub was considered which usually has either four or five blades. The precise hub configurations vary between manufacturers and are dependent on helicopter size; the permutations are large. In this case we were looking to improve on a titanium structure consisting of tubes which flared out in section from each blade root to the centre of the hub. These titanium tubes are discontinuous and have a system of metallic needle bearings to allow torsional displacement for changing the pitch of the blades. The displacement is achieved by a system of push rods which act on each tube as the hub rotates.

These titanium constructions are highly stressed and must never fail in service. They present limitations in terms of;

a) being too heavy
b) create aerodynamic drag
c) require regular maintenance
d) their mechanical bearings are prone to fatigue wear.

These factors lead to short service lives in view of the cyclic fatigue loadings they experience.

It was proposed that a composite solution be sought to overcome these deficiences by;

a) utilising the excellent fatigue properties of composites.
b) eliminating mechanical bearings which can wear.
c) reducing the number of individual components.
d) reducing weight and aerodynamic drag.

This leads to the concept of a bearingless composite rotor hub which will last the life of the helicopter. The exercise concentrated on the section required for one blade with

142

constraints placed on the length between blade root and the centre of the hub. Such a section can be considered a single composite flexure, required to deform elastically in bending and torsion with the necessary strengths for a 10^8 cycle fatigue life. The composite flexure concept is shown in Figure 1.

One of the more obvious requirements was a 200:1 ratio for bending stiffness to torsional stiffness. This implied that fibre reinforcement in circumferential or hoop directions was undesirable, and for the high torsional displacements required a very low modulus matrix was needed. In the first instance a fibre reinforced elastomer construction was sought. This represented a complete new composite materials solution to the problem, rather than a combination of materials and geometry which has been the tendency for evolving helicopter rotor hubs. The proposed elastomeric concept relied on segmented regions of fibre reinforcement with continuous 0° fibres for axial bending stiffness and strength and $\pm45^{\circ}$ fibres for shear support. The construction should be viewed as three dimensional with balanced planes of symmetry on the radii.

A series of models was produced to predict the elastic response of a flexure made in such a manner. These covered bending, shear and torsional stiffnesses for various fibres, elastomers and geometric shapes. The latter being a degree of freedom in terms of transverse cross-section (circular or elliptical) and longitudinal cross-section (parallel or tapered). Further limitations linked to the stiffness of the flexure were specified in the form of natural frequencies in lag and flap.

Concurrent studies into the mechanical properties of carbon fibre reinforced elastomers showed, not surprisingly, that these materials had inferior compressive properties to conventional epoxy based CFRP. The phenomenon being linked to microbuckling of the fibres due to a lack of lateral support by the matrix. This fact necessitated a change in philosophy on the construction, as high compressive stresses would exist in the lower half of the flexure as a result of bending moments. The fibres had therefore to be in a rigid matrix for lateral support, but a solid flexure based on epoxy composite would be far too rigid in torsion. It was shown that if the rigid composite was concentrated in radial segments then as thin long plates they each contributed very little to the torsional stiffness. The segments were in fours, for a symmetrical, balanced flexure, requiring 28 or 32 segments to stay within the torsional stiffness specification; with nothing between the segments.

The next concerns were buckling and elastic instability. The buckling resistance of the segments on the compressive face of the flexure was identified as unacceptably low, but that a very small lateral force was necessary to completely eradicate this occurrence. Very low modulus elastomers were therefore introduced between the segments, to completely fill the gaps. Thus providing more than adequate support whilst minimising any increase in torsional stiffness. The question of elastic instability results from the combined bending moments and torsional distortion. The concern being that high bending loads would trigger an over displacement in torsion as the unit sought a low state of stored energy. A specific modelling package was produced to cover this phenomenon and assess potential designs.

Three modes of helicopter operation were identified as giving the most exacting loading and hence stressing conditions encountered. These were:

a) The Flight Limit Case - a one-off manoeuvre in an emergency which has to be survived, but the rotor can be replaced on landing.

b) Cruise Flight - this represents continuous level flight, necessitating satisfactory fatigue strength, i.e. 10^8 cycles.

c) A Parked Aircraft - here high static bending loads result from the dead weight of the blades.

The tensile and compressive stresses seen in (a) and (c) are the largest, and whilst much smaller in (b) were of equal concern because of their frequency.

In conclusion an all composite flexure is feasible but the required elastic and natural frequency response imposed by the size of the helicopter places limitations on the amount of composite material which can be used, and this limits the strength and hence fatigue life of the construction. The concept of composite bearingless rotor hubs is being pursued for use in the 1990's and later. This study looked at a concept construction and further work is required on aspects such as end connections for load transfer, manufacturing and nondestructive testing.

3 THE USE OF VERY HIGH STRAIN TO FAILURE RESINS TO ENSURE STRUCTURAL INTEGRITY IN GRP SHIPS

As noted in the introduction the brittleness and low strain to failure of thermosetting resins, coupled with low tensile strength, gives laminates with poor transverse tensile properties. Every effort should be made to avoid overstressing composites in this direction or delaminations will result.

In the current generation of Royal Navy minesweepers (MCMV's) which are GRP constructions, such an overstressing situation exists where delaminations could occur. The situation is an unlikely occurrence but the ship has to be designed to accommodate such an incident. In the event of an underwater explosion (mine detonation) the ship's hull will experience shock loading. In this case top-hat stiffeners laminated to the inside of the hull can separate from the hull due to overloading and failure of the brittle polyester resin holding the two together. Total separation of stiffeners from the hull would result in a loss of structural integrity for the ship. Total separation is avoided by installing titanium bolts at regular intervals through the hull and the flanges of the stiffeners. The material and installation costs of this exercise are enormous. Figure 2 shows the sudden change in section that occurs at the heel of the stiffener where it meets the hull. This gives rise to a severe stress concentration as stiffener and hull attempt to separate. The titanium bolts will not stop or inhibit crack formation but arrest crack propagation to prevent total separation.

Early studies[1] on this problem proposed a redistribution of fibre reinforcement around the critical corner. This proved unsuccessful as direct transverse loading of the brittle resin always occurred which dictated failure in the pull-off tests undertaken. It became clear that increased elastic compliance was required in the critical zone to allow redistribution of stresses over larger areas. This could only be achieved by the localised substitution of brittle polyester resin with tough high strain resins. At this time suitable resin systems for ship construction became available through Scott Bader. Table 2 shows the difference between conventional elastic polyester resin and the newer viscoelastic system.

Figure 3 shows the comparative performance, in slow pull-off tests, for top-hat stiffened sections with toughened resin and those with titanium bolts. For the titanium bolted constructions a crack is initiated at a low load and crack damage accumulates to the point at maximum load where the base panel fractures. The toughened resin panels exhibit a different response, whereby the panel deforms elastically - with no crack formation - to the point of failure. The failure loads of toughened resin panels are very high. The high loads combined with the large amount of deformation (displacement) attained suggests the system can store large amounts of elastic energy.

The slow pull-off tests do not provide the same high-strain rate conditions as an underwater shock loading. However, shock trials conducted by the Admiralty Research Establishment, Dunfermline have shown panels containing toughened resins to perform as well, if not better, than titanium bolt reinforced constructions. Toughened resins are to be used in the next generation of Single Role Mine Hunters (SRMH).

Composites can be combinations of many materials, hence their versitility. The term 'toughened resin' has been used in this case study but 'toughened adhesive' is an equally appropriate term. Adhesive bonding is in many ways a preferable joining technique to mechanical fastening. Considerable advances have been made in recent years towards producing adhesives with greater resistance to peel, faster processing times and a reduced need for elaborate surface preparation.

4 COMPOSITE SPACE FRAME CONSTRUCTIONS

Whilst composites are attractive for their weight saving potential, cost is a restriction particularly when considering applications using traditionally low cost materials such as steel. In space frame type constructions which at some time require transportation, e.g. crane sections, foot bridges and temporary structures, both the cost of composite materials and fabrication have to be minimised. An efficient composite design is also required to maximise weight savings.

Jointing is an important aspect of composite constructions in view of the anisotropic and brittle nature of the materials. Fulmer Research Laboratories have developed low cost crimp-bonded joints for composite pultrusions[2]. These enable the full strength of the pultrusions to be utilised. Composite pultrusions represent attractive sections for crane constructions.

In replacing steel in the stiletto and lattice sections of an extendable mobile crane, an allowance is necessary for a composite construction of marginally less overall stiffness than the steel versions - so avoiding the mass use of carbon fibres. Strength, however, is not compromised.

In the 4.5 metre stiletto an elliptical section was sought as offering an efficient geometry in terms of bending stiffness. Calculations showed that the construction would need to be a combination of carbon and glass fibre reinforced sections, with small amounts of CFRP in the outer regions of the ellipse for maximum effect in terms of stiffness. Due to budget restrictions, readily available pultruded composite sections formed the building blocks which were subsequently overwound with glass cloth/polyester resin to complete the construction. The initial design had an optimised weight of 11 kg per metre length to replace a steel section of 35 kg per metre length. However, the range of pultruded sections commercially available are limited and a compromise had to be reached in using a

Table 2

Polymeric System	Initial Tensile Modulus (GPa)	Tensile Strength (MPa)	Strain to Failure (%)
A2785CV Isophthalic polyester resin	3.6	46	∿1.8
Scott Bader System - Unsaturated urethane	∿0.5	∿25	∿120

© IMechE 1986 C16/86

number of over specified pultruded sections; giving a final weight of 18 kg per metre and a saving of 50 per cent.

The lattice construction proved more problematical as the existing steel construction was in its own right a fairly efficient design. An attempt to produce a composite geodesic construction of considerably less weight proved difficult. The composite design was optimised on its compressive performance in terms of crippling and buckling strengths. An idealised design gave a total weight of 90 kg as opposed to 180 kg for the steel version. The composite lattice consisted of four 5 metre pultruded GRP tubes with steel collars at set intervals to accept the interconnecting struts. These struts consisted of pultruded GRP tubes with crimp-bonded aluminium alloy end fittings. It was these struts which proved inefficient as their short overall lengths inhibited weight saving. The metal end fittings became heavy in giving them sufficient bearing strength for the bolted connection to the steel collars and the pultruded lengths virtually disappeared once an allowance for the fitting to pultrusion bonded overlap was made. In having 72 struts the weight rose and the final weight was a little less than the steel original of 180 kg. It is appreciated that aluminium to steel connections should be avoided on grounds of possible corrosion problems.

Subsequent testing of two extended demonstrator lattice/stiletto sections showed agreement on the predicted deformations under load. On increasing to design load premature compressive failure occurred in one of the four main lattice tubes at the point of load transfer to the remainder of the crane. A localised increase in composite sectional thickness would solve this.

A number of conclusions can be drawn from this programme:

a) It has been demonstrated that composite crane constructions can meet the performance of existing steel versions provided a reduction in overall stiffness is allowed.

b) Whilst on paper significant weight savings can be made in going to composites these can only be achieved if the fabrication route is feasible.

c) In assessing the economics of composite lattice and stiletto sections it was concluded that overall combined weight savings of 30% could be easily achieved but at an overall cost two to three times that of steel versions. This was not thought to be a commercially acceptable proposition.

5 CONCLUSIONS

In the use of structural composite materials a successful application requires combined consideration of design, materials and manufacture. On feasibilty studies materials and design are invariably studied first with manufacturing appearing later. This process is satisfactory provided the integration of the manufacturing route is not left too late. Also, any cost premium in proceeding with composites has to be objectively considered in relation to the benefits of their use. These benefits are usually a reduction in weight, reduced servicing or maintenance or acquiring performance not possible with metallic constructions.

Of the case histories presented in this paper those on helicopter and crane constructions were feasibility studies whilst that on minesweepers was an optimisation of a proven product.

The development of helicopter rotor constructions is a continuous process of evolution. A bearingless composite construction is a possibility for the 1990's but as with all dynamic helicopter components strength and fatigue life will always create problems.

The replacement of steel by composites to save weight in highly loaded structures will always be closely governed by cost to improved performance considerations. The application of composites to mobile cranes is difficult to justify. The extensive tonnage use of composites in oil rigs or suspension bridges could be justified on the grounds of excessive self weight for steel with subsequent revenues paying for the initial capital cost of using composites.

The high cost of using GRP for minesweepers is justified on the grounds of a non-magnetic construction giving reduced maintenance. The necessity for sufficient resistance to shock loading introduces complications. The application of toughened resins gives greater damage tolerance whilst reducing manufacturing costs. As an aside, it could be argued that a corrugated hull with integral stiffness is a more elegant solution than a top-hat stiffened structure.

6 ACKNOWLEDGMENTS

The work on composite helicopter bearingless rotors was conducted on behalf of Westland plc (Mr. C.J. Saddler and other members of the Advanced Technology Department) and funded by the British Government through the Procurement Executive, Ministry of Defence (Dr. P.T. Curtis, Royal Aircraft Establishment, Farnborough).

The work on minesweeper constructions was carried out with the support of the Procurement Executive, Ministry of Defence (Dr. C.S. Smith and Mr. J. Bird, Admiralty Research Establishment, Dunfermline).

REFERENCES

1) A.K. Green and W.H. Bowyer. Top-Hat Stiffener Fabrication Methods in GRP Ships. 1st International Conference on Composite Structures edited by I.H. Marshall. Paisley College of Technology. Applied Science 16th -18th September 1981 p. 182-201.

2) A.K. Green and L.N. Phillips. Crimp-bonded
End Fittings for use on Pultruded Composite
Sections. Composites 1982, Volume 13
Number 3. p. 219-224.

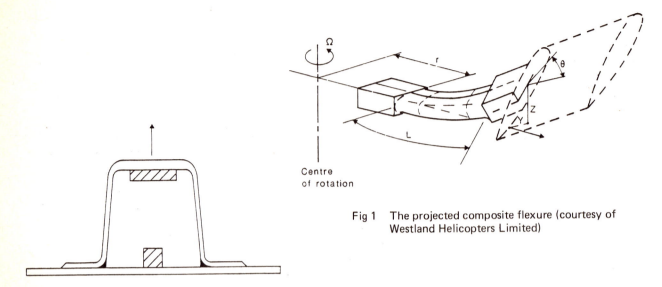

θ → Feathering
Z → Flap
Y → Lag

Fig 1   The projected composite flexure (courtesy of
Westland Helicopters Limited)

Fig 2   Top-hat stiffener section in slow 'pull-off'
configuration with base panel centre clamped

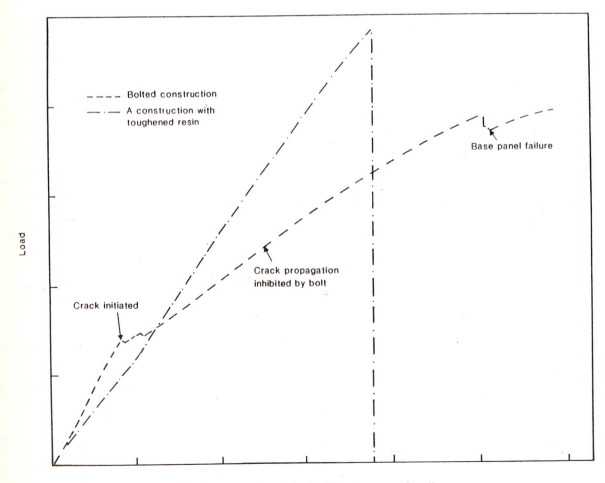

Fig 3   Comparison in 'pull-off' performance for all
composite and bolted constructions

C42/86

Airworthiness of composite structures – some experiences from civil certification

J W BRISTOW
Civil Aviation Authority, Redhill, Surrey

In this paper details are given of the guidance material used in the application of airworthiness requirements to primary structure in composite material in recent years. Some experience gained from the application of these requirements from the viewpoint of an airworthiness authority is also presented.

The approach to the airworthiness of composite structures in civil aircraft adopted by the CAA and the work leading to the evolution of joint European-USA guidance material on the application of such requirements has been the subject of previous papers published recently (ref 1 and 2). This paper will give details of the guidance material produced and will review some of the experience gained in the practical application to the airworthiness certification for a range of civil aircraft structures, as seen by an airworthiness authority. The subject will be dealt with in sections stating the guidance material in bold type taken from references 3 and 4 followed by a commentary on the practical experience to date.

A wide variety of components are presented for certification ranging from simple fairings or access doors, through airship gondolas, propellors, rotor blades up to wing, empennage and fuselage structures. The majority of components have an epoxy matrix reinforced with aramid carbon or glass fibres.

MATERIAL AND FABRICATION DEVELOPMENT. To provide an adequate design data base, environmental effects on the design properties of the material system should be established. Environmental design criteria should be developed that identify the most critical environmental exposures, including humidity and temperature, to which the material in the application under evaluation may be exposed. This is not required where existing data demonstrate that no significant environmental effects, including the effects of temperature and moisture, exist for the material system and construction details, within the bounds of environmental exposure being considered. Experimental evidence should be provided to demonstrate that the material design values or allowables are attained with a high degree of confidence in the appropriate critical environmental exposures to be expected in service. The effect of the service environment on static strength, fatigue and stiffness properties should be determined for the material system through tests; e.g., accelerated environmental tests, or from applicable service data. The effects of environmental cycling (i.e., moisture and temperature) should be evaluated. Existing test data may be used where it can be shown to be directly applicable to the material system.

The material system design values or allowables should be established on the laminate level by either test of the laminate or by test of the lamina in conjunction with a test validated analytical method.

For a specific structural configuration of an individual component, design values may be established which include the effects of appropriate design features (holes,joints,etc).

Impact damage is generally accommodated by limiting the design strain level.

The concept of a data base is not new but when it involves a new material the question is often asked what should be contained in a data base? The structural substantiation of composites is based on a pyramid shape culminating in a full scale structural test at the top, supported by a series of tests on major subcomponents (e.g., a tailplane to fuselage joint). That is in turn supported by a range of detailed tests on simple structural features (e.g., ply changes,fastener holes etc) and in turn that is based on a larger number of material property tests on basic laminates. It is the lowest two tiers of this pyramid that form the data base. Its extent will vary with the project under consideration,but it should encompass all the relevant design properties in the range of environments expected in service. It has been shown in results of recent full scale tests that it would be wise to include total thickness, stacking sequence and impact damage in the range of variables to be considered in such a data base.

Taking two examples from the environmental investigations carried out to date, the maximum temperature to be considered is focused around 70 - 80 $^\circ$C for subsonic transport operations and around 50 - 55°C for gliders and light aircraft with reflective paint schemes.

In the earliest days of carbonfibre composites boiling water immersion was considered as an appropriate definition of the environmental effect. When this was found to be too severe a 1 per cent weight up-take of moisture was assumed. However there is a danger that this 1 per cent figure can be read across to other composite systems, also the definition of moisture uptake is a particularly tricky one in defining the zero uptake position. For these reasons the currently favoured approach is to expose the material to a relative humidity appropriate to world exposure say 84 per cent (after ref 5) until saturation is reached and evaluating properties in that material condition.

In establishing the design values for a structure the same statistical basis is used for composite material as is for metallic material (see ref 6). Although in the early days of a particular material, small sample sizes are allowed provided that the statistics are adjusted for this reduced sample size.

As far as CAA is concerned the establishment of design values of specific material configurations by test is an important approach, either in its own right or to validate analytical approaches that may be used. The significant anisotropy, elastic behaviour and multivariate failure modes of composites make it extremely difficult at present to analyse complex structures reliably for all loading cases to be considered.

PROOF OF STRUCTURE - STATIC   The static strength of the composite design should be demonstrated through a programme of component ultimate load tests in the appropriate environment, unless experience with similar designs, material systems and loadings is available to demonstrate the adequacy of the analysis supported by sub-component tests.

The effects of repeated loading and environmental exposure which may result in material property degradation should be addressed in the static strength evaluation. This can be shown by analysis supported by test evidence, by tests at the coupon, element or subcomponent level, or alternatively by relevant existing data.

Static strength structural substantiation tests should be conducted on new structure unless the critical load conditions are associated with structure that has been subjected to a repeated loading and environemntal exposure. In this case either (1) the static test should be conducted on structure with prior repeated loading and environmental exposure, or (2) coupon/element/subcomponent test data should be provided to assess the possible degradation of static strength after application of repeated loading and environmental exposure, and this degradation accounted for in the static test or in the analysis of the results of the static test of the new structure.

The component static test may be performed in an ambient atmosphere if the effects of the environment are reliably predicted by subcomponent and/or coupon tests and are accounted for in the static test or in the analysis of the results of the static test.

The static test articles should be fabricated and assembled in accordance with production specifications and processes so that the test articles are representative of production structure.

When the material and processing variability of the composite structure is greater than the variability of current metallic structures, the difference should be considered in the static strength substantiation (1) by deriving proper allowables or design values for use in the analysis, and the analysis of the results of supporting tests, or (2) by accounting for it in the static test when static proof of structure is accomplished by component test.

Composite structures that have high static margins of safety may be substantiated by analysis supported by subcomponent, element, and/or coupon testing.

It should be shown that impact damage that can be realistically expected from manufacturing and service, but not more than the established threshold of detectability for the selected inspection procedure, will not reduce the structural strength below ultimate load capability. This can be shown by analysis supported by test evidence, or by tests at the coupon, element or subcomponent level.

The keypoint of the guidelines is a static test to ultimate load considering the effects of environment, cyclic loading and manufacturing variabilities.(Note: In civil aircraft ultimate load is 1.5 x limit load where the limit load approximates to the maximum load expected during the service life).

Aircraft constructors have chosen a variety of ways to meet this requirement. The simplest way of testing is at room temperature without environemntal conditions, using a test factor on loads to cover the environment, cycling and variability.These factors which are specific to the construction used can vary, some examples are:-

For the effect of variability in carbon fibre epoxy laminates compared to conventional aluminium alloy plate a factor of 1.06.

For aramid/phenolic honeycomb sandwich under warm moist conditions a factor of 1.7.

For the hot wet effect in carbon fibre epoxy laminates a factor of 1.31.

As the necessary additional factors are greatest for the matrix dominated properties, such as compression strengths , a room temperature test with large factors necessitates over design of tension elements. Additionally any metallic component in the structure will also have to be designed to cope with the increased test loading.

© IMechE 1986 C42/86

For this reason some constructors elect to carry out structural testing at elevated temperature on a moisture conditioned test specimen. This is a more expensive approach, but eliminates the need for additional structural size (i.e. weight) to cover the increased test load in lieu of environmental exposure.

Another approach to the problem is to establish by extensive subcomponent test and analysis the complete range of allowable strains for the material system in the environmentally degraded state, with a statistical allowance for variability ,then to carry out a fully instrumented structural test up to ultimate load at room temperature without environmental conditions. At design ultimate load if any strain measured in the structure is indicated above the degraded statistical allowable previously derived, the test is not passed.

Major static tests may be carried out on the fatigue test specimen, as the elastic behaviour of typical composites does not lead to the strongly misleading interactive effects seen in conventional metallic structures statically loaded before fatigue testing. It needs to be established however, on small scale specimens, whether prior cyclic loading can have a beneficial effect on the subsequent static strength.

The need for full scale testing to ultimate load has been underlined by the results of a significant number of full scale composite airframe tests in recent years. The majority of which have failed below ultimate load. There have been a variety of reasons for the failures :- inadequate allowance for environmental effects, omission of relevant loading cases in the design, the effect of impact and certain design features. Such testing has enabled the structures to be modified accordingly with subsequent testing proving satisfactory.

PROOF OF STRUCTURE-FATIGUE/DAMAGE TOLERANCE

The evaluation of composite structure should be based on the applicable requirements. The nature and extent of analysis or tests on complete structures and/or portions of the primary structure will depend upon applicable previous fatigue /damage tolerant designs, construction, tests, and service experience on similar structures. In the absence of experience with similar designs, structural development of components,subcomponents, and elements should be performed. The following considerations are unique to the use of composite material systems and should be observed for the method of substantiation selected by the applicant. When selecting the damage tolerance or safe life approach, attention should be given to geometry, inspectability, good design practice, and the type of damage/ degradation of the structure under consideration.

Damage Tolerance (Fail-Safe) Evaluation

Structural details, elements, and subcomponents of critical structural areas should be tested under repeated loads to define the sensitivity of the structure to damage growth. This testing can form the basis for validating a no-growth approach to the damage tolerance requirements. The testing should assess the effect of the

environment on the flaw growth characteristics and the no-growth validation. The environment used should be appropriate to the expected service usage. The repeated loading should be representative of anticipated service usage. The repeated load testing should include damage levels (including impact damage) typical of those that may occur during fabrication, assembly, and in-service, consistent with the inspection techniques employed. The damage tolerance test articles should be fabricated and assembled in accordance with production specifications and processes so that the test articles are representative of production structure.

The extent of initially detectable damage should be established and be consistent with the inspection techniques employed during manufacture and in service. Flaw/damage growth data should be obtained by repeated load cycling of intrinsic flaws or mechanically introduced damage. The number of cycles applied to validate a no-growth concept should be statistically significant, and may be determined by load and/or life considerations. The growth or no growth evaluation should be performed by analysis supported by test evidence or by tests at the coupon, element, or subcomponent level.

The extent of damage for residual strength assessments should be established.Residual strength evaluation by component or subcomponent testing or by analysis supported by test evidence should be performed considering that damage. The evaluation should demonstrate that the residual strength of the structure is equal to or greater than the strength required for the specified design loads (considered as ultimate). It should be shown that stiffness properties have not changed beyond acceptable levels. For the no-growth concept residual strength testing should be performed after repeated load cycling.

An inspection program should be developed consisting of frequency, extent, and methods of inspection for inclusion in the maintenance plan.Inspection intervals should be established such that the damage will be detected between the time it initially becomes detectable and the time at which the extent of damage reaches the limits for required residual strength capability. For the case of no-growth design concept, inspection intervals should be established as part of the maintenance program. In selecting such intervals the residual strength level associated with the assumed damages should be considered.

The structure should be able to withstand static loads (considered as ultimate loads) which are reasonably expected during a completion of the flight on which damage resulting from obvious discrete sources occur (i.e., uncontained engine failures, etc). The extent of damage should be based on a rational assessment of service mission and potential damage relating to each discrete source.

The effects of temperature,humidity, and other environmental factors which may result in material property degradation should be addressed in the damage tolerance evaluation.

Fatigue (Safe-Life) Evaluation. Fatigue substantiation should be accomplished by component fatigue tests or by analysis supported by test evidence, accounting for the effects of the appropriate environment. The test articles should be fabricated and assembled in accordance with production specification and processes so that the test articles are representative of production structure. Sufficient component, subcomponent, element or coupon tests should be performed to establish the fatigue scatter and the environmental effects. Component, subcomponent, and/or element tests may be used to evaluate the fatigue response of structure with impact damage levels typical of those that may occur during fabrication, assembly, and in service, consistent with the inspection procedures employed. The component fatigue test may be performed with an as-manufactured test article if the effects of impact damage are reliably predicted by subcomponent and/or element tests and are accounted for in the fatigue test or in analysis of the results of the fatigue test. It should be demonstrated during the fatigue tests that the stiffness properties have not changed beyond acceptable levels. Replacement lives should be established based on the test results. As appropriate inspection program should be provided.

The relevant requirements determine which approach is to be used for each project. For example large transport aircraft are damage tolerant in design, whereas helicopters have a large number of safe life features.

A number of issues have arisen in the approaches outlined above. The most significant one is something of a dilemma. Some composites although they have a good fatigue strength, have a very flat S.N. curve which leads to high scatter in life. Thus, larger factors than the currently accepted ones for metallic materials would be needed for adequate demonstration, say between 10 and 50. However the factor on stress, such as is typically used in helicopter fatigue testing, can cover the point, but problems may arise if metallic components are also present in the structure and would in consequence be over stressed. The author is aware of two approaches to this dilemma at present. One is to use separate test procedures to validate the composite component and the metallic component independently. The other, is to use a combined life and stress test factor, so that the test does not go on for too long, nor that the structural loads become unrepresentatively high.

Associated with a good fatigue strength the margin of static strength over fatigue strength can be considerably reduced (see figure 1). This may influence the nature of substantiation testing particularly for helicopter components where substantiation for fatigue has taken precedents over static testing for many years, due to a large implied margin.

Environmental interaction with fatigue performance is another issue. Helicopter rotor blades are among the longest serving primary structural components and here the problem is addressed in two ways, using real time component exposure tests and also re-testing of service components. In one approach the constructor has assumed a degradation in performance by increasing his fatigue test loads leading to fixed reduced life. In the other, the results of the long term and component recall tests are used to reassess the fatigue life periodically. In general the results on these thick section glass epoxy structures have shown to be conservative.

The no-growth concept, referred to above, is that for some configurations of composite structures large readily detectable flaws have good residual strength capability and can be demonstrated not to grow under cyclic loading. There is however a need to have an adequate inspection programme of sufficient frequency, such that if the flaw does occur the aircraft is not exposed to extended periods of operation, with the large flaw and hence reduced strength capability.

Impact damage must be considered both in the damage tolerant, and in the safe-life approach. Impact is discussed in a subsequent section of this paper.

PROOF OF STRUCTURE - FLUTTER. The effects of repeated loading and environmental exposure on stiffness, mass and damping properties should be considered in the verification of integrity against flutter and other aeroelastic mechanisms. These effects may be determined by analysis supported by test evidence, or by tests at the coupon, element or subcomponent level.

No fundamental issues in the flutter area has arisen specifically because of the use of composites rather than metallic material. However two examples have come up on the effect of changing the type of composite materials providing the skins to control surfaces. In one of these cases the minor change revealed a mild flutter situation that was inherent in the aircraft but had escaped detection previously, so the composite material could hardly be claimed to be the culprit. In the other a change from glass to aramid fabric associated with other changes to primer, filler and paint, led to water passing through the skins and remaining trapped in the surface, thus affecting the balance of the control surface.

IMPACT DYNAMICS. The present approach in airframe design is to assure that occupants have every reasonable chance of escaping serious injury under realistic and survivable impact conditions. Evaluation may be by test or by analysis supported by test evidence. Test evidence includes but is not limited to element or subcomponent tests and service experience. Analytical comparison to conventional structure may be used where shown to be applicable.

Aircraft structures may experience impact in any of four categories :-
1. Low Energy Impact
2. Bird Strike
3. Discrete Source Damage e.g. turbine disc burst.

4. High Energy Crash Impact

For composites, the first and last cases are of special interest. In the first case particular composites can experience significant strength reductions without visible damage if the impact energy is low. Therefore for the range of energies up to the maximum likely to be experienced it is necessary to demonstrate that the structure can withstand ultimate loads for damage up to the point where it becomes detectable. Furthermore detectable impacts should not reduce the strength below limit value (2/3 ultimate) see the requirement boundary on figure 2.

The definition of maximum likely impact is obviously an important issue and will vary from application to application. Significant energies can be involved. To take a simple example -

a full executive briefcase (weight 10lb) dropped from an overhead locker (height 5ft) can inpart 50ft-1lb to a composite seat structure below.

In the fourth category the elastic behaviour of composite structure up to failure would indicate that the energy absorption characteristics are likely to be different from metallic structure, see the typical line in figure 3, taken from ref 7, showing little energy absorption. However by suitable composite configuration significant energy absorption can be achieved by inducing progressive collapse (see the optimum line on the figure). Furthermore in reality wing leading edge structures have been successfully designed for bird impact in aramid epoxy laminates. Also glass carbon epoxy propeller blades have shown a gentle shredding behaviour during impact after undercarriage collapse.

FLAMMABILITY
LIGHTNING PROTECTION

Fire smoke and toxicity and some electrastatic electramagnetic effects are important in the overall consideration of composites for aircraft structures and furnishings, but are not addressed in this paper which is specific to structures.

PROTECTION OF STRUCTURE

Weathering, abrasion, erosion, ultraviolet radiation, and chemical environment (glycol, hydraulic fluid, fuel, cleaning agents, etc) may cause deterioration in a composite structure. Suitable protection against and/or consideration of degradation in material properties should be provided for and demonstrated by test.

The composites commonly in use in aircraft structure are demonstrating good resistance to most commonly used aeronautical fluids, apart from water, as previously discussed. Erosion shields are important on rotating components, as are reflective coatings to prevent ultraviolet degradation of surfaces exposed to sunlight. There is however one practice in parts of the industry which is a little difficult to understand. That is the use of conventional metallic primers on composite components "to make them look like metal".

This practice cannot help any maintenance task, nor can any use for a chromate primer be seen on a non-metallic structure.

QUALITY CONTROL The overall plan required by the certifying agency should involve all relevant disciplines, i.e., engineering, manufacturing and quality control. This quality control plan should be responsive to special engineering requirements that arise in individual parts or areas as a result of potential failure modes, damage tolerance and flaw growth requirements, loadings, inspect ability, and local sensitivities to manufacture and assembly.

PRODUCTION SPECIFICATIONS Specifications covering material, material processing, and fabrication procedures should be developed to ensure a basis for fabricating reproducible and reliable structure. The discrepancies permitted by the specifications should be substantiated by analysis supported by test evidence, or tests at the coupon, element or subcomponent level.

INSPECTION AND MAINTENANCE Maintenance manuals developed by manufacturers should include appropriate inspection, maintenance and repair procedures for composite structures.

SUBSTANTIATION OF REPAIR When repair procedures are provided, it should be demonstrated by analysis and/or test that methods and techniques of repair will restore the structure to an airworthy condition.

The above topics are of fundamental importance to the production and operation of airworthy composite structures and are the subject of a joint European, USA requirement study project for the near future. They are not discussed here in detail, the reader is referred to reference 2 for a summary of the current CAA approach.

COMMENT IN CONCLUSION

Apart from the necessary expansion in the areas of specification manufacturing and maintenance practices, referred to above, the requirement framework for the airworthiness certification of composite structures appears to be functioning satisfactorily. That is not to say that all the answers to questions posed have yet been completely found, or that all the relevant questions have been unearthed, but engineering was ever like that.

REFERENCES

1. BRISTOW J.W. "Aspects of Certification of Civil Aircraft Structures in Composite Materials". Paper 17 PRI Conference Liverpool 1984.

2. BRISTOW J.W. "Structural Composites - Airworthiness in Civil Aircraft", Paper 39, 6th International SAMPE Conference 1985 Schevingen - The Netherlands

3. Joint Airworthiness Requirements JAR 25, ACJ 25-603 to be issued.

4. Advisory Circular 20-107A "Composite
    Aircraft Structure" Federal Aviation
    Administration April 1984

5. COLLINGS T.A. An Assessment of the Effects
    of Observed Climatic Conditions on the
    Moisture Equilibrum Level of Fibre
    Reinforced Plastics R.A.E. Tech.Memo
    MAT/STR 1034 Feb 1984.

6. DEF STAN 00-932,Metallic Materials Data
    Handbook August 1985.

7. SNOWDON P and HULL D. "Energy Absorption
    of SMC Under Crash Conditions" PRI
    Conference Liverpool April 1984.

ACKNOWLEDGEMENT In addition to the
information obtained from the quote
references the author would like to
thank all his colleagues in the aviation
industry who have contributed to discussions
on the subject. The views thus formed and
set out in this paper are those of the
author and may not necessarily be those of
the Civil Aviation Authority.

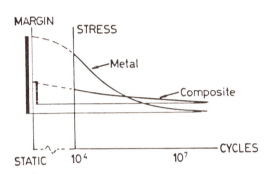

Fig 1   Relationship of static to fatigue strength

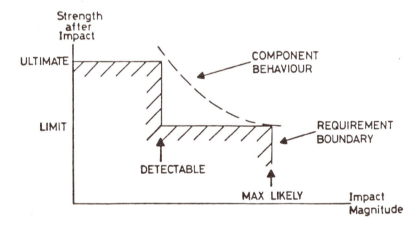

Fig 2   Strength after low energy impact

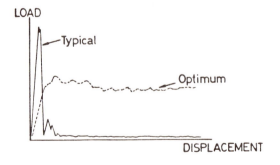

Fig 3   Energy absorption of composites

© IMechE 1986 C42/86

AV-8B/GR Mk 5 airframe composite applications

B L RILEY, BS, MS
McDonnell Douglas Corporation, St Louis, Missouri, USA

SYNOPSIS This paper describes the composite applications on the AV-8B/GR Mk. 5 Harrier II Airframe.
Twenty-five percent of airframe weight is composite materials resulting in a weight savings of 525
pounds per aircraft. Topics covered in this paper include the composite wing, forward fuselage,
and horizontal tail designs, the development test programs, design philosophy, interpretation and
evaluation of test results, impact damage, ballistic damage, lightning, repairs, and manufacturing
defects.

1 THE TEAM

This is the story of the composite structural
development for the AV-8B aircraft. The story
really has two sides, one a people story and
the other a technical development story. First
the people story. It is about what a small group
of people accomplished who were virtually left
alone to dream, visualize, plan, execute, develop
and productionize a new structural technology for
production aircraft. This technology is composite structure, the first major new aerospace
material for general use since the introduction
of titanium in the fifties or high strength
aluminum alloys in the forties.

The unique circumstances of the time and
the right mixture of talent and experience provided the foundation for the start of something
big. On the Government side, Dr Allen Somoroff,
head of Navy aircraft R & D was instrumental in
funding composite research and development in
key structural areas that ultimately led to the
decision for the incorporation of composite
structure for major portions of the AV-8B airframe. On the MCAIR side, L A Smith and E A
Harper, AV-8 Program Management, provided the
key management approval and support needed to
bridge the gap between just another R & D program and the realization that this structure
could someday be on a production AV-8B aircraft.

The composite design team itself consisted
of a small group of design, strength, material
and process, and test support personnel numbering
less than fifteen. The key people had been
through the F-15 boron wing program which had
experienced significant problems from a design,
technical and manufacturing viewpoint. Thus,
the team did not know what the right approach
was for the next project but they knew what
they did not want to do based on the experience
gained on the earlier F-15 boron wing program.

The team was led by K V Stenberg who provided the leadership to obtain the successful
R & D contracts and was instrumental as a coordi-
nator between Company management, AV-8 project
management and Government interests. The design
group was led by J C Watson and M L Huttrop. Jim
Watson was an experienced project designer that
had been involved with composite development at
MCAIR since its earliest days. Marie Huttrop
was a highly skilled and creative designer,

maybe the best that I have seen in my time. Bob
Ahrens, a seasoned veteran of early composite
processing problems, handled the material and
processing development. Victor Padilla headed
up the test lab effort and is in his own right
a very innovative and good investigative engineer.
The writer headed up the technical effort and
brought to fruition a number of key technical
innovations that are commonplace today. Kermit
Kreder provided leadership in the manufacturing
arena and was instrumental in developing pro-
ducible composite manufacturing technology for
tooling and assembly. The remaining members of
the team were a highly motivated and dedicated
group of young engineers each of which today
cherishes the memories that he was a contributor
to the start of something big.

The stage setting for this drama to begin
was the atmosphere of the aerospace community
and the focus of MCAIR's interests at the time.
In the early to mid seventies, a great deal of
skepticism about composites had permeated the
aerospace community. The Air Force had been
instrumental in funding the early development
of composite structure yet their new F-16 fighter
had very little composite structure on it. Issues
such as moisture degradation, impact damage,
repairs, fuel containment, battle damage and
high costs made both government and company
management skeptical about the commitment to
composites for primary aircraft structure. By
the mid seventies, MCAIR had committed itself
to the F-18 with composite applications for
primary structure on the wing skins and empennage.
Thus, with a major new aircraft program, MCAIR's
attention was focused on the development of the
F-18 aircraft. Go-ahead for an AV-8B demonstrator
called the YAV-8B with an all-composite wing
was also given in the mid seventies. Congress-
ional support for a separately developed Marine
aircraft was a controversial issue and continued
funding leading to full scale development and
production of an AV-8B was almost a year to
year debate. Thus, it was paramount that the
first all composite wing design be successful
because any premature structural failures, schedule delays, or cost overruns would severely
endanger the continued survival of the AV-8B,
and for that matter, severely delay the industry
acceptance of composite materials for primary
aircraft structure. Therefore, the stage was

set for the composite team to do its thing almost totally free from the usual constraints placed on aircraft development. The right people in the right place at the right time.

2 THE START OF SOMETHING BIG

On February 1, 1974 work began on a small study contract to determine the weight and cost impact of composite applications on a new version of the AV-8A Harrier. The entire aircraft with a new wing having a super critical airfoil was studied. This work was completed in December of 1974. The study results identified the wing, forward fuselage, and horizontal tail as the most promising candidates weight wise with a projected weight savings of twenty percent. Cost implications tended to be on the high side but the anticipated rapid advancement of composite manufacturing technology was considered to be a likely offset. In most cases this judgement on cost turned out to be correct. The wing was selected as the most promising candidate since a new airfoil was required for long range cruise performance and increased fuel volume. Thus, another piece of the puzzle fell in place for the development of a new version of the Harrier.

3 AIRCRAFT DESCRIPTION

The technology improvements selected for the new V/STOL called the AV-8B are shown on Figure 1. The AV-8B/GR Mk. 5 provides the Marine Corps and Royal Air Force (RAF) with a V/STOL light jet attack air support capability, increased flexibility and a more rapid response to the requirements of ground forces, either for conducting amphibious or land operations.

The AV-8B is a single place, close air support, light attack aircraft with vertical and short takeoff and landing capability. It has a raised cockpit with a one-piece wraparound windshield and shoulder-mounted swept wing with marked negative dihedral. The wing includes a large positive circulation flap and drooped ailerons in the high lift configuration. Conventional aerodynamic controls are used in wing-borne flight and engine bleed air reaction controls are used in jet borne flight with a mix of the two systems being used when transitioning between modes of flight. The aircraft is powered by a Rolls-Royce Pegasus fan jet engine and has two large side inlets and four exhaust nozzles which may be rotated to provide thrust vectoring for V/STOL operations or to enhance in-flight manoeuvering. The landing gear consists of a nose gear and a single main gear mounted in bicycle or tandem arrangement and two outrigger gears located approximately mid-span on each wing between the flap and aileron. Lift Improvement Devices (LIDs) consisting of two longitudinal strakes and a retractable forward fence mounted on the lower fuselage between the nose and main gear, are provided to improve performance in vertical takeoff or landing. The hydraulically operated speedbrake is located on the lower fuselage surface immediately aft of the main landing gear.

The AV-8B has six wing stations and one fuselage station for external stores and may carry an external fuselage mounted 25mm gun. External fuel may be carried on four of the wing stations. The aircraft is equipped with a stability augmentation and attitude hold system (SAAHS) with pilot relief modes for wing borne flight, a Departure Resistance System (DRS) for improved high Angle-of-Attack (AOA) handling qualities, an Angle Rate Bombing System (ARBS) and a Head Up Display (HUD). Other features include a self-contained start system and a Stencel SJU-4/A ejection seat. A retractable in-flight refueling probe may be added to the aircraft if required. The GR Mk. 5 is structurally identical to the AV-8B except for an additional store station on the wing and bird strike beef-ups to the wind screen, nose cone, engine inlet, wing leading edge and a Martin Baker ejection seat.

4 AV-8B/GR MK. 5 COMPOSITE APPLICATIONS

The application of composite materials to the AV-8B airframe is shown in Figures 2 and 3. Primary structural applications are the wing torque box and control surfaces, horizontal tail, and forward fuselage. Secondary applications are the gun and ammo paks, strakes, LIDs fence, ventral fin, rudder, engine bay doors, nose cone, overwing fairing and outrigger fairing. Twenty-five percent of the airframe weight is composite materials resulting in a weight savings of 525 pounds per aircraft (see Figure 4).

5 DESIGN AND FABRICATION OF THE WING

The AV-8B wing is composed of a main torque box, aileron, trailing edge flap, auxiliary flap, overwing fairing, outrigger gear fairing, and a metal leading edge and tip as shown in Figure 5. The use of carbon/epoxy composite material on the wing saved 330 pounds per aircraft. The wing is continuous from tip to tip and attaches to the fuselage at six points as in the original AV-8A Harrier design. There are six pylon stations on the wing for the AV-8B and eight for the GR Mk. 5.

The composite torque box is a multispar monolithic cover design. The upper and lower wing covers are simple one-piece monolithic laminates extending 28 feet from tip to tip. No titanium splice plates are used in the cover design. There are eight sine wave composite spars in the main wing box. Figure 6 shows the sine wave front spar design which is fabricated with woven carbon/epoxy cloth. Eighty percent of the torque box serves as an integral fuel tank. Sealing is accomplished around the tank periphery by injecting a flexible compound into a channel groove moulded into the composite spar caps. All fasteners penetrating the fuel cell are removable and have O-ring seals. The spars are locally reinforced with additional plies or metal fittings for concentrated loads. Holes in the auxiliary spars permit fuel flow within the tank.

Four composite ribs are also of sine wave construction. The centerline rib, the fuel closure rib, and the wing tip rib are metal. These ribs experience high out-of-plane loadings which were more suitable for metal design. Figure 7 shows the metal substructure and fittings for the wing torque box. The torque box is assembled primarily with titanium fasteners. No adhesive bonding is used during the assembly process.

The control surfaces and wing fairings are of unitized design using cocuring to fabricate these composite structures. Cocuring reduces the number of tools needed for fabrication and the number of manhours needed for assembly. Details of the cocuring process will be covered in the discussion of the horizontal tail.

Each wing cover is cured on a one-piece machined steel bonding jig. The wing covers are tooled to the inner mould line to minimize shimming and provide a better fuel sealing interface with the substructure. It is significant to note that the varying height of the machined metal parts require the most shimming and fit-up rework during wing assembly. The wing cover is collated in sections on the bonding tool by flat ply collation. Eight separate ply packs are laid up flat and then transferred to the bonding tool. Plans are underway to automate the flat ply collation packs by use of a tape laying machine. The sine wave spars are cured in aluminum matched metal tools using a reduced resin prepreg woven cloth material. The lay-up process is semi-automated using a machine with rollers to position and compact the plies in the sine wave tool.

The wing design just described was arrived at after conducting a design study to identify the most promising candidates and taking into account the experience gained by the team on the F-15 boron wing research and development program. The competing wing designs that were evaluated are as follows:

o   Multirib honeycomb sandwich cover design
o   Multirib hat stiffened cover design
o   Multispar monolithic cover design

Both the multirib honeycomb and hat stiffened cover designs saved more weight than the multispar monolithic cover design. The team's experience with the difficulties of bonding large areas of honeycomb skin, higher projected manufacturing costs, the concern for fuel leakage into the honeycomb core, and the greater propensity for impact damage on thin honeycomb laminate face sheets resulted in eliminating the multirib honeycomb wing cover design from further consideration. In the mid seventies, cocuring technology had not been developed to the extent that the hat stiffeners could be laid up and cured on the skin with one operation. The difficulty of separately fabricating and secondarily bonding the hat stiffeners to the skin became a major concern to both manufacturing and the team. Again, the team's experience with secondary bonding and hat stiffener terminations on the F-15 boron wing program at major wing ribs played a major role in eliminating the multirib hat stiffened design. The multispar monolithic wing cover design was the natural selection of a team that gave attention to experience gained on a previous program.[1]

This design predicted a weight savings of 20 percent whereas the honeycomb and hat stiffened designs predicted a 25 percent weight savings. The tecnical challenge for the multispar design was the substructure. The super critical airfoil was some fifteen inches deep at the aircraft centre line and with a multispar design, the spar loading index was low, generally less than 500 pounds per inch shear flow. Conventional stiffened web designs were too heavy and honeycomb sandwich was ruled out for the same reasons as for the wing covers. Modest success had been achieved with sine wave spars laid up with tape for the B-1 bomber horizontal tail. The team chose woven cloth as the material form

to overcome wrinkling problems with the tape and set out to develop a simple cost effective tooling concept. A simple tool fabrication method was developed by machining a block of aluminum with the sine wave contour and folding it back on itself and chem milling the spar cavity for a matched metal tool. The wing skins being solid laminates, mostly buckling critical, were now thicker and more damage tolerant than the honeycomb sandwich face sheets or the thinner hat stiffened skin. Also, the additional skin thickness resulted in lower operating strain levels that were easily repaired by traditional bolted repair techniques. The military maintenance community fully endorsed this approach for primary structure since they were not prepared to take up storing perishable materials and training personnel in clean room bonding techniques on a forward basing aircraft that did not have scheduled depot maintenance. However, wet lay-up type repairs are used for ease of repair for secondary structures where a less than perfect bond is more than adequate.

Thus, the die was cast and work began in earnest to design, fabricate, test, and fly the first all composite wing intended for production (see References 1 and 2). The AV-8B Composite Wing Program began on 1 November 1975 to develop the multispar monolithic wing cover torque box design selected for the advanced AV-8B Harrier wing. Figure 8 summarizes the substantial amount of test development work that went into the AV-8B wing development. The entire test program was structured to develop and verify the design through an orderly process in which the end result would be a well proven and tested wing torque box design.

Figures 9, 10, 11 and 12 depict the extent and depth of the composite wing development program. This wing development program featured element test, subcomponent tests, box beams, a full scale prototype wing static test, and a full scale AV-8B production static and fatigue test program. Element tests were conducted to develop the basic design allowables, to determine the effect of environmental degradation on fatigue and static strength, and to determine the effect of ballistic damage and manufacturing imperfections on both fatigue and static strength. Subcomponent tests were conducted to verify the design of critical structural areas. Five major box beam tests of critical design areas on the wing torque box structure were completed prior to fabrication of the first full scale wing structure. The subcomponent and box beam test specimens also served as manufacturing development articles to work out tooling, lay-up procedures, curing operations, and non-destructive testing techniques.

A full scale YAV-8B prototype wing was fabricated and subjected to a complete set of static test conditions and a limited amount of fatigue testing sufficient to clear the YAV-8B prototype for flight testing. The first full scale wing was successfully static tested in August 1978 and the first flight of the AV-8B prototype, the YAV-8B was on 9 November 1978 (see References 3 and 4). During the full scale development program for the AV-8B, some minor

---

[1] see Footnote 1

production improvements were incorporated into the wing design and the production wing successfully passed all of the fatigue and static strength requirements for the AV-8B aircraft. The entire development and certification test program for the composite wing was completed without a failure or any redesign required except for production improvements. The development and certification testing of the composite forward fuselage and horizontal tail was structured in a similar manner to the wing development program just described.

6 FORWARD FUSELAGE DEVELOPMENT

Unlike the wing torque box, the AV-8B/GR Mk. 5 forward fuselage structure is a lightly loaded structure incorporating provisions for crew environment, flight controls, hydraulic systems, nose landing gear, avionics and various equipment installations. Since equipment changes and frequent access for maintenance is common, the structure was designed to be compatible with equipment mounting modifications and for ease of access. The structure also provides for cockpit pressurization and protection from lightning strike.

The composite forward fuselage shown in Figure 13 is a semimonocoque, multiframe, longeron and stiffener design fabricated from carbon epoxy woven cloth (see References 5 and 6). The basic structure is fabricated in five cocured unitized parts. Two half shell side panels, two bulkheads, and the floor are mechanically assembled and attached to the forward centre fuselage. No secondary bonding or assembly bonding is required. On the AV-8A aircraft, the forward fuselage skins, bulkheads, and floor were chem milled aluminum sheet with many small clips and stiffeners attached with rivets to the skin to resist pressure loading. The forward fuselage sidewall skins, bulkheads, and floor have integrally moulded hat stiffeners. This concept greatly reduces the parts required in the assembly and saves appreciable weight and assembly time. The AV-8A had 237 parts with 6440 fasteners compared to AV-8B/GR Mk. 5 with 88 parts and 2450 fasteners.

The composite forward fuselage design saved 58 pounds per aircraft for a weight savings of 25.3 percent (see Figure 14). Of the 171 pounds of structure, 67 percent or 115 pounds was of composite material. The remaining portion consisted of fasteners, small metal fittings, paint, etc. The point here is that the forward fuselage structure is approaching almost an all composite design yet one third of the weight is still of noncomposite materials.

The forward fuselage sidewalls were tooled to the outer mould line since the cocured hat stiffeners were on the inside. Based on cost, electro-formed nickel was selected over numerical controlled machined aluminum for the mould line surface bond jigs. The material cost of the electro-formed nickel is ten times more than aluminum, but fabrication costs increased the total cost of the aluminum jigs to 35 percent more than the nickel concept. In fabricating electro-formed nickel bond jigs, Figure 15, a plastic mould line master model is built. From the master model, a reinforced plastic mould is fabricated. Approximately 0.19 inch of nickel is deposited on the plastic mould surface to create the bond jig shape. A steel frame is added to the nickel tool for stability. The resulting bond jigs are aerodynamically smooth and have excellent heat distribution characteristics. Steel moulding details were fastened to the surface of the bond jigs to mould door sills and recesses within the skin. The integrally cured hat stiffeners were tooled using removable, hollow core silicone rubber mandrels inside the hat stiffeners and a silicone rubber intensifier pressure pad on the outside. Venting the mandrel core to the autoclave atmosphere equalized pressure on the stiffener and provided pressure to the side panel under the stiffeners. Tools for the remaining forward fuselage composite parts were machined aluminum bonding jigs.

The pressurized cockpit area of the forward fuselage is designed to withstand a 100 ka swept stroke lightning restrike. All areas of the forward fuselage have sufficient current carrying capability to withstand the high level of lightning threat without damage, except in the immediate vicinity of the lightning attach point. Damage is greatly reduced by applying a thin metallic coating to spread and dissipate the lightning energy. Therefore, the main concern was to limit possible damage in the immediate vicinity of a restrike such that it would not compromise flight safety or structural integrity and be repairable. Some damage to the skin surface may be tolerated; however, penetration and sudden loss of cockpit pressure are not permitted. Five woven cloth plies (0.070 inch skin thickness) are required to prevent penetration.

Tests were conducted on samples constructed with the AV-8B carbon/epoxy configuration to evaluate several candidate lightning protection schemes. A conductive surface coating of flame-sprayed tin was selected to prevent lightning penetration. The coating carries most of the current preventing degradation of the basic structure. The coating may be vaporized in the immediate vicinity of a restrike, but the basic laminate remains undamaged. Since the forward fuselage is essentially secondary type structure, repair of any laminate damage is accomplished by a local wet lay-up to re-establish the five ply thickness. Figure 16 shows the lightning protection area.

To assure good performance of the on-board avionics systems, extensive testing and analysis were performed to characterize the electromagnetic (EM) behaviour of the composite material (see Reference 7). The first step was to assure electromagnetic compatibility by providing adequate EM shielding and electrical bonding. Second, the antennas mounted on carbon/epoxy must perform satisfactorily.

The conclusions reached from the electromagnetic behaviour test program evaluation were as follows:

o Carbon/epoxy has the inherent shielding capability needed for fighter aircraft design.

o Tests show little difference in shielding between carbon/epoxy and aluminum structures because of joint effects.

o Designs such as finger doublers and/or tin plating are effective in reducing the electromagnetic field leakage through carbon/epoxy joints.

o   Antennas for UHF and L-brand systems
    function properly when using carbon/
    epoxy ground planes.

Thus, carbon/epoxy aircraft structure can
be designed to provide adequate electromagnetic
shielding for avionics and electrical subsystems.

7 HORIZONTAL TAIL STORY

In the late seventies, K V Stenberg and the
writer visited Douglas Aircraft to see a DC-10
cocured rudder that Douglas had developed under
the Aircraft Energy Efficiency Program for NASA.
Douglas had cocured this part using a large
number of internal rubber mandrels that were
intricately assembled during the lay-up and
later extracted after cure through holes in
the closure spar. The idea sparked an interest
and we set out to simplify the tooling concept
yet preserve the best feature and apply this
approach to a unitized cocured design for the
AV-8B horizontal tail. The in-service maintenance problems with core corrosion and our
F-15 boron wing experience convinced us that
if a weight efficient cocured horizontal tail
design could be developed without honeycomb
it would be a very cost effective durable design.
It is significant to note that almost all
horizontal tail structures on high performance
aircraft had been of full depth honeycomb design.
Thus the stage was set for the next significant
development.

The horizontal tail consists of a primary
structural torque box with separate leading
edge, trailing edge and tip. The torque box
is a carbon/epoxy multispar monolithic cover
design. It consists of upper and lower carbon/
epoxy tape skins, woven cloth channel spars,
and metal drive and pivot support ribs. All
spars are integrally cocured with the lower
skin. The upper skin is a separate component
which is bolted to the spars (see Figure 17).
The cocured lower skin has just enough fasteners
through the bond joint to sustain limit load
in the event of impact damage such that loss
of the bond joint could occur (see Reference 8).

The design just described and a full depth
honeycomb sandwich design were studied to
determine the weight and manufacturing impact
of the two competing designs. It was found
that there was very little difference in the
projected weights using a post buckled design.
A distinct cost advantage was projected for
the unproven cocured tooling concept. Since
the cocured design preserved all of the good
design and manufacturing features, ie no core
material, splice plates, or secondary bonding,
work began to prove out the cocured tooling
concept.

The new design generated for the horizontal
tail employed a tooling concept where the skins
and spars are fitted together in their assembled
position and then cured in a single autoclave
operation. The spars are cocured integral to
the lower skin and the upper skin is made
separable by use of a release film to provide
a good fit and provide access for installation
of the separately fabricated parts.

The tooling concept was developed in a
series of steps. The first tools were used
to fabricate a full depth box section and a
quarter scale torque box. Next, a full depth

manufacturing demonstration article was fabricated to prove the tooling concept. The cured
article contained some visible mark off at the
cocured bond joint and some localized wrinkling
in the skins but confidence was established
and production tooling was authorized.

The bond jig consisted of rolled 0.75 inch
steel plate welded to an egg crate support. The
lower skin mould line contours were numerically
machined into the plate. The spar and spacer
mandrels were aluminum. Female ply lay-up tools
were provided for the upper and lower torque
box skins to minimize wrinkling on the bond
jig. The fabrication sequence, Figure 18, is
as follow: The torque box skin and cocured
spar plies are cut with a Gerber cutter and
concurrently laid up on their respective tools.
These assemblies are then prebled. The prebled
lower skin is then transferred to the bond
fixture. The spacer mandrels and spar mandrels
with the prebled spars intact, are assembled
in place over the lower skin (Figure 19). A
mould release film is placed over the assembled
mandrels. The prebled upper skin is now positioned in place over the release sheet, followed
by another release sheet and a caul plate. The
entire assembly is then vacuum bagged and cured
in an autoclave.

The composite cocured horizontal tail had
a weight savings of 20 percent or 47 pounds
per aircraft. Both the upper and lower skins
were allowed to buckle prior to ultimate load.
The extent of post buckling permitted depended
upon the skin thickness and spar spacing. The
governing limitation on post buckling was the
magnitude of the out-of-plane loading induced
by skin buckling and reacted by flatwise tension
or compression loading at the skin-to-spar
juncture. A post buckling allowable curve of
joint strength versus a buckling parameter was
developed for both the mechanically fastened
upper skin-to-spar joint and the cocured lower
skin joint. The data to establish the allowable
was obtained from compression box tests. Considerable difficulty was encountered in the design,
test and post analysis of the box specimens
used for measuring out-of-plane forces on the
substructure due to post buckling. A combination of simple joint test specimens to determine
the effect of moisture and temperature degradation on the bond joint, post buckling analysis
and correlation to measured spar strains for
the out-of-plane forces, and a conservative
linear interaction between shear and tension
was used to establish the post buckling design
allowables. In the main box section at midspan,
the onset of buckling in the upper skin began
at 80 percent Design Limit Load (DLL). Buckling
prior to limit load is not unacceptable for
aerodynamic smoothness as the buckle contour
is barely perceptible in the best of light.
During static test at room temperature dry
conditions, the horizontal tail failed at mid-
span due to post buckling at a load level of
203 percent design limit load. This value is
reduced to 180 percent DLL using the interpretation of test results procedure to account
for environmental degradation. The section
on interpretation of test results will provide
a discussion of the procedure.

8 UNITIZED COCURED DESIGN

As a result of the successful forward fuselage and horizontal tail designs, it was decided that four additional already designed wing components would be redesigned using the unitized cocured design concept developed for the horizontal tail. The four components selected for improvement were the flap, auxiliary flap, aileron and outrigger fairing (see References 9 and 10). The redesign and recertification was justified on the basis of reduced cost and weight savings. The reduced cost is attributable to fewer parts and fasteners resulting in less manhours to complete the final assembly. Figures 20 and 21 show the results of the unitized co-cured redesign on the outrigger fairing and clam shell doors. The redesign saved 231 manhours and 8 pounds per aircraft, a significant cost and weight reduction. Similar savings resulted from redesign of the wing control surfaces.

9 CARBON/BISMALEIMIDE

Several structural components on the AV-8B are fabricated from a higher temperature resistant resin than the 250°F 3501-6 Hercules system. A 450°F bismaleimide system was selected for production of the gun and ammo paks, lower and trailing edge flap skins, deep strakes, and ventral fin (see Figures 2 and 3 and Reference 11). Higher than expected ground operation and V/STOL lift off temperatures experienced during the YAV-8B prototype development necessitated a change from the carbon/epoxy designs. In the early eighties the only available higher temperature resin was the V-378A polymeric system and was available only in limited quantities for development work. Earlier, preliminary evaluations of the V-378A resin had shown it to be susceptible to micro cracking with unidirectional tape. However, when the AV-8B team tried the resin with woven cloth, the micro cracking problem disappeared. Subsequently, the qualification, development, and allowables testing was completed on a parallel basis with the first introduction of carbon/bismaleimide production parts into fleet service.

10 COMPOSITE ISSUES

Prior to and during the development of the wing design, a number of composite issues had to be resolved and a position established. Some of these issues were internal to the design team, others were concerns of the government and aerospace community. Of importance is that none of these issues significantly altered the basic design approach for the composite structure. The following sections discuss these issues.

11 IMPACT DAMAGE

In the mid seventies, a major concern of the aerospace community was the effect that impact damage could have on the static and fatigue strength of composite structure. Impact damage may occur as a result of tool drops, runway stones, asphalt chunks from the runway, hail, bird strike, etc. This damage can vary from no damage, to minor matrix crazing, to interlaminar shear delamination and fibre damage, to punch through of the laminate. The following position was established and a test program structured to generate the substantiating data base:

o   On a walk-around inspection of the aircraft, if you can see damage you take the required action to repair it.

o   If you cannot see any damage, then it is acceptable as is. The test program verified that laminates damaged by impact up to the threshold value had sufficient static and fatigue strength for the 6000 hour design life. The threshold value is defined as the maximum impact energy that can be absorbed and no surface damage detected visually on an unpainted surface from the impacted side. In the vast majority of the test components, the fatigue life exceeded the 12,000 hour test requirement by a factor of two or more (see Reference 12 and 13).

An 0.5 inch diameter impactor was used for impact damage since it was representative of tools and runway stone threats. Results from the test program show that impact damage size is essentially constant for a given impactor and increasing impact energy. Multiple impacts increased the damage but did not result in any additional detrimental damage beyond that obtained from a single impact at a higher energy level. The impact damage test program results show that for carbon/epoxy laminates damaged to the threshold level the fatigue life is near the endurance limit at the limit operating compression strain levels for the AV-8B. Further, the fatigue tests show almost no growth of the impact damaged area. Also, the static strength reduction for damaged versus undamaged is still above the design allowables used for both tension and compression. In conclusion, for the composite designs chosen on the AV-8B airframe, impact damage is not a major concern for reduction in fatigue or static strength.

12 INTERPRETATION OF TEST RESULTS

During the mid to late seventies, the subject of conducting full scale environmental test programs versus utilizing some other means to account for environmental degradation was a much debated issue. The issue is one that is still under some debate today. However, the AV-8B/GR Mk. 5 aircraft were slated for near term production and it was jointly decided by the Navy and AV-8B team to utilize the 'interpretation of test results' methodology for certification of the structural static strength of the composite structures. This procedure utilizes element and subcomponent test results obtained at room temperature, dry conditions and at the critical environmental conditions, together with finite element model data to adjust the measured full scale test strain data to account for environmental degradation. Essentially, the adjustment for identical modes of failure is the percent reduction in strength between the environmental and room temperature dry condition times the full scale test results, so long as there are no nonlinear effects. In theory, this type of interpretation of test results is a simple alternative to a cost prohibitive full scale environment test program. Some of the problems involved with conducting a full scale environmental test program are the major difficulties in bonding tension pads on; formers and/or tension straps would be required, and an environmental conditioning chamber for both hot/wet and cold/dry conditions would be required.

Because, in actual practice, most full scale test failures involve nonlinear failures, ie load redistribution or post buckling, etc, a great deal of postmortem analysis must be conducted on many different areas of the structure before arriving at the adjusted structural strength. The conservative assumptions to reduce analysis complexities which might be satisfactory for initial design, must be re-evaluated when conducting these post-mortem failure analyses.

13 ROLE OF FULLSCALE STATIC & FATIGUE TEST ARTICLES

A fatigue test article is required to verify the fatigue life of the various metal components on the airframe. Since the adjoining composite structure is generally not fatigue critical, it is essentially just a load introduction structure for the metal parts; hence, the fatigue test could be utilized to check out the proposed repair schemes, impact damage tolerance, and battle damage sensitivity of the composite structure.

In July of 1982, the Air Force hosted an industry wide meeting at Dayton, Ohio to cover a number of composite issues and establish recommendations. The following section is a summary of the industry wide consensus and panel recommendations on interpretation of full scale test results:

Air Force Testing/Certification Panel
   Composite Structure
   Testing, Certification
   Recommendations

o One full scale static test article

   Full scale ambient condition static tests recommended to design ultimate loads

o One full scale durability/damage tolerance article

   Full scale fatigue tests on pristine composite structure is required only if element/subcomponent development tests have shown the design to be fatigue sensitive.

   Full scale damage tolerance tests of composite structure is recommended. Damage tolerance to include impact damage, manufacturing defects, etc.

o Environmental effects are determined

by development tests and methods for factoring or interpreting tests have been resolved.

o Use of materials close to or above glass transition temperatures are to be avoided.

These industry wide recommendations on interpretation of full scale test results are essentially identical to the methodology utilized for the certification of the AV-8B composite structure.

14 DESIGN ALLOWABLES/ANALYSIS

During the F-15 boron wing program, design allowables were generated on the unidirectional tape properties and a failure theory was used along with a best guess strength reduction factor to establish the structural strength. The shortcoming of this approach is that the stress analyst was using a best guess approach to establish the strength reduction factor for structural discontinuities such as fastener holes, fastener holes with bearing stress, notches, cutouts, etc. As you might expect, this approach left something to be desired. It was decided during 1974 that design allowables would be generated for the next program based on structure representative of the design chosen. Since the AV-8B composite wing design was a conventional bolted assembly, the design allowables would be based on laminates with fastener holes and varying bearing stress levels tested at both room temperature and the critical environmental condition. In 1975 the preliminary design allowables were established for both tension and compression as shown in Figure 22. Subsequent data base testing and impact damage data have not altered these originally established design allowables.

Also, just prior to generating data for the design allowables, Dr S L Huang from the Naval Air Defense Center in Philadelphia suggested that we change the carbon/epoxy material from the Narmco 5505 system to the Hercules AS-1/3501-6 system. The reason for the suggested change was a lower price on the material, but it also had a more consistent, higher peel strength attributable to the so-called fuzzy fibre. Peel strength is not a property used for analysis and design but for damage tolerance and shop handling it is an extremely important property for durability.

Use of the simple gross strain versus bearing stress design allowable charts (Figure 22) for design permitted a direct read across of principal gross strain data from finite element model data visually displayed in plotted form. This procedure was a quantum jump in efficiency and accuracy from the F-15 boron wing design allowable and analysis methodology. There was one caveat in their use. These simple charts become less accurate in regions of high biaxial loading and in those cases a more sophisticated analysis technique around a hole or discontinuity was used to establish the design. In using a more complex analysis technique and failure theory, it is of paramount importance that a substantial data base be available to check out the procedure before it is used.

15 COMPOSITE REPAIRS

The basic repair approach for the AV-8B primary composite airframe structure is mechanically fastened repairs. The structure is already designed for fastener holes and additional holes for repairs are easily accommodated. This approach was readily acceptable to the Marine maintenance community since the AV-8B is a forward basing aircraft and does not have any scheduled depot maintenance. Bonded or wet lay-up repairs are used for some secondary structure for ease of repair. An automated repair design tool was developed with company funds to provide basic bolted repair designs on a rapid turn around basis to supplement the maintenance handbooks. A test program verified that the bolted repair schemes would restore the original undamaged laminate strength. Before incorporating the repair schemes in the maintenance handbook, the Marine maintenance facility at Cherry Point, North Carolina checked out the repairs by performing trial repair installations on a number of discarded composite test articles (see References 13, 14 and 15).

16 BALLISTIC/HYDRAULIC RAM

An evaluation test program was conducted to determine the battle damage tolerance of the wing torque box. It was concluded that for the type of composite construction utilized for the AV-8B wing, the test results were not much different than what would be expected for metal construction. Simulated fuel tank boxes with water were subjected to ballistic projectile hits with different ullages and projectile velocities. Since the AV-8B is primarily a close air support attack aircraft, the most probable threat is the 14.5mm Armor Piercing Incendiary (API) projectile with a most probable hit range of 1000 feet at a velocity of 2000 feet/second. For this case, there were no hydraulic ram effects generated on any of the simulated wing box tests. The hydraulic ram effects on the composite structure did not occur with the 14.5mm API until the projectile velocity reached 3500 feet/second which was the muzzle velocity of the gun. The AV-8B aircraft specification requires the structure to sustain limit load after multiple ballistic hits and be able to also sustain 55 percent limit load at the time of ballistic impact. It was found that limit load after ballistic impact is just slightly more critical than ballistic impact at 55 percent limit load.

The most predominant damage mechanism for aircraft battle damage is exploding air vapors for both metal and composite fuel tanks. The probability for this damage mechanism is estimated to range from 50 percent for 14.5mm API to 90 percent for 23mm HEI threats with JP4 fuel; JP5 fuel substantially reduces this threat. The threat from missile fragments for fuel tanks is substantially reduced for composite structures compared to metal structures since you do not have the metal-to-metal contact for incineration. In summary, there are obviously some differences in the failure modes and responses between composite and metal structures for ballistic and hydraulic ram threats, but overall there are no overwhelming concerns with the AV-8B type composite applications that are any different than those for metal structures.

17 MANUFACTURING DEFECTS

It has been the experience of our design team that for the AV-8B type composite structure the vast majority of manufacturing defects have only a minor detrimental effect on the static and fatigue strength of the part. Determination of what size and severity of defect is detrimental does require the expenditure of additional test funds, but the money is well spent when up to one third of the total cost of fabrication may be Nondestructive Testing (NDT). In some cases it was found that the effect of the manufacturing defects could be detected by testing and the test results were used to establish NDT requirements. In other cases, such as sine wave spar porosity, testing showed severe porosity to be acceptable for both static and fatigue strength, but unacceptable for leakage at the fuel tank boundaries. Engineering was unwilling to permit parts containing large areas of severe porosity to be used for the interior spars so judgement was used to establish acceptable limits. Some secondary fairing structures on the AV-8B are just given a visual inspection. Others require X-ray and/or C-scan. In almost all cases, NDT requirements have been substantially reduced based on test results or experience with the structure. In summary, if the NDT requirements are truly reflective of what is acceptable and scrappable, then the number of nonconformance reports accepted by engineering will be a small percentage.

18 A LOOK AT THE FUTURE

The AV-8B/GR Mk. 5 is in full production with almost fifty aircraft delivered for fleet service. Service experience from the YAV-8B prototype, the AV-8B full scale development aircraft, and delivered production aircraft to date has not resulted in any serious problems with the use of composite materials. The drawing boards for future aircraft designs feature composite structure as a primary candidate material. Experience has shown that major cost reduction benefits from manufacturing improvements for composite fabrication and assembly will continue to improve the cost picture for composite structure. From my perspective, one of the dangers may like in the fact that since the material is considered state of the art, everyone is assumed to have acquired the technology. It is quite possible that the real risk is on future applications where the budgets and schedules do not permit the thorough development testing of past programs. In the arena of fixed price contracts in the United States, and the less forgiving properties of composite materials, the next generation of program managers for fighter aircraft may well earn their titles. As Dr A Somoroff, former head of Navy aircraft R & D, once stated:

> 'The only thing that worries me about composites is that you guys on the AV-8B made it look so easy, everybody will think they can do it!'

FOOTNOTE 1

The F-15 boron wing design featured a multirib hat stiffened torque box design. Spanwise fibre glass softening strips were used in the boron skin along the spars to increase the design strain compatibility. At the rib locations, additional plies were added to the skin to lower the strain at the rib attachments. Then, a thin layer of honeycomb was added to the skin between the ribs to provide a flat interior surface to secondarily bond the hat stiffeners to the skin. The ribs were carbon/epoxy honey - comb sandwich design. This wing torque box design had a number of undesirable productibility characteristics. This included titanium splice plates in the skin, secondarily bonded honeycomb in the skin and substructure, soften- ing strips in the skin, and secondarily bonded hat stiffeners on the skin.

REFERENCES

1. Weinberger, R A, Somoroff, A R and Riley B L United States Navy Certification of Composite Wings for the F-18 and Advanced Harrier Air- craft. AGARD Report 660, 1977.

2. Huttrop, M L. Composite Wing Substructure Technology on the AV-8B Advanced Aircraft. Fourth Conference on Fibrous Composites in Structural Design, San Diego, CA, 1978.

3. Stenberg, K V and Harvey, F L. YAV-8B Wing Operational Experience. Fifth Conference on Fibrous Composites in Structural Design, New Orleans, Louisiana, 1981.

4. Stenberg, K V. YAV-8B Flight Demonstration Program. AIAA Prototype and Technology Demon- strator Symposium, Dayton OH, 1983.

5. Watson, J C. Preliminary Design Development AV-8B Forward Fuselage Composite Structure. Fourth Conference on Fibrous Composites in Structural Design, San Diego, California, 1978.

6. Watson, J C and Ostrodka, D L. AV-8B Forward Fuselage Development. Fifth Conference on Fibrous Composites in Structural Design, New Orleans, Louisiana, 1981.

7. Skouby, C D and Weinstock, G L. Techniques in the Design of Graphite Epoxy Structures for Electrical Performance and Maintainabi- lity. Fifth Conference on Fibrous Composites in Structural Design, New Orleans, Lousisana, 1981.

8. Moors, G F, Arseneau, A A, Ashford, L W and Holly, M K. AV-8B Composite Horizontal Stabilizer Development. Fifth Conference on Fibrous Composites in Structural Design, New Orleans, Louisiana, 1981.

9. Huttrop, M L and Watson, J B. Production Improvements to the AV-8B Composite Wing Structure. Fifth Conference on Fibrous Compo- sites in Structural Design. New Orleans, Louisiana, 1981.

10. Huttrop, M L. Cost Reduction Through Design for Automation. Seventh Conference on Fibrous Composites in Structural Design. Denver, Colorado, 1985.

11. Groenenboom, J A and Morrs, G F. AV-8B Carbon/Bismaleimide 25mm Gun/Ammo Pak Design Seventh Conference on Fibrous Composites in Structural Design. Denver, Colorado 1985.

12. Padilla, V E. Low Velocity Damage Tolerance of Thin Graphite/Epoxy Laminates. Fifth Conference on Fibrous Composites, New Orleans, Louisiana, 1981.

13. Carrier, W L. Damage Tolerance and Repair of Composite Fuselage Structure. Seventh Conference on Fibrous Composites in Structural Design. Denver, Colorado, 1985.

14. Bohlman, R E, Renieri, G D and Libeskind, M. Bolted Field Repair of Graphite/Epoxy Wing Skin Laminates. ASTM Symposium Composite Materials, Minneapolis, Minnesota, 1980.

15. Bohlman, R E, Renieri, G D and Riley, B L. Bolted Composite Repairs Subjected to Biaxial or Shear Loads. NADC Contract N62269-81-C-0297, Classification Code for public release 22CFR125.11(A)(2).

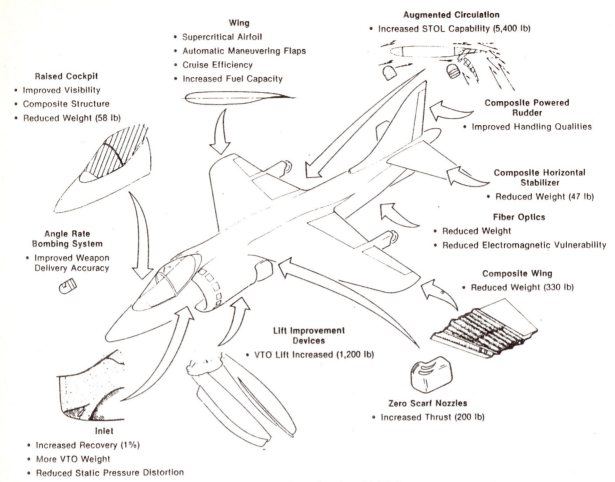

Raised Cockpit
- Improved Visibility
- Composite Structure
- Reduced Weight (58 lb)

Wing
- Supercritical Airfoil
- Automatic Maneuvering Flaps
- Cruise Efficiency
- Increased Fuel Capacity

Augmented Circulation
- Increased STOL Capability (5,400 lb)

Composite Powered Rudder
- Improved Handling Qualities

Composite Horizontal Stabilizer
- Reduced Weight (47 lb)

Fiber Optics
- Reduced Weight
- Reduced Electromagnetic Vulnerability

Angle Rate Bombing System
- Improved Weapon Delivery Accuracy

Composite Wing
- Reduced Weight (330 lb)

Lift Improvement Devices
- VTO Lift Increased (1,200 lb)

Zero Scarf Nozzles
- Increased Thrust (200 lb)

Inlet
- Increased Recovery (1%)
- More VTO Weight
- Reduced Static Pressure Distortion

Fig 1   AV–8B technology highlights

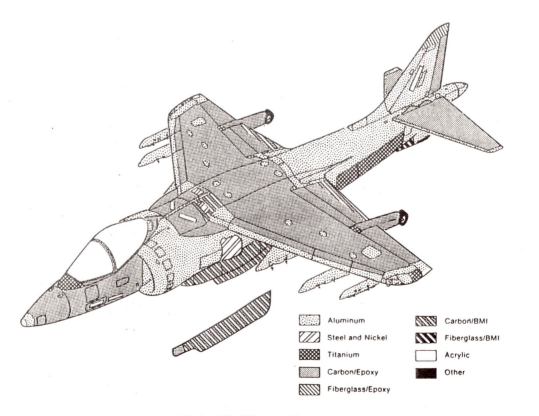

░ Aluminum		▨ Carbon/BMI
▨ Steel and Nickel		◥ Fiberglass/BMI
▓ Titanium		☐ Acrylic
▒ Carbon/Epoxy		■ Other
▨ Fiberglass/Epoxy		

Fig 2   AV–8B materials usage

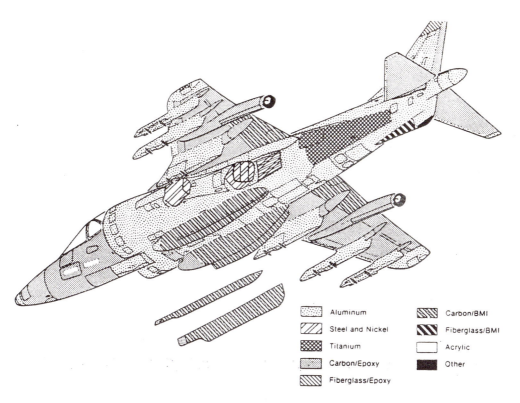

	Aluminum			Carbon/BMI
	Steel and Nickel			Fiberglass/BMI
	Titanium			Acrylic
	Carbon/Epoxy			Other
	Fiberglass/Epoxy			

Fig 3   AV—8B materials usage

Structure	Weight Saving Percent	Composite Percent	Total Weight (lb)	Weight Savings (lb/AC)
Wing				
• Torque Box	19	64	1,078	256
• Flap	17	62	153	32
• Auxiliary Flap	16	68	31	6
• Aileron	17	77	43	9
Forward Fuselage and Nose Cone Structure	23	67	196	58
Overwing Fairing	37	67	27	16
Outrigger Fairing	24	71	75	27
Engine Bay Doors	19	60	42	10
Rudder (Less Balance wt)	20	38	12	3
Horizontal Stabilator	20	73	182	47
LIDs Fence	20	60	8	2
Fuselage Mounted Strakes	13	62	94	14
Ammo Pod	19	42	114	26
Gun Pod	10	30	174	19
			Total	525

Fig 4   AV—8B composite weight summary

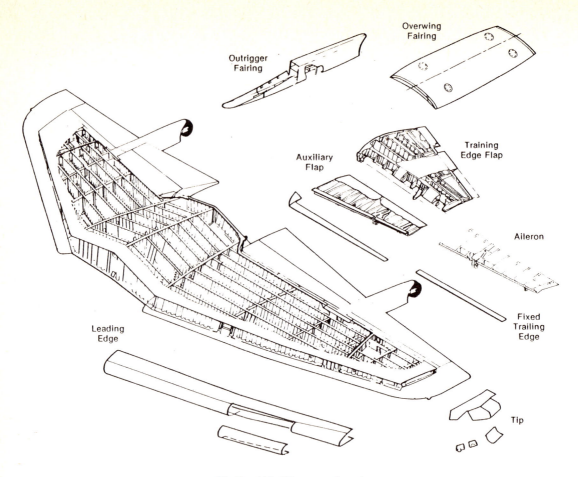

Fig 5   AV–8B composite wing

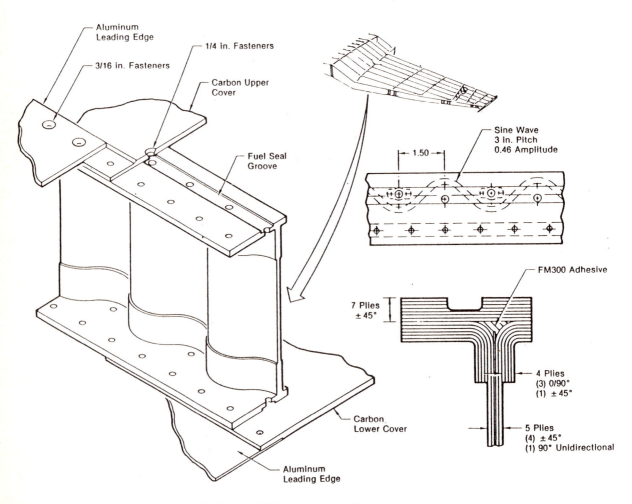

Fig 6   AV–8B front spar — fuel tank area

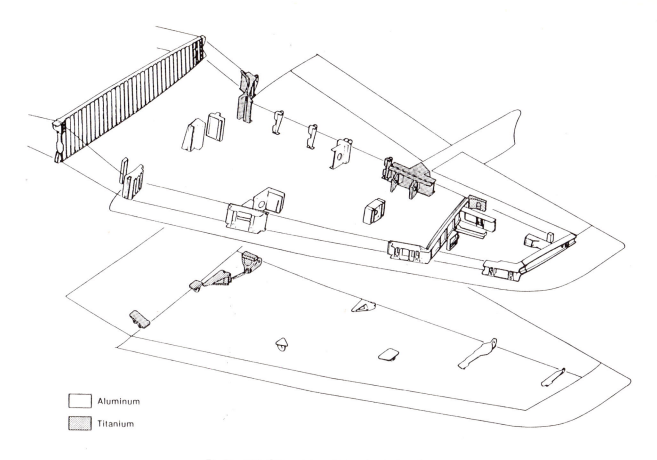

Aluminum

Titanium

Fig 7   AV−8B metal parts in wing torque box

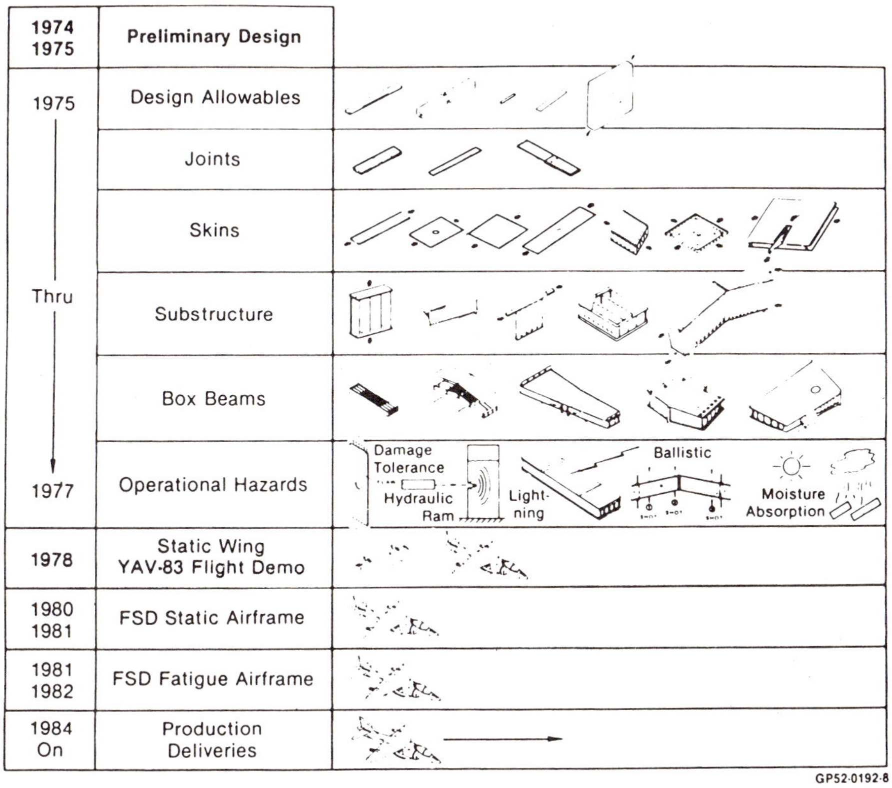

1974 1975	**Preliminary Design**	
1975	Design Allowables	
	Joints	
	Skins	
Thru	Substructure	
	Box Beams	
1977	Operational Hazards	Damage Tolerance Hydraulic Ram Ballistic Light-ning Moisture Absorption
1978	Static Wing YAV-83 Flight Demo	
1980 1981	FSD Static Airframe	
1981 1982	FSD Fatigue Airframe	
1984 On	Production Deliveries	

GP52-0192-8

Fig 8   AV−8B composite wing development

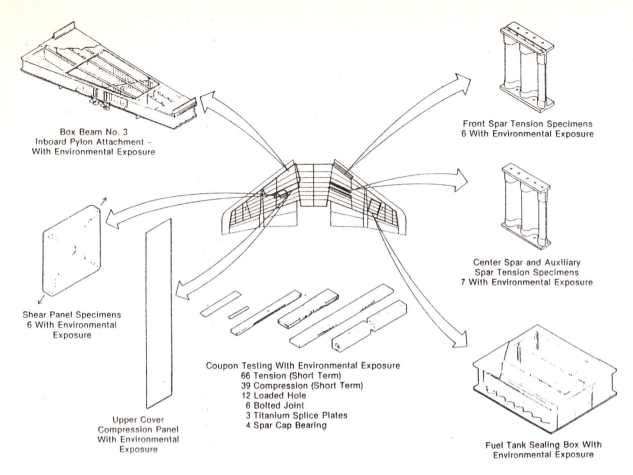

Box Beam No. 3
Inboard Pylon Attachment –
With Environmental Exposure

Front Spar Tension Specimens
6 With Environmental Exposure

Shear Panel Specimens
6 With Environmental
Exposure

Center Spar and Auxiliary
Spar Tension Specimens
7 With Environmental Exposure

Upper Cover
Compression Panel
With Environmental
Exposure

Coupon Testing With Environmental Exposure
66 Tension (Short Term)
39 Compression (Short Term)
12 Loaded Hole
6 Bolted Joint
3 Titanium Splice Plates
4 Spar Cap Bearing

Fuel Tank Sealing Box With
Environmental Exposure

Fig 9   AV–8B environmental testing

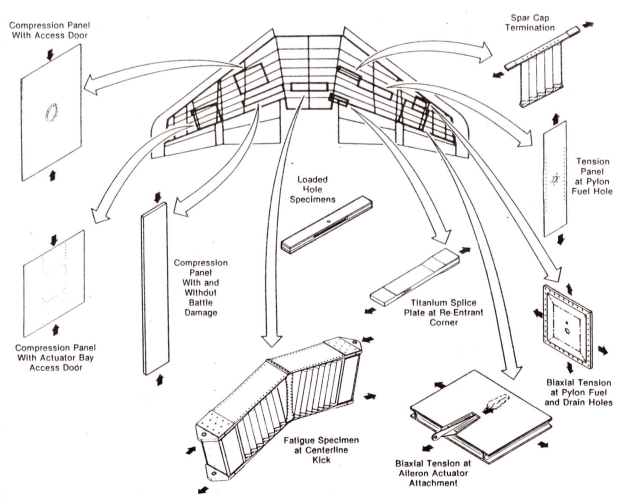

Compression Panel
With Access Door

Spar Cap
Termination

Loaded
Hole
Specimens

Tension
Panel
at Pylon
Fuel Hole

Compression
Panel
With and
Without
Battle
Damage

Compression Panel
With Actuator Bay
Access Door

Titanium Splice
Plate at Re-Entrant
Corner

Biaxial Tension
at Pylon Fuel
and Drain Holes

Fatigue Specimen
at Centerline
Kick

Biaxial Tension at
Aileron Actuator
Attachment

Fig 10 AV–8B cover test specimens

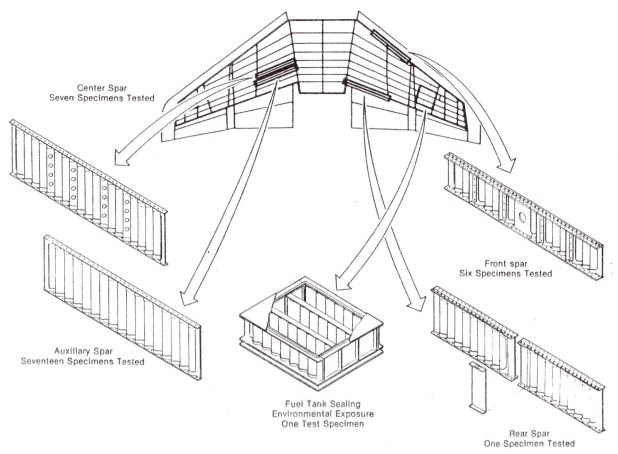

Center Spar
Seven Specimens Tested

Auxiliary Spar
Seventeen Specimens Tested

Front spar
Six Specimens Tested

Fuel Tank Sealing
Environmental Exposure
One Test Specimen

Rear Spar
One Specimen Tested

Fig 11 AV—8B substructure testing

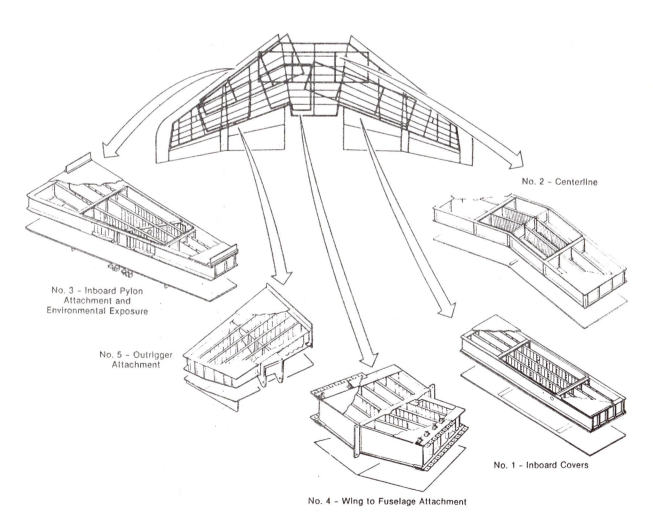

No. 2 – Centerline

No. 3 – Inboard Pylon
Attachment and
Environmental Exposure

No. 5 – Outrigger
Attachment

No. 1 – Inboard Covers

No. 4 – Wing to Fuselage Attachment

Fig 12 AV—8B major box beams

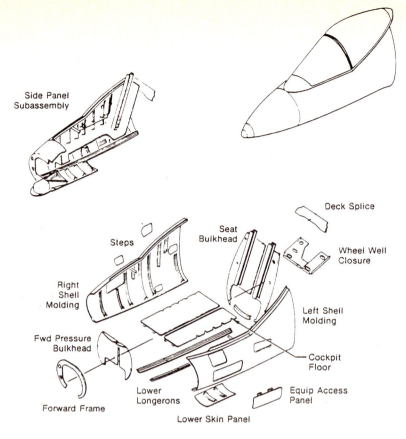

Fig 13 AV—8B forward fuselage structural components

	Metal, (lb)	Composite, (lb)	Savings, (lb)	Savings, (%)
Skin and Doors	88	65	23	26.1
Frames	75	54	21	28.0
Longerons	28	19	9	32.1
Flooring	24	24	0	0
Miscellaneous	14	9	5	35.7
Total	229	171	58	25.3

GP52-0192-14

Fig 14 AV—8B forward fuselage structural weight

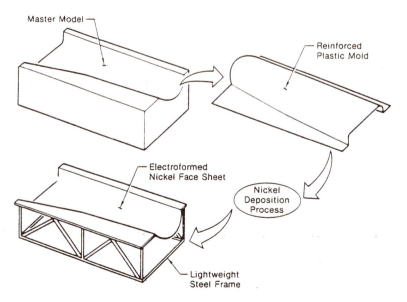

Fig 15 AV—8B electroformed nickel bonding jig fabrication sequence

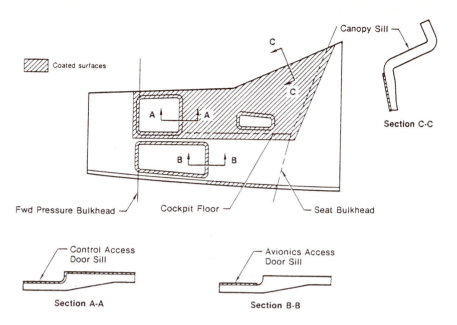

Fig 16 AV—8B forward fuselage lightning protection

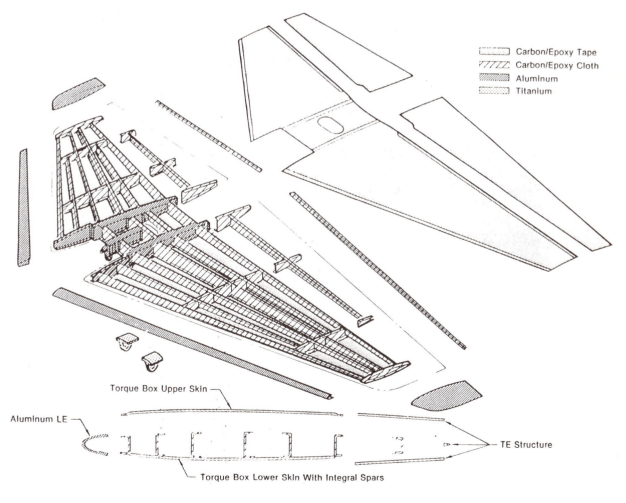

Fig 17 Horizontal stabilator

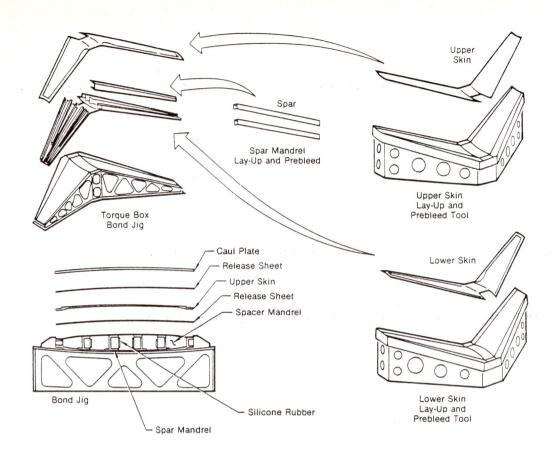

Upper Skin

Spar

Spar Mandrel
Lay-Up and Prebleed

Upper Skin
Lay-Up and
Prebleed Tool

Torque Box
Bond Jig

Caul Plate
Release Sheet
Upper Skin
Release Sheet
Spacer Mandrel

Lower Skin

Bond Jig

Silicone Rubber

Spar Mandrel

Lower Skin
Lay-Up and
Prebleed Tool

Fig 18 AV—8B composite torque box fabrication flow chart

Fig 19 AV—8B spar and spacer mandrels on bond jig

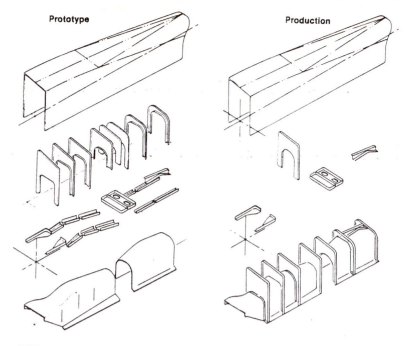

Design Concept		Prototype	Production (Cocured)	Savings	Percent Savings
Parts/Assembly		26	9	17	65
Fasteners/Assembly		929	510	419	45
Weight/Aircraft	(lb)	46	42	4	9
Cost Reduction/Aircraft	(MH)	—	—	164.8	—

Fig 20 AV--8B landing gear fairing — design comparison

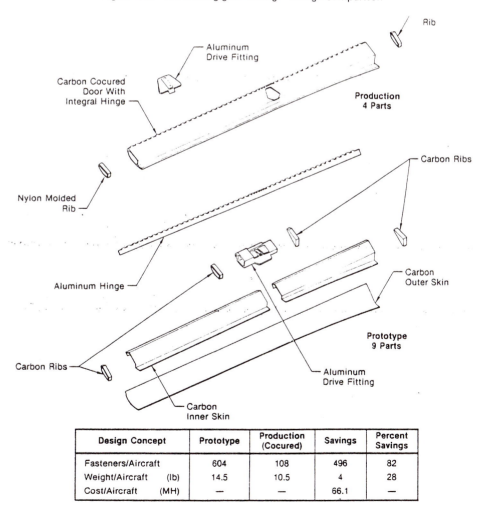

Design Concept		Prototype	Production (Cocured)	Savings	Percent Savings
Fasteners/Aircraft		604	108	496	82
Weight/Aircraft	(lb)	14.5	10.5	4	28
Cost/Aircraft	(MH)	—	—	66.1	—

Fig 21 AV—8B landing gear fairing door — design comparison

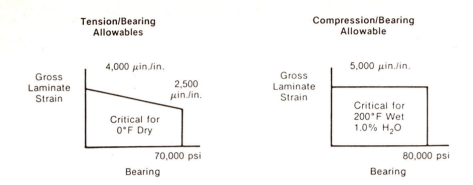

Fig 22 AV—8B laminate strain/bearing design allowables

C28/86

Fatigue growth of transverse ply cracks in (0/90₂)s glass fibre reinforced and (0/90₃)s carbon fibre reinforced plastic laminates

S L OGIN, BSc, PhD, **L BONIFACE**, BSc and **M G BADER**, BSc, CEng, MRAeS
Department of Materials Science and Engineering, University of Surrey, Guildford

SYNOPSIS   The growth of individual transverse ply cracks during tension-tension fatigue of $(0/90_2)_s$ GFRP and $(0/90_3)_s$ CFRP has been observed using transmitted light and penetrant enhanced X-radiography, respectively. The growth rate of cracks in both laminates is independent of crack length but depends on the spacing between the cracks. Very low growth rates occur for small crack spacings or if crack branching occurs. The results are interpreted using a stress intensity factor approach.

1   INTRODUCTION

The complexity of the mechanical response of composite laminates is illustrated by the fact that even the simplest of damage mechanisms, matrix cracking, is still not well characterised for dynamic loading. In polymer based, continuous fibre laminated composites this is usually the initial mode of damage and can lead to large stiffness changes in both angle-ply and crossply laminates. Matrix cracking can also lead to fibre breakage in adjacent plies and is often the precursor of delamination. It is thus important to be able to model and predict this form of damage.

In early work on dynamic loading, Broutman and Sahu (1) showed that fatigue leads to crack accumulation and a consequent stiffness reduction. Reifsnider and co-workers (2,3) demonstrated that the final crack patterns appear very regular when viewed from the laminate edge, with an average crack spacing about the same as the thickness of the transverse ply. A number of authors have derived analytical expressions which relate the density of matrix cracks, accumulated either by monotonic or dynamic loading, to the macroscopic stiffness reduction (e.g. 4-6).

Several attempts have been made to derive damage growth laws for matrix cracking which would enable crack growth rates to be predicted for different loading conditions. These models have, perhaps not surprisingly, been based on fracture mechanics. Wang and co-workers (7,8) based their approach on the strain energy release rate as a flaw grows across the thickness of the transverse ply into a crack (the flaw is assumed to extend initially across the laminate width). Poursartip et al. (9) suggested a relationship between the damage accumulation rate, the current damage state and the cyclic load, and Poursartip (10), using a compliance/crack length relationship, demonstrated a correlation between the stiffness reduction rate (i.e. the rate of damage accumulation) and the peak stress during fatigue loading of $(0/90)_{2s}$ laminates.

An alternative fracture mechanics approach has been proposed by Ogin et al. (11,12) and Ogin and Smith (13). Here, transverse ply cracks are treated as short, enclosed cracks bounded by the 0 plies. Growth of the cracks across the width is governed by a stress intensity factor which is independent of crack length but dependent on crack spacing and transverse ply thickness. This model has been applied to crack development during the fatigue of a $(0/90)_s$ GFRP laminate to predict the stiffness reduction with cycles curve at different load levels. In this work, the relevance of the model to GFRP and CFRP laminates containing thick transverse plies is investigated.

2   EXPERIMENTAL

The $(0/90_2)_s$ GFRP laminates were manufactured by Fothergill Rotorway Composites without the usual 'peel ply' so that the surfaces were smooth and the transverse ply cracks could be observed in transmitted light. Coupons 25 mm wide, nominal ply thickness 0.125 mm, were cycled between various load levels at R=0.1. The tests were interrupted at various stages and the cracks photographed whilst the coupon was under a small load. The CFRP laminates, with a $(0/90_3)_s$ configuration, were laid-up and autoclave-moulded at the RAE, Farnborough from prepreg of cured thickness 0.125 mm containing Ciba-Geigy 'XAS' fibres in 'Fibredux 914' epoxy resin. Coupons 20 mm wide were fatigued with R=0.01. Both sets of tests were at 10 Hz.

Penetrant-enhanced X-radiography, using a zinc iodide penetrant, enabled crack growth in the CFRP specimens to be followed. A small load was applied to the coupons and the zinc iodide solution applied to both edges for at least 30 minutes to ensure complete penetration of the cracks. Crack density measurements in these tests have been reported elsewhere (14,15). Enlarged prints of the radiographs enabled individual crack growth rates to be measured.

3   RESULTS

Photographs of the complete width of a section of a GFRP laminate cycled at $\sigma_{max} = 125$ MPa are shown in Fig. 1. Most transverse ply cracks in the early part of the test initiate at the edges of the specimen. Later in the test, an increasing number of cracks initiate away from the edges. Crack A, for example, which is first seen at cycle 810, initiates well away from the edge and then grows parallel to its length. At cycle 42 000, many cracks which do not extend to the laminate edge can be seen.

Crack growth rate measurements show the growth rates to be independent of crack length. These observations are illustrated by cracks B and C (Table 1). Although crack C is some 3 times longer than crack B at cycle 110, its growth rate is about 1/5 that of crack B. The growth rates are, however, dependent on crack spacing. Fig. 2 shows growth rates against the distance of the growing crack tip to the next nearest crack. The scatter is quite large but, in general, at small crack spacings (<1 mm) the growth rates are small rising rapidly above 1 mm. Cracks D and E illustrate these observations (Table 2). These cracks grow rapidly at spacings of 2.3 mm and 1.2 mm respectively but when the spacing is reduced by crack overlap the growth rate falls to very low or negligible values ('crack arrest').

Another mechanism which can severely reduce crack growth rates is bifurcation or crack branching. Crack F, for example, which is bifurcated at cycle 110, grows very slowly until one branch extends beyond the bifurcation point. The cause of this crack branching is not clear at present.

Results on CFRP are very similar to the GFRP results. Figs. 3 and 6 show the development of the cracking in a laminate tested at σ_{max} = 285 MPa. In the early part of the test, crack growth cross the laminate width occurs very rapidly (i.e. in less than 2000 cycles) so that new cracks appear to span the laminate width immediately (compare cycles 12 000 and 14 000). Later in the test, when the crack spacings are smaller, slow crack growth across the laminate width is observed. All the cracks appear to nucleate at the free edges but this observation must be treated with caution. The X-radiograph technique is dependent on cracks running to a free edge to allow penetration of the dye.

Measurements of individual crack growth rates from the radiographs again show the growth rates to be independent of crack length but to depend on the spacing of the next nearest crack. Table 3 illustrates these results for the five cracks labelled 'a' to 'e' in Figs. 3 and 6. The Table shows the growth rate, the current crack length and the spacing of the next nearest crack. As the crack length increases, the growth rate remains roughly constant unless the crack spacing decreases. Fig. 5 shows crack growth rates against distance to the next nearest crack. The growth rates are negligible for spacings <0.4 mm rising rapidly for larger spacings.

4 DISCUSSION

The results can be interpreted using a stress intensity factor approach. An approximate expression for the stress intensity factor, K,

of a growing transverse ply crack which spans the laminate thickness is (12,13)

$$K = \sigma_t \sqrt{(2d)} \qquad (1)$$

where σ_t is the stress in the transverse ply which acts on the crack tip and 2d is the transverse ply thickness. This expression is independent of crack length, but depends on the crack spacing through σ_t which decreases as the crack spacing decreases. (The stress, σ_t, can be calculated using a shear-lag analysis and the appropriate laminate moduli and dimensions.) The crack is assumed, in deriving eqn. 1, to be flat and to extend across the thickness, but not across the width, of the transverse ply. Away from the crack tip, load is transferred into the 0 plies by shear at the 0/90 interfaces and hence does not build up at the crack tip. The stress intensity factor arises from the localised disturbance at the crack tip which has a characteristic length of about half the transverse ply thickness. The stress intensity factor will be less than $\sigma_t \sqrt{(\pi d)}$ since the crack does not extend through the full thickness of the laminate but is bounded by the 0 plies.

The results can be understood qualitatively using this approach. The stress acting on the crack tip decreases as the crack spacing decreases but is independent of crack length. If a general fatigue crack growth law is assumed, i.e.

$$\frac{dz}{dN} = f(\Delta K, K_{max}) \qquad (2)$$

where dz/dN is the crack growth rate, ΔK is the stress intensity factor range and K_{max} is the maximum stress intensity factor, then the crack growth rate is independent of crack length, as observed. The growth rate depends on the crack spacing however, through σ_t, and decreases if the spacing decreases.

A quantitative test of these ideas for the growth of individual cracks in the CFRP laminate tested at σ_{max} = 285 MPa is shown in Fig. 5. The stress acting on a crack tip consists of a mechanical and thermal contribution both modified by a shear-lag term (16). As a first approximation, the mechanical stress can be taken to be the peak stress given by a shear lag expression for cracks spaced 2s apart. This has been calculated using a parabolic shear lag analysis (6) and the CFRP laminate moduli and dimensions. The thermal strain in the CFRP laminate is about 0.38% coresponding to a thermal stress of 38 MPa. Fig. 5 shows a log-log plot of the expression

$$\frac{dz}{dN} = A K_{max}^m \qquad (3)$$

(results over a single load range are insufficient to distinguish between ΔK and K_{max}) using average values of the crack growth rates for a given crack spacing. The results suggest that a power law expression can be used to describe the growth of individual cracks. Further work is required over different load ranges to establish whether the growth rate is a function of K_{max}, ΔK or of both.

It is interesting to compare the values of K_{max} derived here with an estimate of the fracture toughness, K_c, for matrix cracking. From work on the fatigue growth of delaminations in angle-ply XAS 914 laminates (17), the toughness, G_c, corresponding to the onset of delamination is about 500 J/m^2. Taking the value of G_c for transverse ply crack growth to be the same and assuming the isotropic relation $K_c = \sqrt{(E_2 \cdot G_c)}$ (where E_2 is the modulus of the transverse ply), then $K_c \approx 2.2 MPa \sqrt{m}$. This is slightly above the value of K_{max} for the highest growth rates observed here. The results suggest, then, that the crack growth behaviour can be idealised by the usual sigmoidal fatigue crack growth curve. For $K>K_c$, fast fracture occurs. This corresponds to the early part of the test described above when the crack spacing is large. Slow crack growth occurs for $K<K_c$ and a threshold value for crack growth corresponds to crack arrest (at very small crack spacings). This idealised picture of crack growth will be complicated by variations in the local fracture toughness throughout the laminate caused, perhaps, by varying interfibre spacing and fibre/matrix strengths. The greater scatter of the GFRP results indicates a greater variation in the properties of this laminate.

5 CONCLUSIONS

1 Transverse ply crack growth in $(0/90_3)_s$ CFRP and $(0/90_2)_s$ GFRP is independent of crack length but depends on the spacing to the next nearest crack. Crack growth rates can be described using power law relationships.

2 Crack density measurements in CFRP must be treated with caution. Firstly, observation of the edge alone may include cracks which have not grown across the laminate width. Secondly, X-radiography will not detect cracks which initiate within the laminate but do not extend to the laminate edge.

REFERENCES

1 BROUTMAN, L.J. and SAHU, S. Progressive damage of a glass reinforced plastic during fatigue, Proc. 24th Ann. Techn. Conf., S.P.I., 1969, Section 11-D, 1-12.

2 HIGHSMITH, A.L. and REIFSNIDER, K.L. Stiffness reduction mechanisms in composite laminates. ASTM STP 775, 1982, 103-117.

3 REIFSNIDER, K.L. and TALUG, A. Analysis of fatigue damage in composite laminates. International Journal of Fatigue, 1980, 3-11.

4 TALREJA, R. Fatigue of Composite Materials. Thesis, Technical University of Denmark, 1985.

5 LAWS, N. and DVORAK, G.J. The loss of stiffness of cracked laminates. Proc. Eshelby Memorial Symposium, eds. Bilby B.A., Miller K.J. and Willis J.R., Cambridge University Press, 1985, 119-127.

6 STEIF, P.S. see Appendix to (11).

7 WANG, A.S.D., CHOU, P.C. and LEI, S.C. A stochastic model for the growth of matrix cracks in composite laminate. Journal of Composite Materials, 1984, 18, 239-254

8 WANG, A.S.D., KISHORE, N.N. and LI, C.A. Crack development in graphite-epoxy cross-ply laminates under uniaxial tension. Composites Science and Technology, 1985, 24, 1-31.

9 POURSARTIP, A., ASHBY, M.F. and BEAUMONT, P.W.R. Damage accumulation during fatigue of composites. Proc. 3rd. Riso Int. Symp. on Metall. and Mater. Sci., 1982, 279-284.

10 POURSARTIP, A. Aspects of damage growth in fatigue of composites. Ph.D. Thesis, Cambridge University Engineering Department, 1983

11 OGIN, S.L., SMITH, P.A. and BEAUMONT, P.W.R. Transverse ply crack growth and associated stiffness reduction during the fatigue of a simple crossply laminate. Cambridge University Engineering Department Technical Report, 1984, CUED/C/MATS/TR105.

12 OGIN, S.L., SMITH, P.A. and BEAUMONT, P.W.R. A stress intensity factor approach to the fatigue growth of transverse ply cracks. Composites Science and Technology, 1985, 24, 47-59.

13 OGIN, S.L. and SMITH, P.A. Fast fracture and fatigue growth of transverse ply cracks in composite laminates. Scripta Metallurgica, 1985, 19, 779-784.

14 BADER, M.G. and BONIFACE, L. The assessment of fatigue damage in CFRP laminates. Proc. of Int. Conf. on Testing, Evaluation and Quality Control of Composites, University of Surrey, Guildford, Sept. 1983, Butterworths, 66-75.

15 BONIFACE, L. and BADER M.G. Fatigue damage in cross ply laminates. Proc. European Conference on Composite Materials, ECCM1, Bordeaux, 1985, 57-62.

16 FUKUNAGA, H., CHOU, T-W., PETERS, P.W.M. and SCHULTE, K. Probabilistic failure analyses of graphite/epoxy cross-ply laminates. Journal of Composite Materials, 1984, 18, 339-356

17 POURSARTIP, A. The characterisation of delamination growth in laminates under fatigue loading. ASTM Symposium on Toughened Composites, Houston, Texas, 1985.

ACKNOWLEDGEMENTS

Some of this work was carried out when one of the authors (S.L. Ogin) was doing post-doctoral research at Cambridge University Engineering Department. S.L. Ogin would like to thank the Science and Engineering Research Council and Dr. P.W.R. Beaumont for support. L. Boniface and M.G. Bader would like to acknowledge the support of the Procurement Executive of the Ministry of Defence and the Assistance of Drs. Dorey and Curtis.

Table 1 Growth rates of two cracks with different initial lengths.

crack	length at cycle 110	110–160	160–240	240–310
B	7.6 mm	96	48	11
C	19.2 mm	19	2.4	6.5

Table 2 Growth rates of two cracks showing dependency on the spacing of the next nearest crack

Crack		110–160	160–240	240–310	310–460	460–570	570–610
D	Growth rate ($\times 10^{-3}$ mm/cycle)	147	2.4	3.2	4.7	0	0
	Spacing of next nearest crack (mm)	2.3	0.7	0.7	0.7	0.7	0.7
E	Growth rate ($\times 10^{-3}$ mm/cycle)	78	84	65	0	0	0
	Spacing of next nearest crack (mm)	1.2	1.2	1.2	0.2	0.2	0.2

Table 3 Crack growth rates (CGR), current crack lengths (CL) and spacing of the next nearest crack (SP) for cracks labelled 'a' to 'e' in Figs. 3 and 6.

Crack	a			b			c			d			e		
Cycles ($\times 10^3$)	CGR	SP	CL	CGR	SP	CL	CGR	SP	CL	CGR	SP	CL	CGR	SP	CL
12–14	66.3	0.90	14.3	1.9	0.54	0.5	0.7	0.51	0.5	–	–	0.4	–		0.5
14–16	17.3	" *	17.8	4.1	"	1.3	3.2	"	1.2	15.4	0.77	3.5	0.7	0.77	0.6
16–18	1.3	0.51	18.1	15.1	"	4.4	3.2	"	1.8	9.0	"	5.3	0.4		1.4
18–20	1.9	"	18.4	5.8	"	5.5	2.6	"	2.3	21.1	"	9.5	4.5		2.3
20–25	1.3	"	19.1	1.3	" *	6.1	0.8	"	2.7	9.5	" *	14.2	17.7		11.1
25–40	0.4	"	19.7	0	0.26	6.1	1.0	"	4.2	0.4	0.26	14.9	6.0		20.1
40–100	0.1	"	20.1	0.1	"	6.9	0.5	"	6.9	0.1	"	15.4	–	"	–
100–200	–	–	"	0	"	7.2	0.1	"	8.3	0	"	15.7	–		–
200–500	–	–	"	0	"	7.8	0.1	"	10.8	0	"	15.7	–		–
500–590	–	–	"	0.1	"	8.5	0	"	11.0	0	"	15.7	–		–

* Crack overlap leads to decrease in spacing of next nearest crack

CGR Crack growth rate ($\times 10^{-4}$ mm/cycle)
SP Spacing next nearest crack (mm)
CL Crack length (mm); maximum length = 20.1 mm

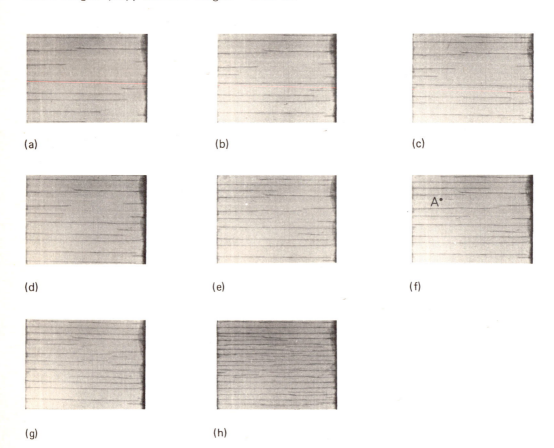

(a)   (b)   (c)

(d)   (e)   (f)

(g)   (h)

Fig 1   (a)   110 cycles   (b)   160 cycles   (c)   240 cycles
(d)   310 cycles   (e)   610 cycles   (f)   810 cycles
(g)   1490 cycles   (h)   42000 cycles

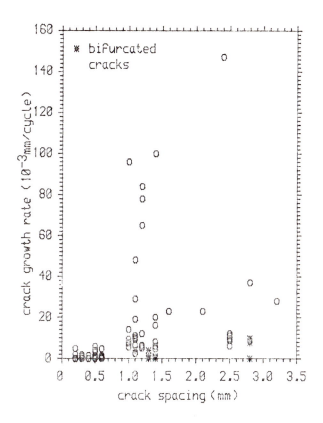

Fig 2

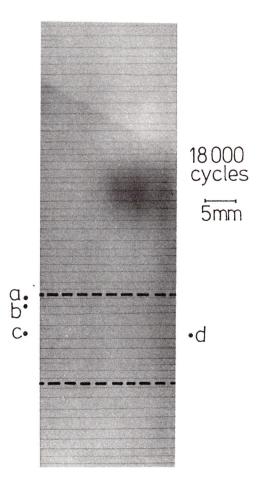

18 000
cycles

5mm

a
b
c

•d

Fig 3

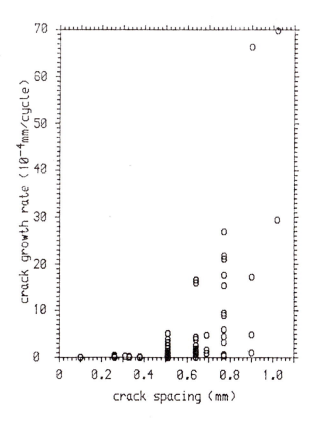

Fig 4

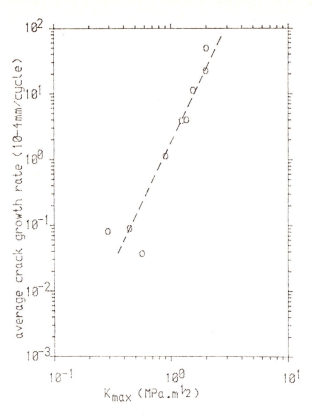

Fig 5

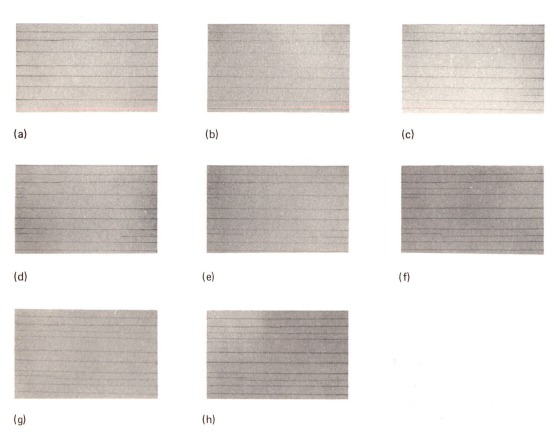

(a)

(b)

(c)

(d)

(e)

(f)

(g)

(h)

Fig 6   (a) 12000 cycles (b) 14000 cycles (c) 16000 cycles
        (d) 18000 cycles (e) 20000 cycles (f) 25000 cycles
        (g) 40000 cycles (h) 100000 cycles

C35/86

The prediction of the energy absorption of composite structural materials

H J BEERMAN, Dr-Ing and H-M THUM
Institute of Transport Technology, Technical University of Brahnschweig, West Germany
G H TIDBURY, BSc, DIC
Cranfield Institute of Technology, Cranfield, Bedford

SYNOPSIS   Tests were made on samples of carbon fibre reinforced composite material with different laminations to determine their elastic constants, strengths and energy absorbing properties. The results are compared with the values given by a preprocessor which computes the elastic constants of complex laminates. A commercially available finite element program is then used to estimate the nonlinear properties of the laminates and the results are again compared with test results. Impact tests on composite tubes are also reported to give some indication of their energy absorbing properties.

1 INTRODUCTION

The three main objectives to be satisfied in vehicle structural design are the achievement of durability under normal operating conditions, the provision of adequate energy absorption in the case of accident and the avoidance structural resonances in the operating speed range of the vehicle. All of these aims have to be achieved with low structural weight. The energy absorption characteristics of sheet metal elements such as engine rails and adjacent wheel arches in the front structure of a passenger car are easier to predict than elements made of composite material because the expected failure mode and the force-deformation behaviour of the former is known approximately from extensive tests eg.(1). Laminated composite materials are commonly used, combined with aluminium honeycomb cores, in the nose cones of racing cars. Fig 1a shows a typical nose cone made of this type of sandwich and Fig 1b the same nose cone after impact by a pendulum. The impacted structure shows different types of failure involving local fractures of the laminated skins. The various types of failure were investigated by testing small rectangular specimens of different laminates of carbon fibre reinforced composite materials. The specimens were chosen to find the effect of fibre orientation in directed fibre laminates and the difference between these and woven fibre laminates.

The tests were used to verify the ability of a preprocessor (LAMANAL) to calculate the elastic constants of the laminates. The elastic constants were used to produce stiffness matrices for a finite element program (LUSAS) which includes special elements of zero dimension but finite extension. These elements were used between the nodes of both thin and thick shell elements which are also available in LUSAS. The special elements were used to simulate the detail fractures in the laminates. They also enabled the program to simulate some degree of non-linearity.

In addition to the static tests on the plane laminates static and dynamic tests were made on tubes made from similar laminates. These tests compared the energy absorption properties of the various laminates for a given deformation.

2 TEST PROCEDURE

Different modes of failure were found in the laminated skins of the crumpled nose cone. They can be characterised as follows:-

    Failures due to peripheral tensile stresses
    Compression failures.
    Failures due to bending or local buckling

The failures were reproduced by testing specimens in tension, compression, bending and buckling separately. All the specimens had the same cross section, namely 20x2mm, with different lengths of unsupported laminate for each type of test - 15mm, for the compression tests, 30mm, for the tensile and buckling test and 42mm, for the bending tests, as shown in Fig 2. Aluminium tags were attached to the ends of the specimens with adhesive where they were clamped in the test machine. Since tensile and compressive stresses are generated in the skins of a sandwich nose cone under impact as shown in Fig 3 these tests were carried out first. The compression tests were made with two configurations, type A with both ends clamped and type B with one free end pressing against the platten of the test machine.

The composite test pieces were made from carbon fibre filaments in a matrix of epoxy resin. The percentage of carbon in the laminates was 60% by volume for the unidirectional and bidirectional samples and 40% by volume for the woven samples. The sequence of the laminates through the thickness of the test pieces is shown in Fig 4. A code was adopted to specify the seven different laminates tested and this is given in table 1.

An Instron test machine was used for the tests which recorded the force-displacement relations. Strain gauges were attached to the unsupported section of the laminate for one tensile test of each type. This enabled a comparison to be made between the measured strain and the strain obtained from the displacement measurement which included the shear deformation of the adhesive between the laminate and the aluminium tags.

Table 1   Laminate identification code.

Code		Description
0° UD	0°	Unidirectional fibres
90° UD	90°	Unidirectional fibres
45° UD	45°	Unidirectional fibres
0°/90° BD	0°/90	Bidirectional fibres
±45° BD	±45°	Bidirectional fibres
0°/90° W	0°/90	Woven fibres
±45° W	±45°	Woven fibres

3   TEST RESULTS

The tensile test results, Fig 5, show the importance of the direction of the fibres, both for the directed and the woven samples. When the load direction is normal to the fibre direction – 90° UD, 45° UD – the laminates have very low stiffness and ultimate strength due to matrix failure at low stress. The ±45° bidirectional and ±45° woven fibre laminates also have low stiffness and fairly low ultimate strength. However these samples had strains of more than 6%, which indicates that they would have reasonably good energy absorbing properties at low force levels. As expected the highest value of tensile strength and stiffness was obtained for the samples with 0° unidirectional fibres. The 0°/90° bidirectional and 0°/90° woven laminates gave nearly the same stiffness, but the woven material had a lower ultimate strength.

A preprocessor LAMANAL,(2), developed at the Cranfield Institute of Technology, was used to calculate Young's modulus and Poisson's ratio for each of the different laminates. The prepropressor requires the unidirectional values of one layer of the laminate as an input and these are usually obtained by testing a sample of the same material. In this case the input values were typical ones for this type of material as there was not time in this limted project to carry out a further series of tests. The agreement between the calculated values for the more complicated composites and those obtained from strain gauge measurements is shown in Table 2. The small difference between the typical values and the actual values for the unidirectional laminates is also shown in this table. It should be noted that only Young's modulus was measured from the tensile tests, while Poisson's ratio was only available from LAMANAL.

The compression tests were carried out using the type A and B samples, Fig 2. As noted earlier these samples were made with only 15 mm unsupported length to avoid buckling. The stiffnesses of the 0° UD, 0°/90° BD and 0°/90° W laminates were practically the same as the stiffnesses obtained from the tensile tests. Rupture failures occured at lower strains than in the tensile tests for type A specimens and at even lower strains for the type B free ended specimens, see Figs 6 and 7. Lower ultimate strengths were also obtained with type B specimens. The rupture mechanisms of the two types are sketched in Fig 8. The samples with free ends (type B) showed large deformations, corresponding to the large strain values at low stress shown in Fig 7. The 0°/90° W (woven) specimens gave the highest energy absorption in both compression test configurations.

Specimens of type C were supported by knife edges 42 mm apart and loaded in the centre by a cylinder 12.7 mm diameter for the bending tests. The tests gave the stress/deflection results shown in Fig 9 from which it can be seen that the ±45° BD and the ±45° W laminates have the greatest deformation while the 0° UD laminate had the highest strength followed by the 0°/90° BD and the 0°/90° W laminates. The very low load carrying capacity of the laminates with laterally disposed fibres was confirmed in the bending tests on the 90° UD and the 45° UD specimens.

Buckling tests were carried out with type A specimens. Using the dimensions in Fig 2 the Euler critical stress can be obtained from the usual formula and compared with the test results shown in Fig 10. It can be seen from Figs 10 and 6 that the maximum compression strength could not be reached as buckling occured at lower stress levels because of the greater length of the unsupported laminate in the type A samples. Table 3 shows the ratio of the measured buckling stress to the Euler critical stress and demonstrates that, except for the 90° UD case, the actual failure stress was less than the calculated value.

Table 2   Elastic constants from test and from LAMANAL

Code	E_{test} N/mm²		$E_{LAMANAL}$	Poisson's ratio
	Start of test	End of test	N/mm²	(LAMANAL)
0° UD	118 000	132 500	130 000	0.28
90° UD	10 400	10 400	9 000	0.0194
45° UD	14 150	12 950	12 390	0.29
0°/90° BD	64 300	72 000	69 790	0.0365
±45° BD	15 300	5 350	16 950	0.766
0°/90° W	54 000	62 400	67 000	0.097
±45° W	17 300	5 100	19 190	0.742

Table 3   Ratio of the experimental buckling stress to the critical Euler stress for the different laminates.

Code	0° UD	0°/90° BD	0°/90° W	±45° BD	±45° W	90° UD
Ratio	0.46	0.48	0.41	0.79	0.50	1.11

4 TESTS ON COMPOSITE TUBES

Quasi-static compression tests and dynamic (pendulum) tests were made on axially loaded carbon fibre reinforced tubes. Because bidirectional stresses are always induced in tubes in compression only 0°/90° BD, 0°/90° W and ±45° W samples were tested. The tubes had an outside diameter of 56 mm and 2 mm thickness with the sequence of laminations as shown in Fig 2. The two types of tubes are shown in Fig 11. The metal end plates were attached to the composite tubes with a suitable adhesive. The free end of the typeII tubes was tapered to ensure that any failure would start at the unsupported face of the tube. The compression tests were made with both force and displacement being measured to find the maximum load and the energy absorbed. The pendulum was set to give an impact speed of 3 m/s with an effective mass of 440 kg.

The results of the static tests are summarised in Table 4 where the maximum load at failure and the energy absorbed to failure are listed, the latter expressed as a proportion of the energy absorbed by the 0°/90° BD composite. The greater load carrying capacity of 0°/90° BD and 0°/90° W laminates compared with the ±45° W laminate is shown in Table 4. However the energy absorbed was essentially the same for all the specimens, especially in the case of the free ended tubes. In the dynamic tests the energy absorbed by the woven fibre reinforced tubes was considerably less than that absorbed by the tubes with bidirectional laminates as shown in Table 5 where the energies are again given as a proportion of the latter. This may well be due to the higher proportion of resin, 60% by volume in the woven laminates compared with 40% for the bidirectional laminates. Resin being inelastic and hence strain rate sensitive it becomes more brittle with increasing impact speed.

5 FINITE ELEMENT ANALYSIS

Most finite element programs are confined to the analysis of linear elastic materials while laminated composites exhibit both non-linear elasticity and anisotropy. The preprocessor LAMANAL calculated the direction dependent elasticity constants which could be used in both a thin shell element with 8 nodes and 32 degrees of freedom and a thick shell element with 20 nodes, each having 3 degrees of freedom. The two elements were used to compile stiffness matrices for use in the FE code LUSAS to analyse the flat plate specimens used for the tests. The matrices were computed for the 0°/90° and the ±45° bidirectional laminates. Between the nodes of adjacent elements special joint elements of zero length but finite deformation, contained in LUSAS, were inserted to enable local failures to be simulated by lowering the stiffness of the special joint elements at predetermined values of stress. In this way bidirectional material non-linearity was approximated while the geometric non-linearity is already included in the program. Fig 12 shows the simple FE meshes used in the analysis. Mesh I uses thin shell elements without the special joint elements, meshes II and III use thin and thick shell elements respectively with the special joint elements and are shown spread out to illustrate their positions.

The comparison between the test and calculated results is shown in Fig 13 for the compression case where it can be seen that the initial slope is the same for both, but only the mesh II system begins to model the nonlinear part of the experimental curve. The results of the comparison between test and theory for the bending case are shown in Fig 14, again the initial slopes are the same and the ±45° BD calculated curve using mesh III shows that a decreasing stress-deformation relation can be obtained with the LUSAS program.

Table 4   Maximum load capacity and relative energy absorption of composite tubes measured in static tests

Tube type	I			II		
Code	0°/90° BD	0°/90° W	±45° W	0°/90° BD	0°/90° BD	±45° W
Max load kN	200	120	70	90	92	50
Energy ratio	1.0	0.84	0.70	1.0	0.97	0.96

Table 5   Relative energy absorbed by composite tubes in impact tests.

Tube type	I			II		
Code	0°/90° BD	0°/90° W	±45° W	0°/90° BD	0°/90° W	±45° W
Energy ratio	1.0	0.39	0.44	1.0	0.67	0.53

6 CONCLUSIONS

The investigation has demonstrated:

a. That the preprocessor LAMANAL is capable of giving correct stiffness matrices for complex laminates.

b. That the finite element program LUSAS with the special node elements can be used to model nonlinear structures although more work is required to make the agreement satisfactory.

In that the work can only be seen as a first step in the prediction of the crash behaviour of composite structures it is hoped that it will lead to further investigations which combine theory and practice.

7 ACKNOWLEDGEMENTS

The work was carried by the second author at Cranfield as a part of the collaboration in research into the problems of automobile structures between Cranfield Institute of Technology and the Technical University of Braunschweig. Guidance on the theoretical work was given by Mr R. Butler of the College of Aeronautics staff who was responsible for the preprossor LAMANAL. The tests were made using the equipment of the Cranfield Impact Centre.

REFERENCES

(1) Beermann, H.J. Behaviour of passenger cars on impact with underride guards. *International Journal of Vehicle Design*, *5*, Nos. 1/2, 86-103.

(2) LAMANAL Handbook, ITC, 1984.

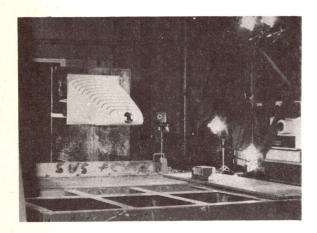

Fig 1a Racing car nose cone

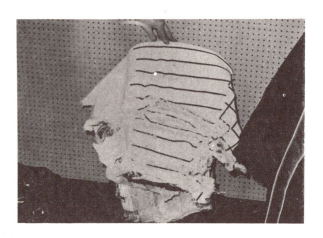

Fig 1b Nose cone after test

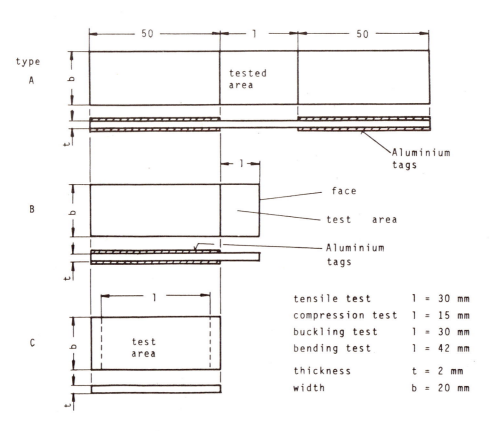

Fig 2   Carbon fibre reinforced composite material test specimens

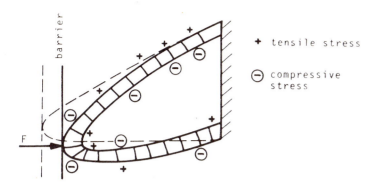

+ tensile stress

⊖ compressive stress

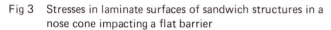

| 0° |
| 90° |
| 0° |
| 90° |
| 90° |
| 0° |
| 90° |
| 0° |

| +45° |
| -45° |
| +45° |
| -45° |
| -45° |
| +45° |
| -45° |
| +45° |

Fig 3   Stresses in laminate surfaces of sandwich structures in a nose cone impacting a flat barrier

Fig 4   Sequences of bidirectional laminates

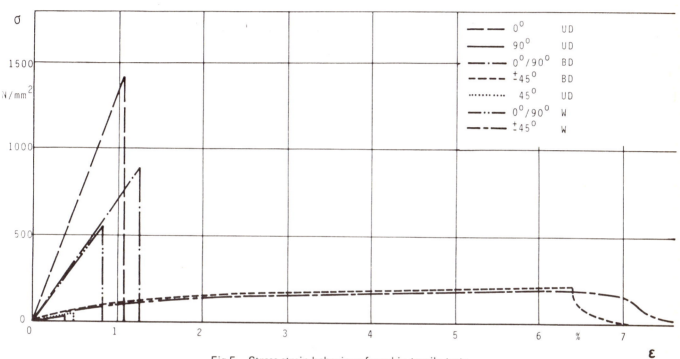

Fig 5   Stress–strain behaviour found in tensile tests

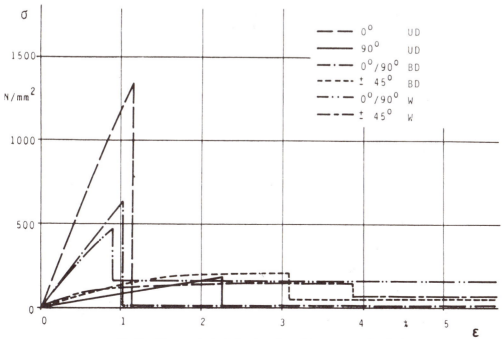

Fig 6   Stress–strain behaviour found in compression tests, type A specimens

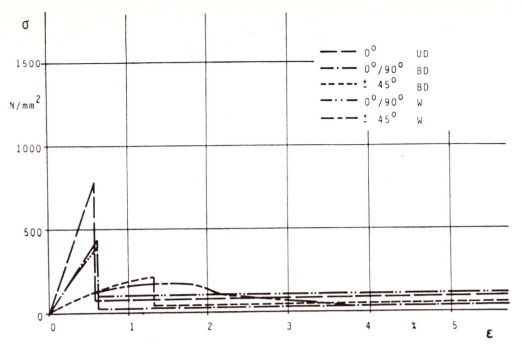

Fig 7   Stress–strain behaviour found in compression tests,
         type B specimens

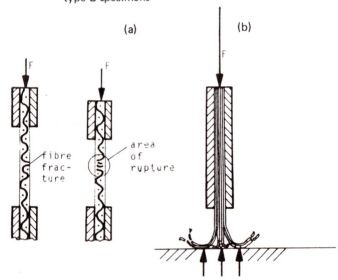

Fig 8   Failure modes in compression tests (a) both ends restrained
         (b) one end restrained, 0°/90° laminate

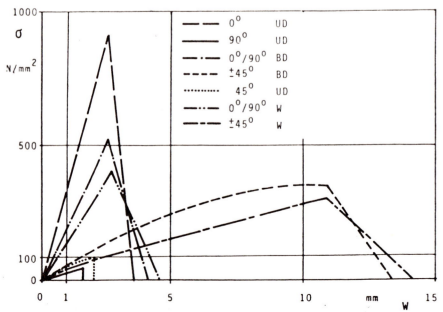

Fig 9   Stress–deflection relationships found in bending tests, type C specimens

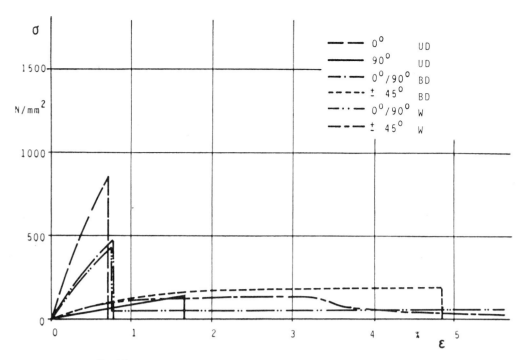

Fig 10 Stress—strain curves of buckling tests, type A specimens

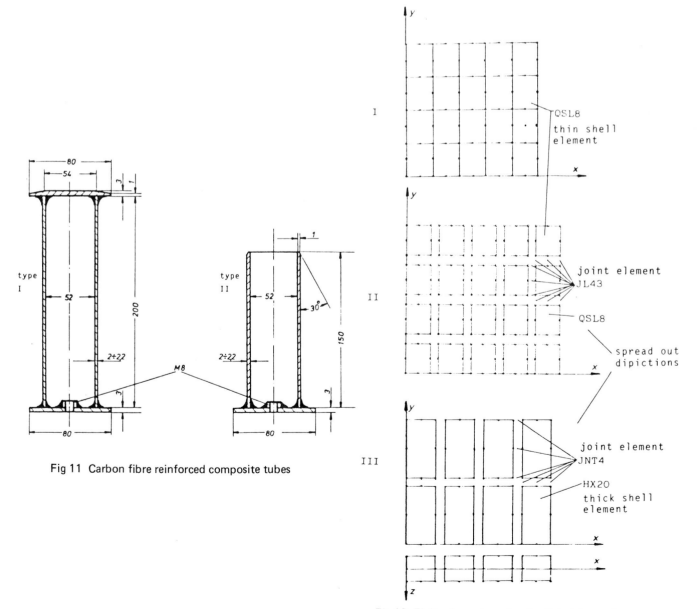

Fig 11 Carbon fibre reinforced composite tubes

Fig 12 Finite element meshes for flat plate specimens

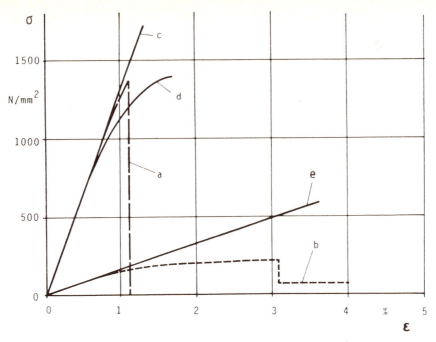

Fig 13 Measured and calculated compression curves
- (a)   0° UD experimental result, see Fig 6
- (b)   ± 45° BD experimental result, see Fig 6
- (c)   0° UD, calculated, FE mesh 1
- (d)   0° UD, calculated, FE mesh 11
- (e)   ± 45° BD, calculated, FE mesh 1

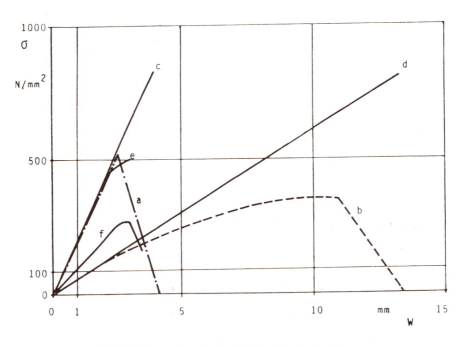

Fig 14 Measured and calculated results for bending
- (a)   0°/90° UD experimantal results, see Fig 9
- (b)   ± 45° BD experimental results, see Fig 9
- (c)   0°/90° UD calculated, FE mesh 1
- (d)   ± 45° BD calculated, FE mesh 1
- (e)   0°/90°UD calculated, FE mesh 111, see Fig 12
- (f)   45° BD calculated, FE mesh 111, see Fig 12

C20/86

Hovercraft skirt design and manufacture

R L WHEELER, MSc(Eng), DIC, CEng, FRAeS, FRINA and A N KEY
British Hovercraft Corporation Limited, Isle of Wight

SYNOPSIS Skirts are the inflated structure which extend around and beneath a hovercraft that allow the craft to operate in rough seas and over uneven ground. The early skirts were designed on the basis of model testing and practical experience, but as the size of craft increased and improved performance was required it became evident that a more sophisticated approach was needed. This was carried out by developing computer programs to design the sectional shapes, investigate the quasi-static behaviour and derive the flat pattern shapes from which the skirt is made. The responsive skirts fitted to large cross-Channel hovercraft, the Super 4, and the latest coastal craft, the AP.1-88/100 were developed using these techniques.

Investigations into the cutting of sheets of non-metallic materials by a high pressure water jet showed that this was ideal for the nylon woven fabric coated with either natural rubber or neoprene from which skirts are made. A cutting table with a NC tape fed controller has been developed with a jet of 0.15 mm (0.006 in) diameter operating at 379 MPa (55000 lbf/in^2). All the processes of producing a skirt, design, detailing with CADAM, part programming and cutting out are now integrated via computer links giving significant savings of time and costs.

1 INTRODUCTION

Before discussing the design and manufacture of hovercraft skirts, it would be useful to give a brief description of the skirt and its function. Figure 1 shows a diagramatic section through a hovercraft, illustrating how the lift air flows through the skirt and into the cushion which supports the craft. The general appearance of a skirt is illustrated on Figure 2.

The development of the fingered skirt has been a major feature in the production of successful hovercraft, raising the hoverheight to a value which enabled the hovercraft to operate in rough seas and over uneven ground.

The design of the fingers is such that the cushion pressure tends to expand the fingers against each other providing a good seal but, being separate, contact with an obstacle or wave enables them to respond easily and drop back into place rapidly after the passage of the disturbance. The air for the cushion is fed to the fingers through holes in the bag, it then passes through the open edges of the fingers into the cushion. Across the rear of the craft, cones are used in place of fingers since these would be forward facing and scoop up water. Each cone is attached to the bag at the leading and aft faces and can move relative to its neighbour. The air is fed to each cone through a number of holes, the total area of these holes determining the cone internal pressure, as the only exit for the air is at the base.

The design process has always relied on considerable support from model tests and now also relies on extensive use of computer programmes.

Arduous operating conditions have required the development of special materials for hovercraft skirts. Many materials have been investigated but the most satisfactory has proved to be a nylon woven fabric coated with either neoprene or natural rubber. Special weaves and chemical treatments have been perfected to give long lives. Methods of fabrication now involve the hot bonding or vulcanising of large sections of the skirt, eliminating many of the mechanical fastenings originally required.

The major activities involved in the design and manufacture of hovercraft skirts are indicated on Figure 3 and are described in more detail below.

2 DERIVATION OF SKIRT CROSS SECTIONS

The sections for the early skirts, up to and including the SR.N4 Mk.2, had been calculated using elementary assumptions, but operational experience had shown that a new approach was required to reduce the amount of tailoring. The aspects of particular relevance were the bow of the craft where the skirt has to be attached to a three dimensional surface and the tapered cushion where the skirt shape varies along the length of the craft.

This new approach was based upon considering a skirt section in equilibrium under static hovering conditions using information from model tests and theoretical analysis. A mathematical model was then developed which was based upon the following:

(a) A unit length of peripheral skirt.

(b) An evaluation of the direction and magnitude of the tensions arising from the air loads in the inner and outer loops of the skirt.

(c) An assessment of the direction and magnitude of the forces and moments from the air loads on the finger.

(d) A definition of the geometry and relationships in the skirt section.

(e) The effect of material weight.

(f) The effect of peripheral tensions acting on sections around the bow.

The computer program that was developed allowed the skirt requirements demanded upon the hull to be determined at an early stage in the overall design.

The experience that had been gained in the operations with the earlier type of skirt suggested that there was scope for improvement in two areas. The first was in the design of bow skirts where, in some cases, it was found that they had a tendency to pull-back under the craft which would then cause a divergent pitch motion called plough-in. This could be avoided by imposing operational limitations on the craft. The other feature that needed to be improved was the general behaviour in waves. There was a lack of response, in that the skirt went through the waves instead of over them, giving a comparatively hard ride and resulting in increased drag due to large wetted areas. To eliminate these adverse features of early skirts, a "fully responsive skirt" was envisaged which would contour the waves. This skirt would absorb the wave impacts to reduce the craft motion and drag would also be reduced because of the reduced wetted area.

To achieve these objectives a new type of skirt was required with the following characteristics:-

(a) A low bag/cushion pressure ratio, of about 1.1, compared with the then currently used values of 1.6. Thus significantly reducing the power required for lift.

(b) A larger finger to bag depth ratio than had been used previously, which model tests had established reduced the drag in rough seas.

(c) Bag sections that could respond upwards, instead of inwards when passing over waves.

(d) Skirt sections which had a high resistance to pull-back when flattened on the water. This would reduce the likelihood of any plough-in tendencies and reduce craft motion in waves.

It was only by using the newly developed

computer program that a skirt with these characteristics could be developed in a reasonable timescale at a reasonable cost.

The Super 4 was the first craft to have the new fully responsive skirt, designed with the computer program. The additional objective of reducing the amount of tailoring needed for skirts was also achieved, neither model nor craft skirt needing special tailoring, and in service in the English Channel the skirt has proved to be very successful. The principles established in the research and development programme have made possible the development of the AP.1-88/80 and a developed version, the AP.1-88/100, see Figure 4.

3 DEVELOPMENT OF THE SKIRT INTO FLAT PATTERNS

After the lines for the skirt have been established, the engineering detail then has to be incorporated. The most significant of these actions is to break down the skirt into a number of panels. It is necessary to have a large number of panels because:-

(a) The skirt material, neoprene coated nylon fabric, is supplied in rolls of around 1.34 m (52 in) in width of which only a 1.27 m (50 in) width is usable. Large items then need to be made from a number of widths. It should be noted that the panels for the Super 4 skirt are about 7.6 m (25 ft) long.

(b) The bow and stern ends of the peripheral skirt have three dimensional surfaces. In general, sheet material can only take-up a single curvature, without creasing at the edges. Although a small amount of stretching will occur in skirt material, a three dimensional surface has to be constructed from a multiple of flat panels.

For the bow skirt, it is important that the profiles of the panels are shaped correctly to ensure that the pressure induced loading of the bag is evenly distributed without any buckling or creasing and peripheral tensions are not built-in which would distort the skirt.

For these objectives to be achieved the inflated panel shapes have to be developed into flat patterns. Prior to the introduction of computers, this had been done manually and required the accurate measurement of the level lines and many repetitive calculations. This was extremely labour intensive and required a high level of skill and concentration. Any changes from the original model lines then required a significant part of the operation to be repeated.

This clearly made a very good case for the flat patterns of skirt components to be produced by computer and the approach used was to adapt the traditional manual method. The program was based on defining the skirt surface by circular arcs and dividing this into a large number of strips, each narrow enough to be treated as a plane surface. The panel shapes were then given x,y co-ordinates which could be plotted. Besides the full-sized versions for lofting, the shapes could also be computer drawn at small scale for the model skirt and to form the basis of the detail drawings.

Computer programs were also written for the flat pattern development of fingers and cones. These together with the programs for the panels for the peripheral skirt and dividers are run on the same computer.

4   MODEL TESTING

All new or revised skirt designs are modelled with representative mass, flexural and longitudinal stiffness and subjected to two-dimensional rig tests. From these experiments the inflated geometry, pressure/volume flow characteristics and resistance to 'pull-back' are established for the given skirt section. Oscillation of the base (floor) of the rig provides dynamic stability and response information.

Further refinements may arise as the three dimensional model skirt system is manufactured and tested on the initial dynamic model.

Craft dynamic models are constructed at the largest practical scale consistent with use in available towing tanks, wind tunnels and for free-flight operation. To ensure dynamic similarity to their full-scale counterparts, special lightweight materials are employed for both structure and skirts and careful attention is paid to weight distribution.

The types of tests that are conducted are:-

(1)   Static hovering tests over a transparent table to determine that the skirt inflation, hover height and cushion stability are satisfactory.

(2)   Towing tank tests that include the measurement of resistance and motion for wide ranges of craft speed and LCG location in calm water, head and following seas. Particular attention is given to craft safety aspects during the towing tank tests.

(3)   Free-flight (radio controlled) model tests in calm water and the highest possible wind speed and sea states to highlight any craft control, handling, performance or safety problems which may not be apparent under the more controlled conditions of the towing tank. Overland tests examine the behaviour during the transition from land to

water and vice-versa and the obstacle clearance capability. Special tests involving the ability to traverse differing types of terrain and features such as gulleys are sometimes carried out.

All the above tests are recorded on video and/or cine film for later analysis.

5   PRODUCTION AND MANUFACTURE

The marking-out of skirt components is done from templates printed on linen backed paper and then the outside profile is cut manually using a cutting knife.

Initial attempts to mechanize the cutting of skirt panel profiles utilised a laser beam with which it was hoped to cut accurately through multiple layers of material and also to eliminate all notches in the cut lines which are liable to be produced by knife cutting. An additional benefit was envisaged in that the laser beam might seal the cut edges of the nylon reinforcing threads and thus avoid the need for subsequent edge sealing operations to prevent sea water wicking through the threads of the skirt material. Preliminary trials showed very good laser cut edges and comparative strength test specimens were produced using laser and knife cutting techniques. The results of the strength tests, however, were disappointing showing a reduction in static strength of 15 per cent and a reduction in fatigue strength of approximately 20 per cent attributable to the laser cutting process.

Meanwhile investigations were being conducted into advanced techniques for the cutting of materials as part of the Company manufacturing research programme. One of the techniques examined was the cutting of sheet material by a water jet.

This used a thin, 0.15 mm (0.006 in) diameter, jet of water at a high pressure of 379 MPa (55000 lbf/in^2) to cut through sheets of non-metallic material. This was of particular interest for the cutting of skirt materials, fibre glass and other laminates. The first practical use of the water jet was for the cutting of small components using templates and a LINATROL optical line follower. A number of parts could be cut at one loading, either by stacking the required layers of material or by producing a template with nested patterns. However there are practical difficulties in using templates for large components, such as skirt panels and the current technique is to employ an NC process. This utilises a Hancock Falcon-S cutting table, 2.4 m by 10 m (8 ft by 33 ft) and the guidance of the jet is performed with a BURNY IV paper tape fed controller. A picture of the machine is shown in Figure 5.

A chart showing the various steps towards the preparation of a NC tape are shown in Figure 6. This tape contains all the information required by the water jet table, viz the profile to be cut, doubler locations, bonding areas and part or identification numbers.

The information for the parts can be taken into Graphical Numerical Control (GNC) in various ways, as follows:

(a) Detail Drawings, for this the data is typed in at a CAM workstation. This is used for components that are fully dimensioned.

(b) Digitiser Tablet, linked to the VAX computer via a Tektronix 4054, an intelligent graphics terminal. This is used for transferring drawings with complex profiles.

(c) Line Tracker. This is a facility on the water jet table whereby existing templates can be copied into the controller and thence to GNC for any necessary processing to produce the production tapes.

(d) CADAM. This is used for all current work and is discussed below.

The shape produced by the skirt computer programs defines the datum lines for each panel with x,y co-ordinates. A considerable amount of engineering detail is then added by CADAM. The interchange of data to and from CADAM is handled by the specially written software packages that convert the files into a compatible format.

The use of the water jet cutting facility almost eliminates the use of paper templates. A few templates will still be required for parts that are larger than the table and for interchangeability purposes. These are drawn on a plotter and then printed on to a suitable material.

Development tests on the water jet have shown that it is suitable for cutting, with a good finish, a wide range of sheet materials as well as reinforced neoprene, these include:-

Glass reinforced polyester and epoxy cured laminates.

Carbon reinforced, epoxy, peek and poly-amide cured laminates.

Kevlar reinforced polyester and epoxy cured laminates.

Composite non-metallic honeycomb sandwich panels.

Tufnol.

The maximum thickness and cutting speed will depend upon the nature of the material.

The skirts on BHC craft are normally made using a combination of bonding (both cold and hot), bolting, riveting and stapling. For minimum weight and susceptibility to self inflicted damage, bonded joints would be preferable throughout, but for various reasons this is not practicable. To begin with many components need to be readily replaceable 'in the field', and it is desirable to break the skirt down into units which can be easily handled, see Figure 7. Also bonded joints are not particularly suitable for joints which are subjected to 'peeling' loads.

The other major area of production development concerns the introduction of a steam heated autoclave for the hot bonding and vulcanisation of flexible skirt components. This accommodates two flat tables approximately 7.5 m x 3 m, see Figure 8. Layed-up components are stacked three deep on each table, covered with a rubber blanket and subject to a vacuum before loading into the autoclave. Curing takes place at a steam temperature of 160°C with a corresponding pressure of 517 kPa (75 lbf/in^2).

Previously hot bonding was performed as individual operations using steam presses of various shapes and sizes. It was not uncommon for a skirt panel incorporating three or four pieces of 1.27 m (50 in) wide calendered material laid edge to edge and reinforced with edge and feed hole doublers to be subjected to twelve separate steam press operations often on several different presses.

Considerable benefits have resulted from the change to autoclave curing. It is estimated that one full loading of the autoclave achieves the workload equivalent to 200 individual press loadings. Shrinkage of components, which was always a problem when local areas of panels were heated in the individual press operations, is constant through the component and can be accurately predicted for autoclave cured items.

6 CONCLUDING REMARKS

The development of the Super 4 from the SR.N4 Mk.2 required a major step in the technology of skirt system design which would not have been possible without the extensive use of computer techniques. In simple terms the advent of the low pressure ratio deep fingered skirt allowed the N4 payload to be doubled for the same performance at the same power, coupled with improvements in ride comfort and performance in higher sea states. Clearly these improvements in skirt efficiency could be utilised in a different way, namely by sacrificing the potential payload increase for the opportunity of using heavier, but considerably cheaper, power plants and structure. This is the approach that has been adopted in the design of the AP.1-88, which employs welded structure and is powered by diesel engines to reduce operating costs and build costs to a minimum. The developments in skirt manufacture described above contribute in no small way to the reduction of the build costs. The result is that the AP.1-88 is competitive in terms of overall economics with comparable hydrofoils and sidewall hovercraft without sacrificing the performance, ride comfort and operational advantages of amphibious hovercraft.

KEY TO ABBREVIATIONS

CADAM   Computer Graphics Augmented Design and Manufacturing, a registered trademark of CADAM Inc.

GNC   Graphical Numerical Control, a software package supplied by Kongsberg Systems Technology Ltd.

NC   Numerical Control.

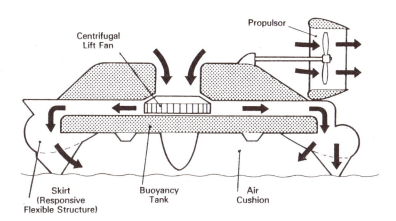

Fig 1   Diagrammatic section through a hovercraft

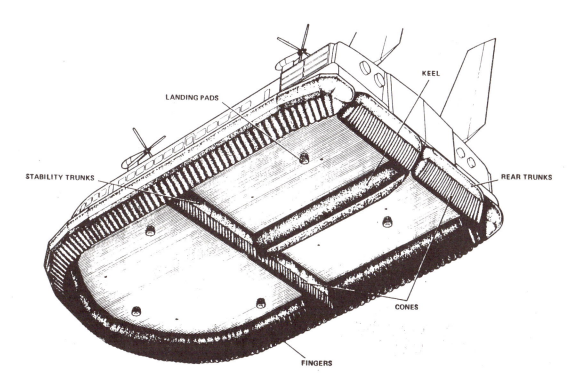

Fig 2   Typical skirt arrangement

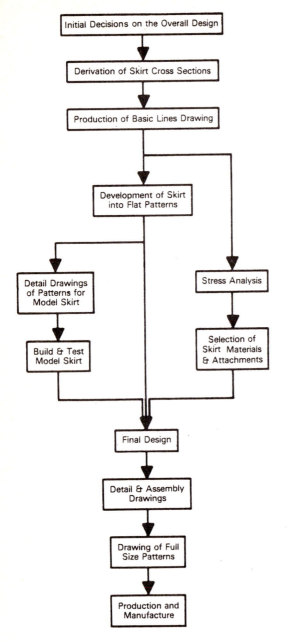

Initial Decisions on the Overall Design

Derivation of Skirt Cross Sections

Production of Basic Lines Drawing

Development of Skirt into Flat Patterns

Detail Drawings of Patterns for Model Skirt

Stress Analysis

Build & Test Model Skirt

Selection of Skirt Materials & Attachments

Final Design

Detail & Assembly Drawings

Drawing of Full Size Patterns

Production and Manufacture

Fig 3   Stages in design and manufacture of hovercraft skirts

Fig 4   The AP.1-88/100 in service over broken ice

Fig 5   Water jet table

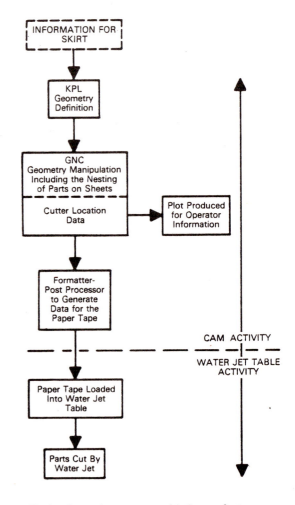

INFORMATION FOR SKIRT

KPL Geometry Definition

GNC Geometry Manipulation Including the Nesting of Parts on Sheets

Cutter Location Data

Plot Produced for Operator Information

Formatter-Post Processor to Generate Data for the Paper Tape

CAM ACTIVITY

WATER JET TABLE ACTIVITY

Paper Tape Loaded Into Water Jet Table

Parts Cut By Water Jet

Fig 6   Stages in computer aided manufacture

Fig 7   Super 4 rear skirt suspended during manufacture

Fig 8   Skirt components being loaded into the autoclave

Prospects for new materials development using biotechnology

J A LIGHT, PhD, and **T R JARMAN**, PhD
PA Technology, Melbourn, Royston, Hertfordshire

SYNOPSIS Some biological materials have an extensive history of service to man. Examples include wood, leather, rubber, cotton, wool, silk. Other biostructural materials have not been available commercially in the past, due to inaccessibility or intractibility of source.

The new capabilities in Biotechnology enable us to develop novel commercial materials. Genetic engineering now allows us to produce hitherto unavailable biostructural materials in commercial quantities: advances in process engineering can allow cost-effective production of biomolecules: protein engineering could allow us to design and produce totally novel structural proteins; and enzyme technology enables us to modify the properties of both natural and synthetic materials in novel ways. Our understanding of micro- and macro-molecular interactions, and their relationship to physical properties of biological materials, though limited, is increasing.

We can now begin to learn from the ways in which materials in nature are constructed (via self-assembly, in elegant composites both between organic, and between organic and inorganic, molecules). The availability of new biostructural materials and an increased ability to formulate them into useful structures will open up new horizons in materials science. We discuss opportunities created by these developments, and look at both short-term and long-term prospects for commercial implementation.

1. INTRODUCTION

In this paper we explore ways in which man can utilise the skills now available to biologists, in making novel fibres and composites.

We present no experimental data; rather, we concentrate on presenting a view which we hope will be thought-provoking. We, the authors, are biologists, not engineers, and although we work in a multidisciplinary environment at PA Technology, may not yet have learnt to present our thoughts in the most appropriate way for engineers to appreciate them. Nevertheless, we hope to stimulate some interest and discussion amongst members of the audience.

2. PERSPECTIVES

PA Technology is investigating ways in which materials found in nature, especially those having a biostructural role, can be exploited via biotechnology. The biological kingdom admirably demonstrates the wide range of properties and functions which might potentially be developed for commercial materials uses. Our new capabilities in genetic engineering and in bio-processing enable us to broaden our horizons in considering routes to exploring the possibilities.

Our fundamental understanding of the relationship between molecular structure, inter-molecular association and resultant mechanical properties is, however, in its infancy.

We therefore advocate an 'evolutionary' approach, first exploiting the known mechanical properties of existing biostructural materials and then, perhaps modifying them for specific end uses. As our understanding increases, the potential for the de novo design of biomolecules, particularly proteins for structural uses could become a real and exciting prospect.

3. BIOSTRUCTURAL MATERIALS

Nature has provided man through past centuries with the majority of his materials needs. Materials such as wool, leather, cotton, wood and bone come readily to mind. These materials are associated with traditional uses and mature markets. Demands for more specialist materials are increasing in areas as diverse as aerospace and healthcare. These demands are usually filled by materials such as new metal alloys, plastics and ceramics, or by composites containing carbon or aramid fibres. In this paper we explore the potential for harnessing the rapidly developing capabilities of bio-technology to provide new materials.

We look at materials which perform structural jobs in nature and which might be exploited for their unique properties, improved performance or cost effectiveness. The large range of materials used in nature is based on a few common themes, with many subtle variations. Not only does a complex organism need many different materials to perform different functions, but each type of animal and plant has structures which enabled it to adapt to a specific environmental niche. Structural biomaterials are not just static supports for animals and plants, they also support locomotion, resist stresses and minimise energy wastage. Such materials are highly engineered to near to their physical limits, and are finely tuned to their specific functions.

In seeking to exploit the properties of biostructural materials we must, of course, recognise that their functioning in nature in many instances requires continuous replacement through growth and metabolism and to the provision of a suitable physiological environment. Nevertheless, this does not preclude the exploitation of similar molecules and molecular associations in totally artificial situations.

4.   COMPOSITES IN NATURE

The structural materials of nature are generally based on polymers, the majority of which are proteins (eg collagen, elastin, keratin) or polysaccharides (eg cellulose, chitin, alginate). Although these might be utilised in modes based on a single molecular type they are more generally found as composites, often consisting of a fibrous structure in a space-filling matrix, Examples of composites include:

(a)   protein-protein (keratin),
(b)   protein-polysaccharide (chitin, cartilage)
(c)   polysaccharide-lignin (wood)
(d)   protein-inorganic compounds (teeth and bone, mollusc shells)

The molecular and macromolecular structures found in such composite materials are interesting because of the resulting properties of the materials. We may find that it would be useful to study the physical and mechanical properties of such materials and to correlate these with structural features.

Examples of interest to industry might be: flexible, or elastic composites and very strong, light-weight composites. We might be interested in mimicking such structures, either at the molecular level, using analogous components, or mimicking the macromolecular shapes and forms, using different components.

We may also hope to learn from the way in which complex, composite materials are synthesised and assembled in nature.

Composites found in nature might be compared with the range of composites known to the materials scientist, such as fibre-reinforced plastics and rubber reinforced with carbon black.

The range and sophistication of composite properties obtained with biological molecules is potentially very great. The protein keratin, for example, is the base component in structures such as hair, hoof, horn and feathers. When relaxed, the major portion of mammalian keratin is in the form of alpha-helices, grouped together in a fibrillar structure. The fibres so formed are embedded in, and crosslinked to a matrix of high-sulphur proteins to form a two-phase composite. It would appear that subtle changes in protein structure and molecular interaction dictate the quite distinct properties characteristic of the structures involved.

5.   PROTEINS

Proteins are of particular interest to us, because of the new scope for producing and manipulating them using biotechnology. 'Protein'

molecules have fixed lengths and exact compositions, determined by the genetic material.

Proteins are highly unusual molecules when compared with analogous synthetic polymers such as the polyamide nylon-2. The synthetic polymer chemist is more familiar with random statistical distribution of molecular components and molecular weights.

There are more than twenty component monomers of proteins, and the precise order of these monomers in each type of molecule is also determined by the genetic material. Each monomer (amino acid) has different properties, for example reactivity of free side groups such as $-CH_3$, $-OH$. This variety of molecular composition of proteins is reflected in the variety of molecular (3-dimensional) structures, and of functions, eg proteins can be structural materials, antibodies, enzymes (biological catalysts). Thus the molecular composition defines the molecular shape, and function, under a given set of physical conditions.

We are interested here in the behaviour of proteins as structural materials. Molecular bonds between and within molecules can be highly specific (hence unique 3-dimensional structures). These molecular bonds are not densely distributed; proteins in general have considerably fewer reactive side groups than polysaccharides (proteins could never show properties like carbon fibre) But the molecular bonds between molecules can be highly specific.

As discussed below, we now have the possibility of designing protein structural materials de novo to fit the industrial requirements of man.

We can also think about proteins as enzymes (biological catalysts) to modify properties of other structural polymers (natural and synthetic). Enzymes can have very high specificity, carrying out particular molecular reactions after the recognition of special structural motifs. Using protein engineering techniques this specificity can be altered to perform desirable reactions (within certain constraints). This technology for modification of materials could be utilised in, for example, altering surface properties of materials, or in altering the interaction between molecular components of fibres and matrices.

6.   BIOTECHNOLOGY ROUTES TO BIOLOGICALLY-
      DERIVED MATERIALS

The advances being made in biotechnology have predominantly been focused on high value products for pharmaceuticals, animal health and nutrition, agronomy and fine chemical markets. We believe that if biotechnology is to have its predicted broad commercial impact it must 'expand' its horizons into other industry sectors through an innovative, multidisciplinary approach. In this context we can look at making interesting polymers, found in nature in small amounts or in inaccessible forms, available commercially via genetic engineering. We can also begin to consider designing new polymers (proteins) with properties needed for industrial materials.

Our abilities to clone genes for proteins of interest into prospective production organisms are developing rapidly. In addition, advances in process engineering and recovery give the prospect of manufacturing products and recovering them from aqueous solution for an acceptable cost.

Some of the technology for exploiting novel biomaterials is therefore in place. If we can identify attractive market areas and learn how to formulate and fabricate biomolecules to give desired mechanical properties, then we have a high chance of making a real commercial impact on the materials world.

Considering various routes and potential materials for reaching these goals:

(a) Microbial products

Microbes, increasingly easy to manipulate to produce desirable products, themselves have structural components of interest.

Compounds already produced by microbes might be exploited, through strain development or process development to give efficient production processes. It is often necessary to manipulate an organism to enhance production of a material normally only produced in small amounts. In other cases it is necessary to "tame" the microorganism, to manipulate it and adapt it to the available process technology, or perhaps to adapt process technology to it. The already established microbial polysaccharide, xanthan gum and the newer commercial microbially-derived polysaccharides could be included in this context though these are usually used as water-thickening agents rather than in a strictly structural sense. Dextran, however, which is a well established commercial microbial polysaccharide has recently been applied as a fibrous non-woven textile in speciality uses such as wound-dressings.

Perhaps a more topical subject is the microbial storage compound poly-beta-hydroxybutyrate (PHB). Though it was known for many years that this compound is accumulated in substantial quantities by a range of bacteria it is only recently that attempts at its commercial exploitation have been made. Imperial Chemical Industries are producing developmental quantities of PHB under the name 'Biopol' which is targeted as a plastic in such applications as specialist coatings and medical appliances.

Other examples of microbial materials include fungal chitosan and bacterial cellulose. These materials, chitosan and cellulose, can be extracted much more cheaply from animal (shell-fish) and plant sources respectively. Process and strain improvements could decrease the costs of the fermentation route, and make these polysaccharides, which are more uniform in structure than the animal/plant derived analogues, viable commercial materials particularly for specialist end-uses.

Another material for possible future production using biotechnology is natural rubber. Some fungi have been reported to produce small quantities of rubber. Through strain improvement and the targeted enhancement of the isoprenoid biosynthetic pathway it is conceivable that a viable microbial production process for this strategically important material could be developed. In this context we note with interest that the US Naval Air Systems Command recently invited research proposals to clone rubber transferase, the native isoprenoid polymerising enzyme, from plants into microbes.

Although new microbial products (including proteins) will presumably continue to be identified and exploited, the new biotechnology allows us to broaden our horizons to consider microbial production of non-native biostructural materials.

(b) Microbial production of structural proteins using genetic engineering

Genes for animal or plant proteins with potential as new commercial materials could be transferred into suitable microbes and produced in substantial quantities. Examples of suitable proteins include collagen, fibrin, the silk protein fibroin, elastin, resilin (which stores strain energy in insect wing hinges) and abductin (from the hinge ligament of bivalves). Some of these proteins have not been commercially available to man, being found in very small amounts in intractable sources. Others, although available in large amounts in a mature form, might be more exploitable if produced microbially in an immature form in which they can be more easily manipulated and formulated for special end-uses. To date it would appear that little serious effort has gone into such developments.

An interesting example of a family of proteins which could be exploited in this way is the silks. Silk from the moth Bombyx mori has long been a commercial textile fibre, prized for its unique properties in terms of lustre, feel, and dyeability as well as its prestigious image. Silks with a variety of different properties are produced by other arthropods, including spiders. Some spiders use up to 5 different silks for their webs, and for cocoons. Each silk is adapted to its particular function, whether as a dragline (like a climber's safety rope), or as the sticky spiral web to ensnare the prey. Such different functions for silks require different physical properties, in terms of strength, extensibility, and visco-elasticity. The differences in physical properties of these silks reflect, at least in part, differences in amino acid composition and their distribution.

Bombyx mori silk fibres consist of two twisted strands of protein fibres (fibroin), coated and held together by another protein (sericin) which is removed for commercial use of silk. Fibroin has a complex structure, but a primary component is a very large polypeptide consisting of tandem arrays of highly crystalline regions interspersed and flanked with more amorphous regions which might be considered as a matrix for the crystalline structure. The amino acid sequence in the crystalline regions is simple and highly repetitive, consisting mainly of glycine, alanine and serine, with alanine and serine alternating with glycine, so that antiparallel beta sheets are formed, stabilised by unstrained H-bonds and stacked. In contrast, the amorphous regions contain amino acids with bulky side chains which destabilise regular structures. Interesting correlations between amino acid composition and distributions, and physical properties can be identified, although the processes involved in silk manufacture presumably are also important in determining physical properties.

Some of the silks found in nature have desirable properties for specialist industrial applications, and beyond this may provide clues for the design of novel synthetic protein fibres for specific requirements. Since silk from spiders' webs is not readily available commercially, the production of specially designed silk-like materials by microbial cells could be a viable commercial proposition for speciality markets.

Biostructural proteins produced using biotechnology could have extended ranges of end-uses if incorporated into composites with other organic (synthetic or natural) or inorganic materials. The formulation and testing of such composites require an understanding not only of the biochemistry and biophysics of the biological components but also an understanding of materials sciences. Mimicking a natural biological composite, such as tooth enamel is a further step away, requiring a deep understanding of the biological system. It has recently been reported that one of the proteins of tooth enamel (amalogenin) has been cloned into a bacterium though much further work is clearly necessary before a useful composite between such proteins and hydroxyapatite, as in the natural system is achieved.

(c) De novo design of new biological materials

De novo design of new structural proteins for specific end-uses is the most exciting prospect for biotechnology to contribute to the development of new materials. Instead of relying on biological materials developed throughout evolution to perform a particular structural role in vivo we can begin to envisage the design of biopolymers to suit special requirements of man.

The technology exists to produce totally novel proteins via synthetic DNA and microbial production. The only limitation, apart from the acceptability of the design to the chosen host organism, is our current lack of understanding of structure-function relationships. We cannot predict the 3-dimensional structure of a protein from its primary sequence, nor can we predict its behaviour from a knowledge of its 3-dimensional structure. Nor do we understand how molecular structure and properties relate to macromolecular structure and mechanics. Advances in molecular modelling and protein chemistry are likely to be made rapidly, enhancing our ability to predict tertiary structures, but we are a long way from predicting how a protein behaves, in terms of interaction with other molecules (other polymers, or solvents) and particularly the dynamics of such interactions under different conditions.

The scope for a molecular biologist to design new protein molecules is almost infinitely broad, but at present he is limited to testing the effects of changing small sections of existing protein molecules at one time, on specific features of protein function. Until he can, with some degree of confidence, productively design molecules to fit the material scientist's dream, we will have to rely on the 'evolutionary' route, adapting proteins found in nature, in small steps, to produce desirable properties.

7. OUTLOOK FOR THE PRODUCTION OF NEW BIOLOGICAL MATERIALS COMPONENTS

We have suggested in this paper that biotechnology could enable new biostructural materials to be exploited by man, but what are the real commercial prospects? In attempting to answer this question we should address issues of relative cost of "biotech" materials to current commercial materials in the light of desirability of their respective properties for particular end-uses.

The variable cost for production of one tonne of a biopolymer could be of the order of $1000-5000, depending on efficiency of production and recovery. On a large scale this cost is achievable only for a microbial product which is stably produced. There are a few examples today of such processes, but no recombinant DNA product has been developed to a very large scale. This is partly because of technical problems (process stability and recovery) but also because most recombinant DNA products to date are aiming at lower volume, higher value markets.

The cost of developing new microbial biopolymers, and the capital cost of equipment required are both high. If these costs are written off over a 10 year period total production costs escalate to around $10,000 per tonne for a 1,000 tonne per annum scale, and up to $10,000,000 per tonne on a 100 kilogram per annum scale.

Although costs of production decrease rapidly with increasing production scale, even high-volume products are likely to be more expensive than their synthetic analogues, and must have better properties to compete. Low-volume, high-value products which fill specific market needs by virtue of a unique combination of properties are more likely to be successful in the near future.

In the longer term we can hope that, through increased understanding of the structure-function relationships of biopolymers, both on micro- and macro-molecular scales, and through an increased commitment to a multidisciplinary approach to development of new materials, horizons will become wider. New biopolymers with desirable properties will be designed, and formulated into new types of composites.

Such developments will initially be market-driven, fitting needs for new materials with defined properties, but could potentially provide "enabling technology", stimulating the growth of new markets.

8. CONCLUSION

In conclusion, we hope that, despite the differences in background and attitude between biologists and engineers, we have managed to open up discussions on a subject where inter-disciplinary communication, though difficult, may prove to be fruitful. Within a university setting, at Reading, Drs Julian Vincent (a biologist) and George Jeronimidis (an engineer), have to some extent blazed a trail in their joint approach to understanding structural materials systems used in biology. They, and others are exploring areas not only of academic interest but also of industrial relevance. Our input, high-lighting the relevant new techniques available in biotechnology, may help to stimulate further interest in this area.

We hope that we have excited the interest of engineers in the materials structures used by nature, and in means of investigating and exploiting them. Such interest may lead to new collaborative multidisciplinary efforts in both academic and industrial settings.

The development of balance tubes for Dowty Rotol composite bladed propellers

R D TIMMS
Courtaulds Research, Coventry

Dowty Rotol Ltd of Gloucester are manufacturers of aeroplane propellers and they supply customers throughout the world.

The propeller blades are made of carbon fibre spars which run the full length of the blade and these are encased in a glass fibre epoxy resin aerofoil section envelope, the core being filled with a polyurethane foam.

When the individual blades are mounted on the hub to form the propeller assembly, Fig. 1A, due to manufacturing differences between blades, the assembly may be out of balance. To allow balancing to take place, a hole 1.25 inch diameter is bored into the polyurethane foam core at the base of each blade and a Balance Tube is bonded into position. The assembly is then balanced by packing the required amount of lead wool into the Balance Tube bores and fitting retaining plugs, Fig. 1B. In the past Balance Tubes have been made from an aluminium alloy and comprise two items, the tube and a loose flange, weighing together 197 grams, Fig. 2 left.

The loose flange is necessary to prevent the tube being forced into the foam core by centrifugal force when the propeller is rotating at speed in service.

Following a request from Dowty to investigate the possibility of producing a one-piece Balance Tube as a Grafil carbon fibre reinforced thermoplastic (CFRTP) injection moulding, the design in Fig. 2 right, using a Grafil RG40 material (40% Grafil c.f./Nylon 66), was agreed. This has a hemispherical ended tube with an integral flange supported by 8 ribs, weight 74 grams. The initial tubes were made using an aluminium mould tool, Fig. 3, in a simple plunger type injection moulding machine, the Manumold Mk 2. This machine allows the use of simple tooling to keep development cost for prototyping to a minimum. There is a strength requirement for the Balance Tube, viz an axial load applied to the hemispherical end of the bore should exceed 15 KN (3370 lbf) before failure.

Tubes were tested to failure using a steel bolster bored to locate the test specimen vertically and supporting the flange. The load was applied to the hemispherical end of the bore through a small amount of steel wool by a round ended plunger, the purpose of the steel wool being to distribute the load around the end of the bore and represent the lead wool which would produce the normal service load, Fig. 4.

Early test failures gave disappointingly low results in the 8 KN-11 KN (1800-2470 lbf) range. These premature failures were caused by voids at the nose end of the moulding, this portion of the component being the farthest from the feed gate.

To eliminate the voids, a number of tool modifications were carried out; larger venting slots and a cold slug well were added. It was hoped to carry any entrapped air into the tab formed by the well, which was subsequently removed.

A gradual improvement in the strength of the components was made and failing loads in the range 11.6 KN - 16.5 KN (2600 - 3500 lbf) obtained.

The tool was also temporarily modified to produce flat ended Balance Tubes, but no significant increase in strength resulted.

Due to the limitations of the Manumold machine with its low injection speeds and pressures, it was not possible, however, to produce components which gave a consistent failing load above the specified value. It was then decided to adapt the mould tool for use on a Daniels 550/170 (170 ton lock) screw type injection moulding machine. The existing tool was mounted onto standard bolster plates with additional ejector and locating pins for the cores, etc., Fig. 5.

The components produced now showed a marked improvement in strength, the test failing load ranging from 18.6 KN - 23.5 KN (4170 - 5270 lbs) over 8 specimens from the first batch of 25. Another batch of 25 was supplied to Dowty Rotol Ltd and a random selection of 5 was tested by them, giving a failure range of 21.9 KN - 23.3 KN (4910 - 5230 lbf), the remainder being acceptable.

Following the successful completion of the development of the single pocket Balance Tube, Dowty Rotol requested mould tool and component costs for the production in quantity of an improved type of component. This has a similar central tube but with 8 additional small pockets around the flange to allow balancing of the blade about its two cross-sectional axes, Fig. 6. The costs quoted were accepted and the steel tool for moulding the components on the Daniels injection moulder was made, Fig. 7.

The complexity of the multi-pocket component necessitated the sprue bush (where the molten material is injected into the cavity) to be located at the nose end of the component, Fig. 6 left, whereas the sprue feed into the single pocket component was at the flange. However, based on the experience gained in developing the single pocket component, it was confidently expected that the multi-pocket component would meet the specified test load of 12.5 KN (2800 lbf).

Test values for the first batch of components produced gave failing loads in the range 23.6 KN - 26.5 KN (5300 - 5950 lbf) and the average value for the first production run was 25.0 KN (5610 lbf) with a standard deviation of 1.275 KN and a coefficient of variation of 0.051.

The successful development of a CFRTP Balance Tube has led to the production of other types and 3 variants are now made for different size propellers, Fig. 8.

This case history is one of a growing number where a suitably designed CFRTP component can successfully and cost effectively replace a conventional metal part, achieving weight reduction, parts consolidation, process savings and other specific benefits.

The author wishes to acknowledge the assistance given by Dowty Rotol Ltd in the preparation of this presentation.

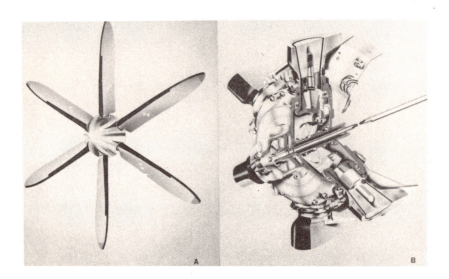

Fig 1

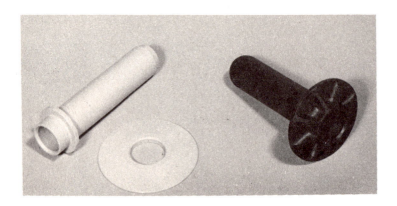

Fig 2

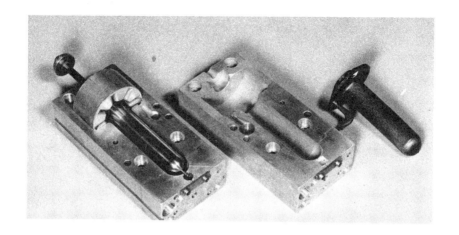

Fig 3

Fig 4

Fig 5

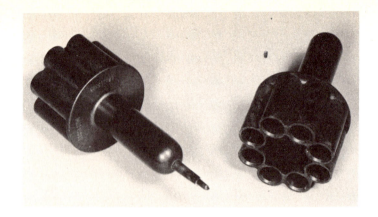

Fig 6

Fig 7

Fig 8

C44/86

Expanding core moulding of composite components

R G WALKER, BSc(Eng), ACGI, MSc, MBL
RK Technologies Limited, Stockport, Cheshire

The moulding of composite components using a filled epoxy resin expanding core wrapped in
unidirectional carbon fibre - epoxy resin prepreg tape is reviewed. Process control and component
design aspects are outlined and typical internal mould pressures generated by the expanding core
during the cure cycle are presented. The economics of moulding of fibre reinforced composite
components using expanding core to mould the component and to provide structural support after
curing are outlined.

1. INTRODUCTION

Fibre reinforced epoxy resin materials are
usually convertedinto a rigid component by the
application of pressure and temperature during
a cure cycle. In many applications several
layers of fibre reinforced prepreg material are
placed on a rigid lightweight honeycomb or
foam core and then cured. The rigid core is
used to improve the performance of the
component in bending by displacing the
laminated skin material away from the neutral
axis. The core also helps to stabilise the thin
skins from buckling in compression.

Pressure needs to be applied to the layers of
preimpregnated material during the cure cycle
in order to eliminate voids caused by trapped
air and volatiles. Pressure can be applied by
a vacuum bag, a vacuum together with positive
pressure in an autoclave or by the application
of pressure in a press. In these cases the
pressure is applied externally to compact the
layers of material together to form a laminate
and to bond the laminate onto the rigid core.

Expanding core moulding makes use of an
internal pressure and expansion generated by
the core during the cure cycle which serves
to compact the layers of material together
to form a laminate and pushes the laminate
into the shaped cavity walls of an external
mould before the resins start to cure.
Expanding core moulds the surrounding
prepreg into a void free laminate and cures
into a lightweight rigid sandwich core to
stabilise the surrounding cured laminate.

2. THE EXPANDING CORE MOULDING PROCESS

We use RK2036 epoxy foam core moulding
compound in this process. RK2036 is an
intumescent epoxy core filled with low
density granulated materials. It is
supplied as a one pack premixed dough.
RK2036 can be readily shaped and wrapped
with prepreg at room or slightly elevated
temperature of 16 to 38 C (60 - 100F). The
uncured composite is inserted into a loose
fitting mould cavity. The mould is closed
and heated to 120 to 177 C (250-350F). The
RK2036 core viscosity reduces rapidly with
temperature and expands (before the gel
point of the prepreg resin is reached) and
pushes the prepreg out to the exact shape of
the mould cavity.

The epoxy resin in the expanded core and
the prepreg then cure in 30 minutes at 177 C
(350F) or in 1 hour at 120 C (250F). The
resulting composite component is a high
strength lightweight integral skin-web-core
structure.

RK 2036 is designed to be compatible with
fibre reinforced epoxy prepegs that cure in
the 120 - 177 C (250 - 350F) range. RK2036
has been used successfully with RK120/177
unidrectional carbon fibre epoxy novolak
prepreg and with several commercially
available prepreg tapes from other
manufacturers.

This expanding core moulding process is
illustrated in FIGS 1 and 2 and a specific
example of moulding a guitar neck is shown
in several slides. This process has also
been used successfully to make tennis and
squash racquets, rotor bows and is
undergoing trials for use in helicoptor
blades, air brakes and missile fins.

3. EXPANDING CORE MOULDING PROCESS CONTROL

The quality of composite components made by expanding core moulding can be influenced by the following variables:-

a. Incoming quality control of the prepreg and expanding core. Regular flow and expansion tests can be carried out on the prepreg and core respectively.

b. The dimensions and weight of the shaped core pieces used.

c. The dimensions and weight of the fibre reinforced prepreg shapes used.

d. The layup sequence of core and prepreg together with prepreg orientation. Positioning of the constituents in the mould cavity is also important.

e. The state of the mould tool cavity and split line. The mould tool is kept closed by torque tightened bolts.

f. The temperature and pressure inside the mould cavity during the cure cycle.

Each mould is filled with a preweighed and checked kit of prepreg and core parts. The individual core pieces may be prewrapped in prepreg in order to generate internal webs on curing. In some cases the skin/web formed by the laminated prepreg may be thickened up by rolling or folding prepreg into a spill or a thick reinforcing strip.

The constituents are loaded into the mould so that they are a loose fit when the mould is closed. This is done in order to prevent excessive movement of the constituents before and during the cure cycle. Movement could cause variation of the internal structure of the component. The constituents may be loaded into a mould so that two thirds of the cavity volume is filled before expansion. This loading makes full use of the 50% expansion of the RK2036 core and will result in the lowest overall bulk density.

We have found that it is better to fill the cavity so that the constituents are a loose fit in the closed mould representing a fill of about 80% of the cavity volume. Raw material quality control and weight control of the kit of parts makes this loading process and the resulting component reproducible. Metal inserts can be located in the mould and become moulded in place during the expansion and cure of the surrounding materials.

During the cure cycle the expanding core is sufficiently liquid to behave like a hydraulic fluid in generating an internal pressure in all directions coupled with an expansion which fills the remaining voids in the cavity by pushing air out through the mould split line.

Fig 3 shows a typical temperature and internal pressure cure cycle with constituents loaded into an aluminium mould and inserted into an air heated oven. Fig 4 shows a typical temperature and internal pressure cure cycle with constituents in an aluminium mould inserted into a fluidised bed to increase the heat transfer into the mould and thereby shorten the cure cycle.

The internal pressure and expansion forces the trapped air out through the mould split line and, depending on the tightness of the split line, a flash of resin is produced. Inspection of the resin flash and the cured laminate indicate an absence of voids. The method of generating internal expansion and pressure in the core prevents any volatiles or gases being exuded from the core and thus porosity in the laminate is prevented.

4. COMPONENT DESIGN WITH EXPANDING CORE

The use of expanding core enables more design freedom to place the laminated fibre reinforced material in the optimum position to resist the design loads. Varying laminate thickness, internal webs and local reinforcement around metal inserts can be readily achieved.

Initial design calculations of deflection and strength of the composite component ignore the contribution of the expanding core except where skin buckling in compression is a significant factor. Strength and deflection tests on prototype moulded components verify the design, and depending on results, fibre reinforced prepreg can be added or removed from particular locations in order to further optimise the performance.

No mould modifications are required to make components of varying strength and deflection, since the component modifications are internal. However the mould must be designed to withstand at least 150 psi at the cure temperature.

The density of the expanding core can vary from 0.4 to 0.6 g/cc depending on the average expansion that has taken place during the cure cycle. This produces a composite component with an average density similar to a hard wood. In some applications this results in too high a component weight. To reduce the component weight two procedures have been developed.

Fig 5 shows how a piece of low density rigid foam core (e.g. Rohacell) can be inserted into the mould cavity in order to reduce the component weight. Overall component densities of 0.3b/cc have been achieved with this technique.

Fig 6 indicates how a removable internal mandrel can be used to reduce the component weight.

In the examples shown in Figs 5 and 6 we have successfully made components using RK2036 sheet thicknesses down to 3 mm in a tightly filled mould cavity.

These procedures enable high strength to weight ratio skin-web-core-hollow components to be made with performances similar to those of conventional aerospace components.

Another application of RK2036 expanding core is as a manufacturing aid. Layers of prepreg can placed in a mould covered with release film and then the cavity is levelled off with RK2036. Expansion of the RK2036 compacts the laminate into the shape of the mould.

This is illustrated in Fig 7. This process is more effective than using cured silicone rubber pre-moulded to the shape required and then inserted in a press to compact the curved layers of prepreg onto the mould surface.

Since RK2036 becomes a fluid during the expansion process, it is capable of applying pressure equally in all directions and can therefore compact laminates of more complex shapes.

5. THE ECONOMICS OF EXPANDING CORE MOULDING

Composite components can require varying sizes of mould, amount and type of material and complication of layup. This makes it difficult to generalise cost information for an expanding core moulding process. However our experience has shown that:-

A. A typical two piece aluminium mould capable of withstanding expanding core pressures that contains a machined cavity for making a component weighing in the range of 200 g to 1 Kg costs about £2,000 to manufacture.

B. The raw materials used to make an expanded core fibre reinforced composite component can vary from £10/Kg with glass reinforcement to £25/Kg with carbon fibre reinforcement.

C. Typical labour costs for cutting and shaping raw materials, loading and cleaning moulds including an overhead recovery amount to £15/Kg of component weight.

D. Using a fluidised bed enables one mould to be cycled 3 to 5 times per 8 hour shift. The fluidised bed is set to the cure temperature and a cold mould is lowered into the bed. A 15 minute heat up to 165 C,30 minute hold followed by a water quench produced a cure cycle time of 45 minutes for typical 200 g components.

E. Mould cleaning, loading and closing times for a typical 200 g component are 45 minutes. One mould can therefore be cycled 5 times per 8 hour shift to make 200 g components.

F. For heavier components the heat up time, cleaning and loading times are higher reducing the number of mould cycles per 8 hour shift to 3 for a 1 kg component.

Allowing for cost of mould release agent, mould amortisation over 1000 components and equipment depreciation over 3 years results in component costs, made by the expanding core process, of:-

200 g component
(5 per 8 hour day from 1 mould)

(Carbon Fibre Prepreg) (Glass Fibre Prepreg)

Component cost	£12.50	£9.50

1000 g component
(3 per 8 hour day from 1 mould)

(Carbon Fibre Prepreg) (Glass Fibre Prepreg)

Component cost	£46.00	£31.00

The relatively low cost of the low pressure mould coupled with the reasonable component cost makes this process attractive for launching new components. The conversion cost of making the component from the raw materials is reasonable since the total component cost is between 2 and 3 times the raw materials cost. As market demand increases more moulds can be employed to increase the output and further reduce costs by repetitive batch time savings.

6. CONCLUSIONS

Expanding core moulding of fibre reinforced composite components is a new process aimed at relatively low cost manufacture of components. It is suited to volume manufactureusing more moulds as demand increases. The conversion process is reproducible and canproduce components at a cost of £30 - £60/Kg depending on the cost of glass or carbon fibre prepreg used.

The conversion cost is 2 to 3 times the raw material cost which makes this process attractive for volume manufacture of components for a variety of industrial applications. The process is more expensive than SMC moulding but is cheaper than resin injection, autoclaving and wet lay up processes. The cost of a low pressure mould makes the process attractive for new product start up applications.

As work continues on developing expanding cores with a range of densities and with thermoplastic resins, a wider range of expanding cores that can be used with thermoplastic and thermosetting prepregs will enable a wider range of components to be produced competitively. Moulding composite components by expanding core that becomes an integral part of the structure after cure is expected to increase in usage because of attractive manufacturing economics compared with other moulding processes.

7. ACKNOWLEDGEMENTS

I would like to thank everybody at our Eccles facility for adapting this process to make such a variety of commercially successful components.

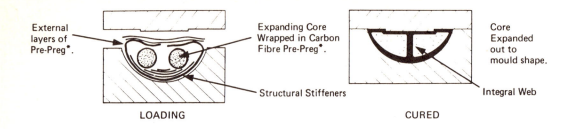

External layers of Pre-Preg*.

Expanding Core Wrapped in Carbon Fibre Pre-Preg*.

Structural Stiffeners

Core Expanded out to mould shape.

Integral Web

LOADING

CURED

Fig 1   Expanding core technique

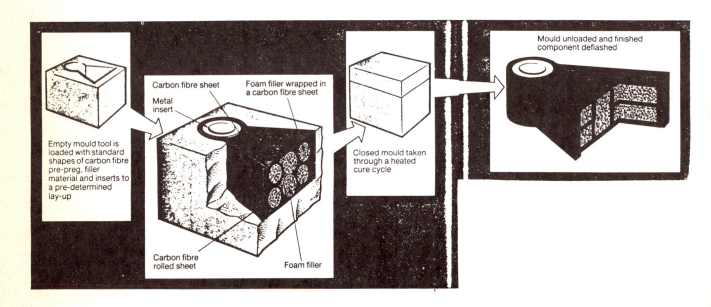

Empty mould tool is loaded with standard shapes of carbon fibre pre-preg, filler material and inserts to a pre-determined lay-up

Carbon fibre sheet

Metal insert

Foam filler wrapped in a carbon fibre sheet

Carbon fibre rolled sheet

Foam filler

Closed mould taken through a heated cure cycle

Mould unloaded and finished component deflashed

Fig 2   Expanding core moulding process

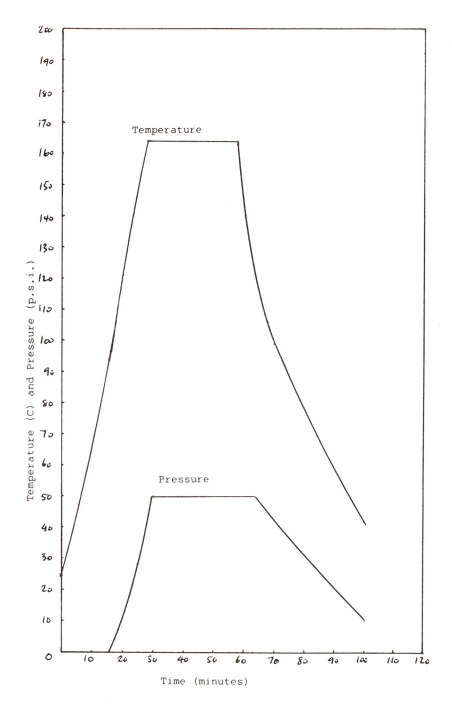

Fig 3   Temperature/pressure cure cycle with aluminium mould in air
heated oven followed by air cooling

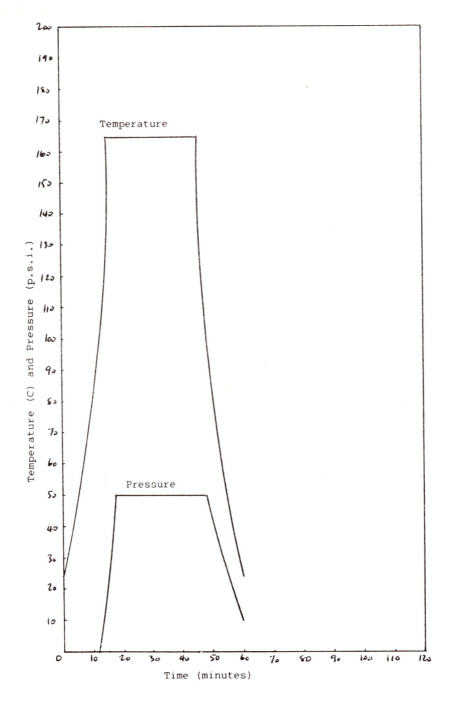

Fig 4   Temperature/pressure cure cycle with aluminium mould in a
fluid bed followed by a water quench

© IMechE 1986 C44/86

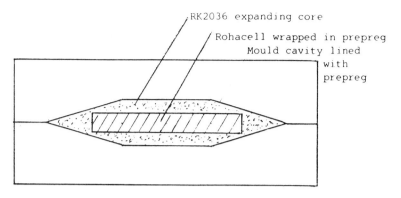

Fig 5   Use of a rigid core insert to reduce overall density of
         composite component

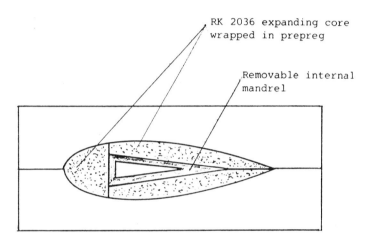

Fig 6   Use of a removable internal mandrel to reduce weight
         of composite component

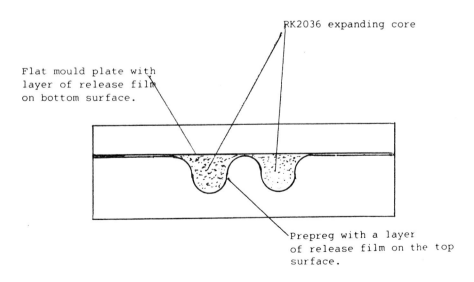

Fig 7   Use of expanding core as a moulding aid to mould a
         sine wave laminate

C45/86

Fatigue and damage in reinforced nylon

W S CARSWELL, BSc, PhD, MInstP
National Engineering Laboratory, East Kilbride, Glasgow

SYNOPSIS   Fatigue tests with a range of mean loads and at two temperatures have been carried out on
short glass-fibre nylon and short graphite-fibre reinforced nylon. The changes occurring during
these tests have been monitored by noting the peak deflection. The maximum and minimum deflections
provide indications of the form and position of the hysteresis loop in the material. In these tests
both the range of deflections and the peak deflection increase, indicating a rotation of the hysteresis loop and creep in the direction of maximum load. The range of deflections at both room temperature and at elevated temperature provide a reasonable indication of endurance although there are differences in performance at elevated temperature. The extent of such changes necessary for failure is
obscure at this stage.

1   INTRODUCTION

It is now generally accepted that under cyclic
loading fibre-reinforced plastics composites
exhibit accelerated or premature failure at
loads well below those required to produce static
failure. Failure may not be defined necessarily
by fracture or inability to sustain the applied
load but can be described by the change in properties which occur in the course of a test, for
example a decrease in modulus.

Such changes can be related to the concept
of damage which is initiated at the beginning of
the test and grows until failure. In physical
terms the damage can be considered as cracking
in the matrix or debonding, or generally as the
reduction in the effectiveness of the reinforcement.

In fibre-reinforced composite materials
this damage is essentially a bulk effect occurring progressively throughout the whole stressed
volume of the material. Fracture, on the other
hand, is a local event. The problem then is to
relate such bulk changes to some failure criterion, particularly where fracture is the
failure mode.

Much of the work in damage mechanics has
been concerned with laminates and continuous
fibre-reinforced thermosetting materials (1, 2).
Fibre-reinforced thermoplastic materials are now
readily available and are of interest for their
toughness and ability to be easily formed.

The reinforcing fibres may be discontinuous
in order to permit ease of moulding. In these
materials fatigue may be considered a more
severe problem than in continuous fibre-
reinforced thermosetting materials due to the
reduced load-carrying capabilities.

Consideration is also being given now to
the properties of the materials at elevated
temperatures for some applications. In order to
predict performance or endurance it is necessary
to be able to apply the concept of damage mechanics over all these parameters.

2   DAMAGE MEASUREMENTS

The deflections occurring during cyclic loading
have been used in the programme as a measure of
the condition of material or the damage. Using
deflection transducers in a servo-hydraulic
fatigue machine operating under load control,
the maximum and minimum deflection occurring in
the load cycle can be recorded progressively
during the test without interruption to the
test.

The fatigue tests were carried out over a
range of cyclic amplitude and mean loads. The
frequency of cycling was 5 Hz. Tests were
carried out at 100°C and room temperature.

The form of the testpiece is generally a
strip of parallel section approximately 75 mm
long. For testing at elevated temperatures the
specimen is contained in an oven. The deflections can be converted to strain by comparing
the strain as measured from an extensometer or
strain gauges with the deflections measured by a
linear displacement transducer in a static test.
Hysteresis loops may be obtained in a similar
manner.

3   MATERIALS

The results which are presented and discussed in
this paper were obtained on two fibre-reinforced
thermoplastic materials:

a   Nylon 66 reinforced with 41 per cent by
weight discontinuous graphite fibres, and

b   Nylon 66 reinforced with 53 per cent by
weight discontinuous glass fibres.

The length of the fibres was approximately 10 mm
and the moulding of the testpieces produced
substantial orientation of the fibres in the
longitudinal axis of the testpiece.

The tensile properties of the materials at
room temperature are given in Table 1.

Table 1   Mechanical properties of materials

Designation	Weight fraction %	Tensile strength MN/m^2	Tensile modulus GN/m^2
Graphite/Nylon	41	265	13.7
Glass/Nylon	53	260	7.38

4   RESULTS AND DISCUSSION

The general form of the hysteresis loop in tension is shown in Fig. 1 for the glass-filled nylon. During the course of a fatigue test under constant load conditions both the maximum and minimum deflections change as shown in Fig 2. The change in maximum deflection is greater than that of the minimum, indicating a rotation of the hysteresis loop as well as a movement or drift of the loop in the tensile direction.

The rotation, as measured by the range of deflection, $\Delta\delta$, in the cycle, and the difference between maximum and minimum deflection, $\delta_{max} - \delta_{min}$, is an indication of the change in stiffness or modulus of the material. The drift can be described approximately by either the maximum deflection, δ_{max}, or the mean deflection in the cycle, $\frac{1}{2}(\delta_{max} - \delta_{min})$.

The general course of the changes in $\Delta\delta$ and δ_{max} during the course of a test are shown in Fig 3. In both parameters three stages can be distinguished (3, 4)

a   an initial stage in which generally there is a rapid change;

b   a second stage of slow change; and

c   a third stage of rapid change just prior to failure.

In some cases there is a transition region between the first and second stages in which there is no change in $\Delta\delta$ and even a decrease as the material is conditioned to the testing conditions.

During the second stage, the rate of change of both $\Delta\delta$ and δ_{max} tends to decrease as the test continues.

The maximum deflection δ_{max} shows the same changes as $\Delta\delta$ and is essentially an exaggeration of $\Delta\delta$. The changes are similar for both temperatures and for the graphite fibre-reinforced nylon as well as the glass fibre-reinforced nylon.

The results of the room temperature tests on glass-reinforced nylon are given in Table 2. The initial and final range of deflections are given with an approximate rate of change of $\Delta\delta$ at approximately mid-life, $d\Delta\delta/dN \times 10^3$. Also included is the change in δ_{max} in the course of a test to failure.

The results of the tests at 100°C on glass-reinforced nylon are given in Table 3. The corresponding results for graphite fibre-reinforced nylon are given in Tables 4 and 5.

In both materials and at both temperatures the range of deflection and the rate of change of range are functions of both the cyclic load and the mean load. The maximum ranges of deflection occur around a repeated stress cycle where the stres ratio, R, is approximately zero. The mean load or maximum load appears to have a considerable effect on the rate of change of the range of deflection.

For each material, the load range to produce a range of deflection is reduced as the temperature increases but at both temperatures the relation between range of deflection and endurance is very similar (5).

In continuous fibre-reinforced composites the stiffness of the composite in the damaged state, E, has been related to a damage parameter D by the expression (3)

$$E/E_o = 1 - cD \qquad (1)$$

where E_o is the original stiffness of the undamaged composite and c is a constant.

The same damage parameter D may be introduced in the concept of effective stress $\tilde{\sigma}$ (6) in the expression

$$\tilde{\sigma} = \sigma/(1 - cD). \qquad (2)$$

From expression (1)

$$cD = 1 - E/E_o \quad \text{or} \quad = \Delta E/E_o \qquad (3)$$

where $\qquad \Delta E = E_o - E.$

In terms of deflection $cD = (\delta - \delta_o)/\delta_o$.

It has been reported (7) that the value of $\Delta E/E_o$ at failure increases as the cyclic amplitude is increased.

There is a trend in this direction in glass-reinforced nylon but it is very small. Similarly the total change in deflection does not appear to be very dependent on the stress or provide a good indication of impending failure.

Microscopic examination of the material around the failure region has not produced any evidence of resin cracking.

In the case of discontinuous fibre-reinforcement the damage is probably in the form of debonding around the fibres, leading to a reduction in both the effective length of the fibres and in their effective volume fraction. This reduction will lead, in load control test conditions, to increased load on the matrix.

Failure will occur when the local load in the matrix is sufficient to initiate a crack and the reinforcement is degraded to the extent that a crack can propagate across the testpiece.

The damage under these conditions, D, might be described as a function of ℓ/ℓ_o, where ℓ is the effective fibre length at any time and ℓ_o is the original fibre length.

In all conditions, the cyclic load appears to have a more significant effect on rate of change of range and deflection than the mean or peak load.

The effects are much greater in glass-reinforced nylon than graphite-filled nylon, reflecting the relation between the moduli of the reinforcing fibre and the different load-transfer characteristics.

CONCLUSIONS

Deflection measurements taken during load-controlled fatigue tests on glass-fibre and graphite-fibre reinforced nylon have permitted the course of damage produced by cyclic testing to be followed.

The range of deflection necessary for any endurance is almost independent of temperature. The changes occurring in the range of deflection during any test are greater in glass-reinforced nylon than in graphite-reinforced nylon, although at 10^6 cycles the initial range is approximately the same.

In addition to changes in range of deflection there is significant creep or an increase in peak deflection.

The changes are much more sensitive to cyclic loads than to mean or peak loads.

The total changes occurring during the course of a test to failure are not very sensitive to peak load and do not provide a useful indication of impending failure.

ACKNOWLEDGEMENTS

This paper is published with permission of the Director, National Engineering Laboratory, Department of Trade and Industry. It is Crown copyright. The work was supported by the Mechanical and Electrical Enginering Requirements Board of the Department of Trade and Industry.

REFERENCES

(1) Highsmith, A. L. and Reifsnider, K. L. Stiffness reduction mechanisms in composite laminates. Damage in Composite Materials, ASTM STP 775, 1982, pp 103-117.

(2) Ogin, S. L., Smith, P. A. and Beaumont, P. W. R. Matrix cracking and stiffness reduction during the fatigue of (0/90)s laminate. Composite Sci. and Techn., 1985, 22(1), 43-74.

(3) Poursartip, A., Ashby, M. F. and Beaumont, P. W. R. Fatigue and Creep of Composite Materials, 3rd Riso Int. Symposium, Denmark, 1982, pp 279-284.

(4) Roylance, M. E., Houghton, W. W., Fuley, G. E., Sauford, R. J. and Thomas, G. R. Characterisation of cumulative damage in composites during service, AGARD Conf. Proc. N355, Neuilly sur Seine, 1983, Paper 7.

(5) Gotham, K. V, and Hough, M. C. The durability of high temperature thermo-plastics. R.A.P.R.A., Shawbury, England, 1984.

(6) Sidoroff, F. Damage mechanics and its application to composite materials. European Mechanics Colloquium, 182, Brussels, 1984, pp 21-35.

(7) Dody, M., Oytana, C., Pierre, M. and Varchon, D. Mechanical properties of a material reinforced with glass cloth (in French). First European Conf. on Composite Materials, Bordeaux, 1985, pp 108-113.

Table 2   Test details of glass/nylon: range of deflection - room temperature

Test-piece	Mean stress MN/m^2	Cyclic stress MN/m^2	$\Delta\delta_o$ mm	$d\Delta\delta/dN \times 10^3$ mm/c	$\Delta\delta_{end}$ mm	$(\delta_o - \delta_{end})$ mm	Endurance cycles
1	77.5	77.5	1.19	0.4	1.50	0.23	652
4	32.5	77.5	0.91	0.28	1.20	0.20	833
13	90.4	64.5	0.90	0.20	1.08	0.50	922
2	64.5	64.5	0.93	0.08	1.10	0.26	2 275
5	38.75	64.5	0.85	0.06	1.03	0.23	2 901
7	12.90	64.5	0.73	0.05	0.91	0.26	3 361
9	103.3	51.6	0.58	0.04	0.76	0.48	2 237
11	77.5	51.6	0.64	0.015	0.75	0.38	6 339
6	51.6	51.6	0.67	0.008	0.71	0.32	14 490
8	25.8	51.6	0.62	0.006	0.74	0.23	87 206
10	116.25	38.75	0.50	0.01	0.64	0.52	7 877
14	90.35	38.75	0.50	0.002	0.55	0.51	21 467
12	64.50	38.75	0.50	0.001	0.517	0.28	269 858
3	38.75	38.75	0.40	-	-	-	>10^6 unbroken

Table 3   Test details for glass/nylon: range of deflection – 100°C

Test-piece	Mean stress MN/m^2	Cyclic stress MN/m^2	$\Delta\delta_o$ mm	$d\Delta\delta/dN$ $\times10^3$ mm/c	$\Delta\delta_{end}$ mm	$(\delta_o - \delta_{end})$ mm	Endurance cycles
8	37.5	37.5	0.93	0.02	1.1	0.32	2 824
4	25.0	37.5	0.75	0.01	0.96	0.16	8 022
5	12.5	37.5	1.11	0.01	1.45	0.21	16 650
2	31.25	31.25	0.84	0.002	1.04	0.53	16 500
3	37.5	25.0	0.56	0.012	0.63	0.11	50 500
1	25.0	25.0	0.60	0.002	0.63	0.20	65 000

Table 4   Test details of graphite/nylon: range of deflection – room temperature

Test-piece	Mean stress MN/m^2	Cyclic stress MN/m^2	$\Delta\delta_o$ mm	$d\Delta\delta/dN$ $\times10^3$ mm/c	$\Delta\delta_{end}$ mm	$(\delta_o - \delta_{end})$ mm	Endurance cycles
1	103.3	103.3	0.08	0.08	0.881	0.14	337
4	51.66	103.3	0.06	0.06	0.84	0.08	1 298
5	25.8	103.3	0.02	0.02	0.90	0.17	2 361
2	84.0	84.0	0.005	0.005	0.664	0.07	5 796
6	38.75	90.4	0.01	0.01	0.78	0.12	9 476
8	12.9	90.4	0.005	0.005	0.68	0.04	9 283
9	129.2	77.5	–	–	0.50	0.07	1 745
10	103.3	77.5	0.0004	0.0004	0.63	0.08	3 490
12	103.3	77.5	–	–	0.58	0.02	1 965
13	77.5	77.5	0.001	0.001	0.671	0.05	10 709
11	116.25	64.5	–	–	0.51	0.10	30 611
14	90.4	64.5	–	–	0.537	0.07	13 919
3	64.5	64.5	–	–	0.51	0.14	321 015

Table 5   Test details for graphite/nylon: range of deflection – 100°C

Test-piece	Mean stress MN/m^2	Cyclic stress MN/m^2	$\Delta\delta_o$ mm	$d\Delta\delta/dN$ $\times10^3$ mm/c	$\Delta\delta_{end}$ mm	$(\delta_o - \delta_{end})$ mm	Endurance cycles
1	50	50	0.77	0.01	0.79	0.17	3 884
3	25	50	0.70	0.005	0.78	0.27	34 000
4	0	50	0.65	0.001	0.68	0.23	174 000
10	56.25	43.75	0.61	–	0.61	0.06	8 010
7	43.75	43.75	0.61	0.001	0.63	0.11	9 800
9	31.25	43.75	0.57	–	0.60	–	192 000
5	62.5	37.5	0.46	–	0.46	0.57	$>10^6$ unbroken
2	37.5	37.5	0.55	–	0.56	0.10	256 000

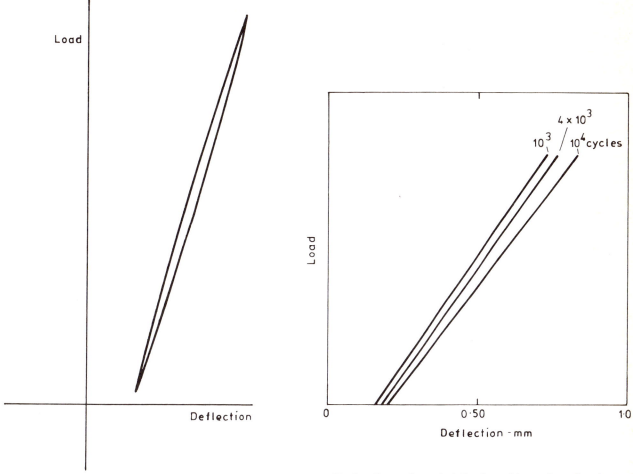

Fig 1   Hysteresis loop in tension only

Fig 2   Change in peak deflection with number of cycles

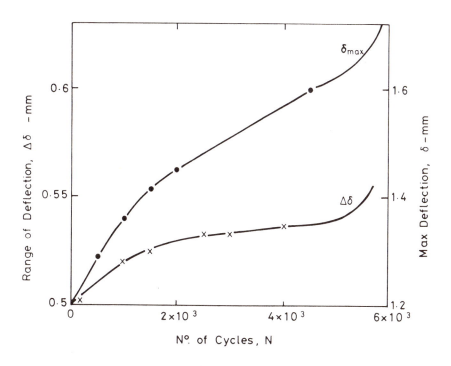

Fig 3   Form of changes in deflection and range of deflection
with number of cycles

C45/86

C46/86

Future use of advanced composites in bridge construction engineering

U R S MEIER, Dipl-Ing
Swiss Federal Laboratories for Materials Testing and Research, Duebendorf, Switzerland

SYNOPSIS   The potential application of advanced composites in bridge construction engineering is described based on two examples. The first deals with the strengthening of existing structures, the
second with a proposal for a CFRP Bridge with a main span of 8400 m across the strait of Gibraltar at
its narrowest site.

1   INTRODUCTION

Advanced composites will never be able to compete with traditional materials such as concrete
and steel for the building of commonplace structures. Where then does the potential for the
structural application of advanced composites
lie? One possibility is in special cases where
lightweight construction is demanded due to local conditions. Another is for structures having
very large spans. Both of these applications
will be discussed based upon two examples.

2   STRENGTHENING OF EXISTING STRUCTURES

Systematic investigations have been carried out
at the EMPA since 1970 [1] with the purpose of
developing a new method to strengthen structures. In this method, steel elements in the
form of plates or segments are bonded to the exterior of steel or reinforced concrete structures. Thus, the structure can be subsequently
reinforced. Such measures are necessary when
existing structures are called upon to accept
greater loads than they were originally designed
for.

A recent analysis showed that the employment of
advanced composites (e.g. carbon fibre reinforced epoxy laminates) permits considerable cost
savings for the scaffolding in the subsequent
strengthening of bridges. Since reinforcement
plates made of advanced composites are extremely
light, scaffolding is not necessary. The plates
are bonded from the working platform of a hydraulic lift and pressed on with a vacuum bag
during the hardening process, as shown in
Fig 1. Despite the much higher price of advanced composites (approx. 40 times higher
than steel), the total cost for this type of
strengthening can be reduced by about 20 per
cent compared to the method with steel plates.

First investigations started with bending tests
on reinforced concrete beams of rectangular
cross section (150 mm x 250 mm) and a free span
of 2.0 m, which were additionally strengthened
with a unidirectional carbon fibre reinforced
epoxy laminate of 200 mm width and only 0.2 mm
thickness on the outside. These tests were intended to supply information on the best form of
the anchoring zone. This zone was always outside
the support lines. A typical load/deflection
diagram for such a test is shown in Fig. 2.

3   PROPOSAL FOR A CFRP BRIDGE ACROSS THE STRAIT OF GIBRALTAR AT ITS NARROWEST SITE

3.1 Background

The idea of joining Europe and Africa together
with a fixed link has interested politicians and
scientists for centuries. The most suitable zone
for such a structure is the Strait of Gibraltar. For the possible implementation of this
structure, an agreement on technical and scientific cooperation between the Spanish and Moroc-
can governments was formulated in 1979. Following a technical colloquium in Tangier in October, 1980 and a large general international colloquium in Madrid in November, 1982 on the feasibility of a fixed link across the Strait of
Gibraltar, the following possibilities have cry-
stallized: A tunnel in the sea floor, constructed with conventional mining techniques or a
bridge mounted on rigid towers.

The alignment for the crossing of the Strait of
Gibraltar was determined by deep sea measurements and the distance between the two shores.
Analysis of the deep sea map suggests two crossing locations:

(a) An axis which connects the two continents
    through the shortest possible path, i.e.
    between Funta Cirès and Oliveros, having a
    length of approximately 15 km. Along this
    axis the sea depth ranges up to 900 meters.

(b) A connection between the Capes of Malabata
    and Palomas, taking advantage of an almost
    straight-line elevation of the sea floor.

The length of this connection amounts to about 26 km. The sea depth does not exceed 300 m.

Considering the **existing** state of the art of building technique, the alternative (b) appears to be the most feasible. Taking into account the considerable sea depth (300 m) at this location and the small allowable grade of 1.5 to 2.0 per cent, a railroad tunnel would have a length of about 50 km of which approx. 28 km would lie under the sea floor. A highway tunnel would be shorter, but would present much more difficult ventilation and exhaust problems. At present, a bridge employing **conventional** building materials and having a length of about 30 km is under study. The central portion of this bridge, where the sea is deepest is to be spanned with a suspension bridge having a length of 22 km. This will be composed of 10 adjacent spans, each of 2000 m length and two end spans, each of 1000 m length. Connection to the two shores is to be accomplished with foreland bridges of 1800 m resp. 6000 m length. The cost for the tunnel solution is estimated at approx. 5000 million US dollars. For the above-described 'conventional' bridge, the cost would be about double so much. The bridge solution has the additional disadvantage that ship passages through the Strait would be permanently hindered by a minimum of 10 towers separated from each other by only 2000 m. A fixed link at the narrowest site (Alignment (a)) is not possible with buiding materials currently employed, since the theoretical limiting span for steel bridges lies in the range of 5000 or maximum 7000 m. In a practical case, this permits a span of at most 3500 to 4000 m. Since tower foundations deeper than 350 m are not possible, a minimum main spain of about 8400 m is necessary for a bridge utilizing alignment (a). Such spans can only be accomplished by employing advanced composites.

3.2 Multiplication of the limiting span of suspended bridges through the use of advanced composites

The use of advanced composites would allow the doubling or tripling of the limiting span in comparison to steel structures. In Fig 3 the specific design loads versus the centre span for the classical form of suspension bridges made of steel are compared with those made of glass fibre reinforced (GRP) or carbon fibre reinforced (CFRP) plastics. The specific design load is defined as the dead load w_{st} of the superstructure compared to the load $w_{Deck} + w_{Live}$ (dead load of the deck and live load). Calculation of the specific design load as a function of the span ℓ is as follows:

$$\frac{w_{Structure}}{w_{Deck} + w_{Live}} = \frac{\ell}{\dfrac{\sigma_{allowable}}{\alpha\,\gamma} - \ell} \qquad \text{Eq. (1)}$$

whereas α (assumed as 1.66 in the case of Fig 3) represents a system coefficient and γ the density. The limiting span ℓ_{lim} was calculated as follows:

$$\ell_{lim} = \frac{\sigma_{allowable}}{\alpha\,\gamma} \qquad \text{Eq. (2)}$$

The uni-axially loaded cables and hangers are mainly responsible for determining the allowable stress in Eq. (1). This allows the use of the allowable stress for unidirectional fibre reinforced plastics.

The assumptions used for Fig 3 can to some extent be questioned. However, this would not have any fundamental influence on the qualitative statement of Fig 3.

The application of glass fibre reinforced plastics would permit a doubling of the limiting span ℓ_{lim} in comparison to steel, and through the use of carbon fibre reinforced plastics, even a three-fold increase is possible.

In Fig 4 the price relation of CFRP to steel, resp. GRP to steel per unit mass is plotted as a function of the centre span of a suspension bridge. The assumptions made are the same as in Fig 3. The diagram in Fig 4 demonstrates that the 'break-even-span' for CFRP in 1985 is 4170 m and for GRP 4150 m.

Assuming a further downward development of prices for carbon fibres and a stable price for cable steel this 'break-even span' for CFRP may be in the range of 3900 m in the future. This implies that only superstructures with main spans in the range of 4000 m and greater will be the domain of composites.

Due to the relatively low modulus of elasticity of GRP, as well as other disadvantages (and only very small economical advantages; Fig 4) the further discussion will only deal with CFRP.

Carbon fibre reinforced plastics (CFRP) have been available since the early 1960's. These composites have been the subject of numerous studies and reports, funded principally by government agencies that have recognised that the intrinsic qualities of the materials were valuable to aerospace applications. The modulus-to-density ratio of CFRP was the highest available for structural materials and, hence, CFRP are found where maximum stiffness and light weight are essential. Other benefits also were observed such as low order of thermal expansion coefficients (much lower than steel) under temperature extremes, enhanced fatigue life, and the relative inertness of CFRP to corrosion.

At first sight these perspectives open enormous new possibilities in the construction of very long span bridges. However, these expectations are somewhat restricted, as a large number of constructive and material problems must be resolved before such a long span structure can become reality. These seem solvable, but require a considerable investment of time and money.

3.3 Choice of the superstructure

To bridge a main span of 8400 to 10 000 m, as required for the crossing location (a) of the Strait of Gibraltar, only two types of super-structures can be considered suspension and cable-stayed bridges made of advanced composites. During the past two decades, cable-stayed bridges have found widespread application especially in Western Europe, but recently in other parts of the world as well. Nowadays the cable-stayed bridge competes against the classical suspension bridge where very long spans are required. It appears easier to adapt a cable-stayed bridge to certain boundary conditions than a suspension bridge.

The renewal of interest in the cable-stayed system in modern bridge engineering is due to the tendancy of bridge engineers [2] to strive toward optimum structural performance. With respect to the proposed CFRP bridge across the Strait of Gibraltar, optimum structural performance is of primary importance, due to the high cost of advanced composites. In the following discussion, the suspension bridge solution will be mentioned only for comparison purposes, since, as will be shown, the cable-stayed bridge is superior.

Depending on the longitudinal cable arrangement, cable-stayed bridges will here be classified as fan and harp systems. In the fan system, all cables lead to the top of the tower. Structurally, this arrangement is perhaps the best since by directing all cables to the tower top (Fig 5), the maximum inclination to the horizontal can be achieved. Consequently, the smallest quantity of CFRP is required. The cables carry the maximum component of the dead and live load forces, and the axial component of the deck structure is a minimum. This is important in a design using CFRP. These cables can easily be built up of several hundred parallel CFRP wires. Each individual wire having a diameter of about 6 mm is pulltruded from about 500 000 continuous carbon fibres. Such cables are excellent applications of CFRP since they take advantage of the outstanding unidirectional properties (strength, stiffness) of these composites.

In the harp system, the cables are connected to the tower at different heights and placed parallel to each other. This system is preferred from an aesthetic viewpoint. The quantity of CFRP required for a harp-shaped cable arrangement is slightly higher than for a fan-shaped arrangement.

To compare the quantities of material required for the cables of suspension and cable-stayed bridges, calculations were performed in accordance with [2]. The results are shown in Fig 6. The end spans are assumed to be 0.4 times the center span for each of the three bridge types. Furthermore, the following assumptions were made:

Dead load plus live load (including girder and deck) for CFRP structure: 0.17 MN/m[1]

Dead load plus live load (including girder and deck) for steel structure: 0.28 MN/m[1]

Bridge width according to Fig 7 with 6 road lanes and 2 railroad tracks:
- Truss type (2 decks)   28 m
- Girder type (single deck)   40 m

Cable unit weight:
- CFRP 0.015 MN/m^3
- Steel 0.078 MN/m^3

Allowable cable stress:
- CFRP 380 MN/m^2
- Steel 720 MN/m^2

Ratio of tower height to center span length h/ℓ:
- Suspension type   0.10
- Stayed type   0.20

It is important to note that the material weight of the cables increases as the square of the span length for the suspension bridge as well as for the cable-stayed bridge. Therefore the saving of material with the cable-stayed bridge remains in the same order of magnitude (Fig 6) for long spans; the criterion also applies for its superior stiffness. This undoubtely proves that compared to suspension bridges, cable-stayed bridges are superior for all spans above approximately 200 m.

For the Gibraltar bridge proposals the following total cable weights were calculated:

Cable-stayed type with a main span of 8400 m (Fig 5) for CFRP Cables: 1026 MN (= 104 600 metric tons)

Suspension type with a main span of 10 000 m for CFRP Cables: 2879 MN (= 293 500 metric tons)

For the girders and decks the total CFRP quantity is in both cases approximately 454 MN (= 46 240 metric tons).

Finally the total CFRP weight of a superstructure as shown in Fig. 5 would amount to 1480 MN (= 150 840 metric tons).

3.4 Feasibility Considerations

The proposal described above for the bridging of the Strait of Gibraltar at its narrowest site with a cable-stayed bridge of CFRP would be a highly challenging civil engineering task. This is true not only for the superstructure but also for the towers, the foundations and the anchorages. Employing building materials currently used for superstructures, this challenge cannot be met. The development step from previously designed main spans (approx. 1400 m) utilizing steel to the proposed span of 8400 m (Fig 5) or greater can only be achieved with advanced composites. Indeed, composites seem predestinated, above all for the cables which comprise approx. 70 per cent of the weight of the superstructure, since the outstanding properties of strength and stiffness of unidirectional fibre composites can be used to full advantage.

The fabrication of cables from 6 to 10 mm wires does not present any technical problems. The wires themselves would be manufactured using pulltrusion. In addition, the anchoring of the individual wires should pose no problems since sufficient anchoring length is available.

On the other hand, the fabrication of the CFRP girders appears to present some problems. The alternative of Fig 7 a represents an aerodynamically optimized cross-section presently employed for steel bridges. Such CFRP cross-sections cannot be fabricated with currently available machines. It would be necessary to produce huge filament-winding or pulltrusion machines to enable their fabrication. Furthermore, a great difficulty lies in the design of the joints for the individual pre-fabricated girder sections.

Based on the current state of the art, the truss arrangement sketched in Fig 7 should be considerably more simple to achieve using available means. The principle bracing members of the truss would be of box-section design. The manufacture of box beams from composites is described in [3]. The truss diagonals would be fabricated as filament-wound tubes.

Although a great numer of technical details remain to be solved, the bridging of the Strait of Gibraltar at its narrowest site with a CFRP bridge appears possible from the technical standpoint within the next 20 to 30 years.

From the cost point of view, a CFRP bridge at the narrowest site (alignment (a)) would certainly be less expensive than a conventional bridge at alignment (b). Whether the CFRP bridge at alignment (a) would also be competitive with the tunnel alternative at alignment (b) (and could eventually be carried out) depends, above all, on future developments in the price of carbon fibres.

Another obstacle is the financing of a long span bridge made of fibre reinforced plastics. Surely no contractor is willing to build such an object, without being able to estimate the risk. This is only possible after years of practical experience on an object of medium span. Bridges of medium span however will never be built economically from fibre reinforced plastics as shown in Fig 4.

But what can be done to develop confidence for CFRP in civil engineering applications? At the Swiss Federal Laboratories for Materials Testing and Research (EMPA) a research program is going on since 1975 on GRP box beams which could be used for the girder construction.

Six FRP-box beams loaded in bending, each having a length of 3 m, a height of 188 mm and a width of 118 mm, were tested under static and fatigue loading. In the short-time tests, these beams yielded similar failure loads to those of a HEA 200 steel beam. Two beams were tested under sinusoidal load-controlled conditions at a frequency of 2 Hz. The minimum load in each case was 4.8 per cent of the failure load, while the maximum load was 19.1 per cent in one case and 28.6 per cent in the other case. The beam fatigue tested at 19.1 per cent of the failure load did not develop any detectable damage before the test was terminated at 10^8 cycles (1.59 years of loading). The bending stiffness of this beam after 10^8 cycles was identical to that before testing. The beam tested at 28.6 per cent of the failure load showed no damage for the first $4 \cdot 10^6$ cycles; at this point matrix cracks initiated on the matrix-rich surface of the beam on the tensile side. Even at 10^8 cycles, the observed damage caused a decrease in bending stiffness of only 2 per cent [3]. This investigation is still going on.

Even more important are the cables. What can be done in this case? In a first step EMPA started 1984 with laboratory tests on single CFRP-wires. This program will be continued with investigations on cables on large test rigs. In a second step the replacement of one or two steel cables by CFRP cables in an existing cable-stayed bridge is planned. As shown within the following section such applications could alredy be quite interesting in todays bridge engineering for the longest cables.

The stiffness of a cable-stayed bridge (Fig 5) depends largely upon the tensile stiffness of the stay cables. The displacement of the end of a free hanging stay cable under an axial load depends not only the cross-sectional area and the modulus of elasticity of a cable but to a certain extent on the cable sag, as described in Ref. [4].

An equivalent modulus of elasticity E_i of a cable is defined as:

$$E_i = \frac{E}{1 + [(\gamma \ell)^2 / 12 \sigma^3] E} \qquad \text{Eq. (3)}$$

where

E_i = equivalent modulus of inclined stay-cable having sag

E = modulus of elasticity of vertically tested straight cable

γ = specific weight of cable

ℓ = horizontal span of stay cable

σ = tensile stress in the cable

In Fig 8 the relative equivalent modulus of elasticity of CFRP and steel cables are compared and the good performance for CFRP-cables is demonstrated. The actual modulus of elasticity of CFRP wires is the same as for steel wires or even slightly higher.

The proposed field tests appear to be feasible and will help very much developing confidence in CFRP for civil engineering.

Although a great numer of problems remain to be solved, the bridging of the Strait of Gibraltar at its narrowest site with a CFRP bridge appears possible from the technical standpoint within the next 20 to 30 years.

REFERENCES

[1] Ladner M., Weder Ch.: Concrete Structures with Bonded External Reinforcement. EMPA Report 206, Dübendorf/Switzerland.

[2] Leonhard F. and Zellner W.: Comparative Investigations between Suspension Bridges and Cable-stayed Bridges for Spans Exceeding 600 m. International Association for Bridge and Structure Engineering, Publication 32-1, 1972.

[3] Meier U., Müller R. and Puck A.: FRP-Box Beams under Static and Fatigue Loading. Proceedings of the International Conference on Testing Evaluation and Quality Control of Composites, Guildford, Editor T. Feest, Butterworth 1983.

[4] Ernst H.-J.: Der E-Modul von Seilen unter Berücksichtigung des Durchhanges. Der Bauingenieur, 1965, 40, 52-55.

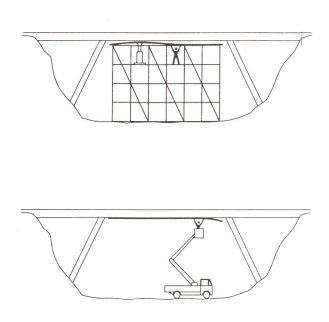

Fig 1 Strengthening of existing bridges. Since reinforcement plates made of advanced composites are extremely light, costly scaffolding is not necessary. The plates are bonded from the working platform of a hydraulic lift and pressed on with a vacuum bag during the hardening process

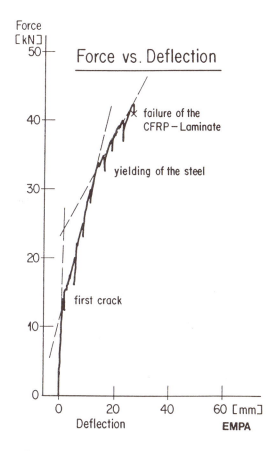

Fig 2 Experimental loading curve of a concrete beam strengthened with a 0.2 mm thick carbon fibre reinforced plastic (CRFP) plate

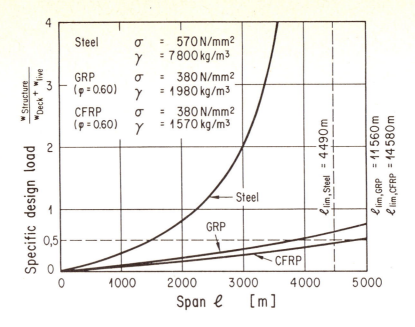

Fig 3   Specific design load versus the main span for the classical
form of suspension bridges. ℓ_{lim} = limiting span

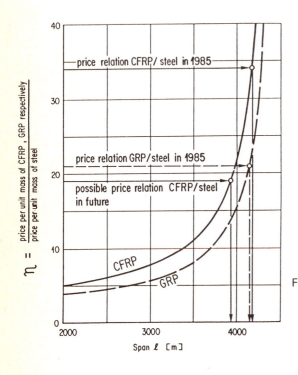

Fig 4   Price relation of CFRP to steel, respectively glass fibre
reinforced plastic (GRP) to steel per unit mass as a function of
the main span. These calculations are based on the following
prices for 1985. CFRP $ 21.00 per Kg, GRP $ 13.00 per Kg,
cable steel $ 0.61 per Kg and for the future: CFRP $ 12.00
per Kg

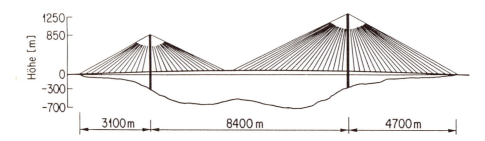

Fig 5   Proposal for a CFRP bridge across the Strait of Gibraltar at its narrowest site. Since tower foundations
deeper than 350m are not possible, a maximum main span of about 8400 m is necessary

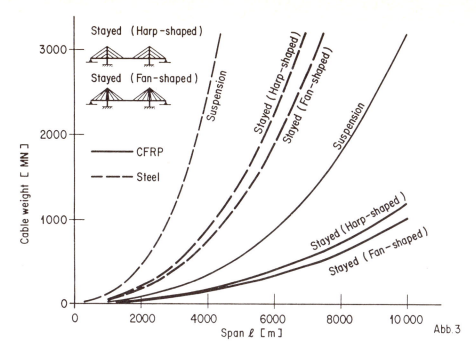

Fig 6 Comparison of the quantities of material required for the cables of suspension and cable-stayed bridges of steel of CFRP as a function of the main span ℓ. The end spans are assumed to be 0.4 times the main span for each of the three bridge types

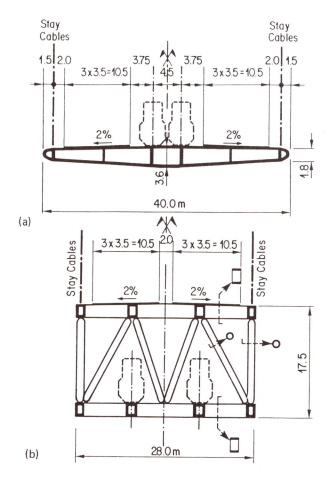

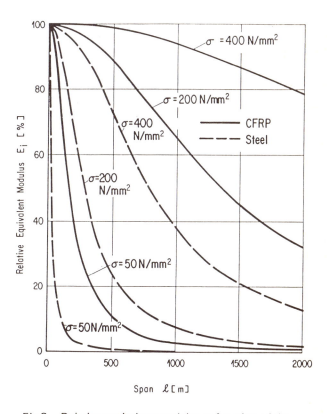

Fig 7 Possible girder cross-sections
(a) Aerodynamically optimized cross-section with six road lanes and two railway tracks on one deck
(b) Truss type cross-section with six road lanes on the first and two railway tracks on the second deck

Fig 8 Relative equivalent modulus as function of the horizontal span ℓ of the cable for CFRP and steel

C47/86

Tensile behaviour of unidirectional glass/carbon hybrid laminates

G KRETSIS, MSc, ACGI, DIC, F L MATTHEWS, BSc(Eng), ACGI, CEng, MRAeS, FPRI,
J MORTON, MA, DPhil, CEng, MIM and G A O DAVIES, BEng, PhD, DCAe, MRAeS
Department of Aeronautics, Imperial College of Science and Technology, London SW7

SYNOPSIS Tensile tests carried out on 8- and 16-ply unidirectional carbon/glass/epoxy laminates confirmed that the stiffness of hybrids is predictable using the rule of mixtures. They also showed a small positive hybrid effect (i.e. enhancement in the initial carbon fibre failure strain) which was greater for decreasing proportions of carbon and which was reasonably well represented by calculations accounting for statistical effects and thermal stresses.

1   INTRODUCTION

Composites containing more than one type of fibre material are commonly known as 'hybrid composites'. The term 'hybrid' is generally used to denote the incorporation of two different types of material into one single material, and the level of mixing can be either on a small scale (fibres, tows) or on a large scale (layers, pultrusions, ribs). The purpose of hybridisation is to construct a new material that will retain the advantages of its constituents but not their disadvantages. However for most of the properties the rule of mixtures (i.e. the volume-weighted sum of the constituents' properties, according to the composition) is only an upper bound. For example, in the case of tensile strength, the material with the lower failure strain will fail first and therefore the hybrid is weaker than both its constituents. However, there are other factors like cost, weight, post-failure behaviour, fatigue performance, impact resistance, etc, that sometimes lead the designer to the use of hybridisation in order to tailor the material to the needs of the structure under design.

2   EXPERIMENTAL PROCEDURE

Unidirectional hybrid composites were fabricated by stacking pre-impregnated sheets of either XAS-carbon or E-glass fibres in Ciba-Geigy 913 epoxy resin (Ref (1)). Both 8-ply and 16-ply symmetric laminates were produced with stacking sequences which spanned the entire range of hybrid compositions (see Table 1). These laminates were then used in a study of the effects of thickness, carbon/glass content and stacking sequence on the tensile performance.

The tensile tests were carried out on specimens which were about 200 mm long and 20 mm wide, with a gauge length of 100 mm. It was found that tapered aluminium alloy end plates gave the most reliable results. During testing, load, strain and acoustic emission activity were monitored. The tests were carried out under both displacement control and load control.

A distinction was made between initial and ultimate failures, initial failure being defined as the first detectable drop in the stress-strain curve, although one was not observed in every test. The 'hybrid effect' was defined as the enhancement of the initial failure strain of the carbon layers in the hybrid, as compared with the initial failure strain of the all-carbon composite. However, for those specimens in which initial failure was not observed, the strain at ultimate stress was used for the calculation of the hybrid effect. Furthermore, the datum value for the initial failure strain of an all-carbon composite was calculated from specimens of exactly the same dimensions as for the hybrid specimens. This is an important detail since thickness, edge (width), and statistical (volume) effects are all quite significant, bearing in mind that the aim of the work was to investigate the hybrid effect, which was only expected to be a few per cent above or below the simple theoretical prediction based on the failure strain of carbon.

In order to compare the strength of the various stacking sequences, the nominal ultimate stress was used (instead of strain), since there are errors associated with strain measurements once initial damage has occurred.

3   EXPERIMENTAL RESULTS

The tensile Young's modulus in the fibre direction was shown to obey the rule of mixtures, i.e. the linear relationship between the all-carbon and the all-glass values, to within ±5 per cent. It was also found that scatter in the experimental results increased slightly with the amount of carbon in the hybrid, values ranging between ± 2 and ±6 per cent.

The modes of failure that have been observed with displacement-controlled loading are the following:

a)   In single phase CFRP some of the fibres failed under, or near the end plates, while many small bundles failed within the gauge length. A large number of longitudinal cracks that ran from one end to the other divided the specimen into bundles of typically $1 mm^2$ in cross-section. Interlaminar cracks were also observed but did not extend lengthwise as far as the longitudinal through-thickness ones. In most tests two main cracks were observed, one within the gauge length and one under the end plates, as shown in Fig 1. It is believed, because of the catastrophic nature of the failure, that the latter was produced by the stress wave of the former. Mean strain to failure was 1.38%.

b) In single phase GFRP, where the bond between fibre and matrix is known to be weaker than in CFRP, all the specimens failed in the characteristic mode of excessive longitudinal and interlaminar cracking. Fibres broke in bundles at various places along the gauge length and under the end plates (see Fig 2). These bundles were typically 1 mm wide and one or two layers thick (1/8 or 1/4 mm). Also, GFRP specimens, unlike CFRP ones, did not fail in a catastrophic fashion. The mean strain to failure was 2.22%.

c) In hybrids, the carbon layers failed first in a way similar to that described above for single-phase CFRP. However, in the cases where the surface layers were not of carbon, longitudinal splitting was greatly reduced (see Figs 3 and 4). In most cases, all-carbon layers failed across the same section. In some instances, however, multiple cracking along the specimen's length produced crack patterns that were not symmetric about the mid-plane (see Fig 5). Such multiple cracking can occur when the quantity of glass in a hybrid is sufficient to withstand the stress concentration produced when a crack is formed in a carbon layer. A short delamination is created around that crack, and provided the interlaminar strength of the composite is high enough to stop the delamination from growing, another crack may appear in that same carbon layer.

Once all the carbon layers had failed, they delaminated from the remaining glass layers, which were virtually undamaged. The glass layers in turn failed in the characteristic failure mode of GFRP. In certain cases however, the catastrophic failure of the carbon layers precipitated complete failure of the glass layers at the same instant. Those cases were the carbon-rich stacking sequences (75 per cent carbon), and the alternate layer stacking sequences (numbers 7 and 19).

In the load-controlled tests the modes of failure were similar to those observed in displacement-controlled tests but, as expected. a catastrophic failure occurred upon failure of the carbon layers, since the remaining glass layers were not strong enough to withstand the load attained at that moment. The single exception was stacking sequence number 13 (gggggggcggggggg), for which the carbon content was too low to produce immediate failure of the glass layers upon failure of the carbon layer.

The initial failure strains of the laminates were recorded and the hybrid effect (as defined above) was calculated. The results for the 8-ply laminates are presented in Fig 6, where it can be seen that the general trend is for higher failure strain as the relative carbon content is decreased. Also note that when the carbon layers were placed on the outside, the hybrid effect was practically zero. However, such a trend was not observed with the 16-ply laminates. Furthermore, the standard deviation associated with the above results was roughly ±6 per cent on average, and the single negative value of hybrid effect that was observed (stacking sequence number 5) was probably due to

this scatter. The number of specimens tested for each hybrid stacking sequence was between 5 and 8, rising to 12 for the single-phase laminates.

It should also be noted that the above results have been obtained from the complete set of specimens tested, including a few that failed under the end plates. It was found, however, that using only the results from specimens which failed within the gauge length did not alter the overall picture significantly. In fact, it was not possible to say whether the gauge failures produced higher values of strength or not.

Values for the hybrid effect found in the present work are in good agreement with values quoted by Phillips (2), Zweben (3), and Manders and Bader (4), both in magnitude and trend, despite the fact that different materials were used in each work. Furthermore, the above researchers have attributed the hybrid effect to factors similar to those outlined in the following section (i.e. thermal, statistical, etc.), although the approach to quantifying it has been slightly different in each case.

The results of ultimate strength for the 16-ply laminates are presented in Fig 7, where it can be seen that the strength of hybrids generally falls just above the simple theoretical prediction, but still below the rule of mixtures line connecting the strength of all-carbon to that of all-glass composites. Moreover, in the small region to the left of the cross-over point A, failure of carbon should not lead to ultimate failure since the residual strength of the glass would still be sufficient to take some extra load, and hence the ultimate strength is expected to lie on the line A C of Fig 7. However, stacking sequence number 13 did not reach the expected strength level, possibly because of early damage caused in the glass layers at a microscopic level arising from the failure of the carbon layer. On the other hand, in load-controlled tests, similar hybrid specimens yielded a value of ultimate strength that was 14 per cent higher than that obtained with the corresponding displacement-controlled test. Similarly, single-phase GFRP specimens were 11 per cent stronger under load-controlled than under displacement-controlled loading. This increase in ultimate strength may be due to the significantly higher strain rates associated with load-controlled loading. It is not known, however, whether a similar increase exists in the initial strength of the above laminates.

Acoustic emission monitoring equipment was also used, for two main reasons: firstly to pinpoint the start of the failure process, and secondly to deduce at what amplitude of emission the various processes take place: e.g. carbon fibres breaking, matrix cracking, etc. The results showed that there usually is an increase in activity at around 85 per cent of the initial failure strain, but there was no sign of a particular failure process happening at a specific emission amplitude. It is believed that the single probe used for monitoring the emission also picked up noise arising from the adhesive under the end plates, and therefore the task of analysing the results became very difficult. Hence, unless two probes are used to filter out unwanted signals, acoustic emission monitoring is not a very useful method of evaluating damage in similar test coupons.

4   THEORETICAL ANALYSIS

The aim of the current work has been to understand

and quantify the hybrid effect. It was known that this enhancement of the failure strain was partly due to thermal stresses present in the laminate, caused during the curing procedure as a result of the differential contraction of the glass and carbon plies. It was believed however, that some enhancement was also due to statistical effects, test procedure effects, and material stacking sequence effects. Indeed, a simple calculation showed that thermal stresses accounted for only a small part (about 15 per cent) of the measured hybrid effect. Therefore, the tensile test specimen, including end plates, was modelled using finite elements, so that the stress field near the gripping points and the dependence of these stresses on stacking sequence were established and stored in computer memory. The specimen was then represented by a large population of bundle-elements (not finite elements) each being assigned a certain value of direct strain at failure selected at random from a two-parameter Weibull distribution. Suitable values for the two parameters were found in the literature (Refs (5) and (6)), in particular a Weibull modulus of 20 was used. The weakest element was identified by the ratio of the strength of the bundle to the local value of stress, and failure commenced there. This constituted the initial failure of the composite and the theoretical hybrid effect was computed on that value. The complete chain of elements linked to the failed one along the length of the specimen was then taken out of the failure process and the next weakest element was identified.

The model assumed the delamination around a broken bundle to be so long that no stress concentration was imposed on neighbouring bundles. This technique was adopted because interest was focussed on the initial failure strain and not in the process of crack growth until ultimate failure. Thermal stresses were also included and the results obtained with this model are compared to experimental data in Fig 8 (8-ply laminates only). It can be seen that the effect of the end plates on the theoretical model was not very important since the predicted hybrid effect was very similar whether the carbon layers were placed on the outside or on the inside. Furthermore, the agreement between theory and experiment was much better in the cases where the carbon layers were placed on the inside, implying that certain factors affecting the lay-ups with the carbon layers on the outside had not been taken into account.

5   CONCLUSIONS

a)   The axial stiffness of 8- and 16-ply unidirectional glass/carbon epoxy laminates was found to obey the rule of mixtures.

b)   There was found to be a small positive hybrid effect, i.e. an enhancement of the initial failure strain of the carbon fibres.

c)   The strength was lower, for a given glass/carbon ratio, when the carbon layers were on the outside of the specimen. This may have been related to the end plates used for gripping.

d)   A simple statistical model showed that the hybrid effect was mainly due to strength variations throughout the volume. Thermal strains had a minor influence.

ACKNOWLEDGEMENT

This work has been carried out with the support of Procurement Executive, Ministry of Defence, UK.

Table 1   Stacking sequences tested

All laminates unidirectional
c = carbon,   g = glass
Single layer thickness = 0.127 mm when cured
Fibre volume fraction = 0.60 when cured

Stacking sequence number	Material stacking sequence	CFRP by volume (%)
1	$(gggg)_s$	0
2	$(cggg)_s$	25
3	$(gggc)_s$	25
4	$(gcgg)_s$	25
5	$(ccgg)_s$	50
6	$(ggcc)_s$	50
7	$(cgcg)_s$	50
8	$(gccg)_s$	50
9	$(cccg)_s$	75
10	$(gccc)_s$	75
11	$(cccc)_s$	100
12	$(gggggggg)_s$	0
13	$(ggggggggc)_s$	6.7
14	$(gcgggggg)_s$	12.5
15	$(ccgggggg)_s$	25
16	$(gccggggg)_s$	25
17	$(gccgcggg)_s$	37.5
18	$(ccccgggg)_s$	50
19	$(gcgcgcgc)_s$	50
20	$(gccgccgg)_s$	50
21	$(ccccccgg)_s$	75
22	$(ggcccccc)_s$	75
23	$(cccccccc)_s$	100

REFERENCES

(1) KRETSIS, G., MATTHEWS, F.L., MORTON, J. and DAVIES, G.A.O. Basic technology of hybrid composites. Third Progress Report, MoD(PE) Agreement No 2037/0245 XR/STR, Department of Aeronautics, Imperial College, June 1985.

(2) PHILLIPS, L.N. On the usefulness of glass-fibre-carbon hybrids. Paper 21, British Plastics Federation Congress, Brighton, November, 1976.

(3) ZWEBEN, C. Tensile strength of hybrid composites. Journal of Materials Science, 1977, 12, 1325-1337.

(4) MANDERS, P.W. and BADER, M.G. The strength of hybrid glass/carbon fibre composites: Part 1, Failure strain enhancement and failure mode. Journal of Materials Science, 1981, 16, 2233-2245.

(5) MANDERS, P.W. and BADER, M.G. The strength of hybrid glass/carbon fibre composites: Part 2, A statistical model. Journal of Materials Science, 1981, 16, 2246-2256.

(6) BADER, M.G. and PRIEST, A.M. Statistical aspects of fibre and bundle strength in hybrid composites, Progress in Science and Engineering of Composites, T. Hayashi et al, Eds., Proc. International Conference on Composite Materials IV, Tokyo 1982, 2, 1129-1136.

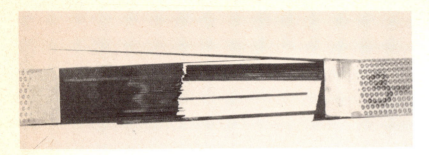

Fig 1 Typical failure of carbon fibre reinforced plastic (CFRP) (stacking sequence number 11)

Fig 2 Typical failure of glass fibre reinforced plastic (GFRP) (stacking sequence number 1)

Fig 3 Hybrid specimen, carbon outside (stacking sequence number 2)

Fig 4 Hybrid specimen, carbon inside (the outside glass layers have been cut away for better viewing — stacking sequence number 6)

Fig 5 Double failure in the carbon layers of stacking sequence number 16 (the outside glass layer has been cut away so that the cracks can be easily observed). The arrow indicates the position of the crack in the carbon layers of the opposite side.

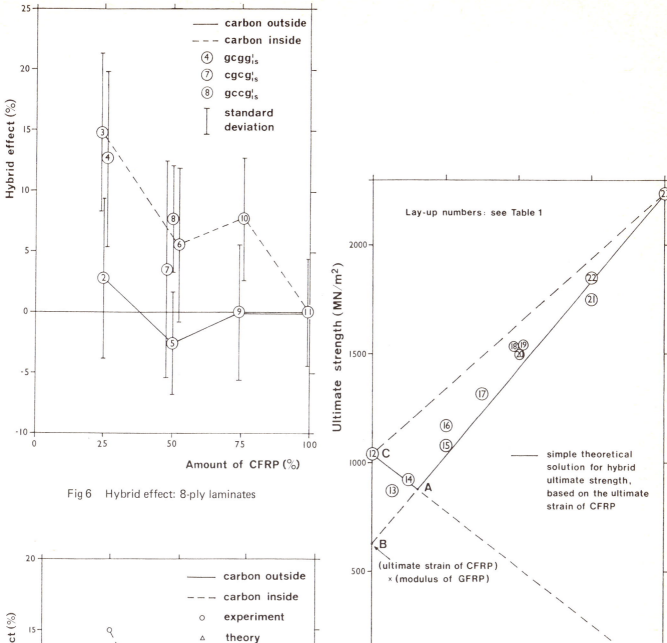

Fig 6   Hybrid effect: 8-ply laminates

Fig 7   Ultimate strength: 16-ply laminates

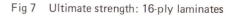

Fig 8   Comparison between experimental and
computational results (hybrid effect)

The durability of adhesive bonds

A P DAVISON, BSc
BL Technology, Lighthorne, Warwick

Abstract For correct selection of adhesive bonding systems the principles involved must be clearly understood. The most crucial factor in structural adhesive bonding is good bond durability. This is critically dependent on the correct pretreatment of the substrate to be bonded.

This paper outlines some of the general concepts involved in adhesive bonding and then goes on to examine how the use of suitable pretreatments greatly enhances bond durability.

Introduction

The expansion in the use of adhesives is due to the continually improving range of properties offered by adhesives as well as the increasing recognition of their advantages which include:-

a. joining of dissimilar materials

b. a more uniform distribution of stress across the joint

c. a increased fatigue life compared to mechanical fastening

d. corrosion between dissimilar materials can be prevented or reduced

e. weight savings can be obtained

Adhesives are used in practically all industries, the following being of particular importance:- aircraft, cars, boats, shoes and packaging. In certain applications the bonding requirements are modest and the use of a high performance adhesive is pointless. However, in applications where a high demand is placed in the adhesive joint attention must be paid to long term integrity of the bond.

The main requirements needed to achieve a good adhesive bond are:-

a. good wettability of the substrate to be bonded

b. absence of weak boundary layers

c. avoidance of stress concentrations which can lead to disbonding

The fulfilment of these requirements in itself is not the complete answer. These requirements have to be met in such a way that not only does the bond exhibit the necessary bond strength but that this bond strength should be maintained for the service life of the application. It is clearly of no benefit to have a component that is twice as strong as necessary on construction yet fails prematurely in service. What is needed is the making of a joint that not only fulfils the desired characteristics but retains those properties throughout its service life.

This obviously leads us on to how is the best way to achieve the necessary level of bond durability. It is first useful to outline some of the techniques used to determine the durability of an adhesive bonding system.

Testing Adhesive Joints

In practice, bonded joints are subjected to various stresses and environmental effects. For both the purpose of simplicity and relevance the most popular methods of testing adhesive joints are the single lap and T-peel joints, as illustrated in Figure 1.

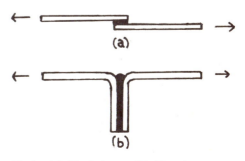

Fig 1   (a) Single lap   (b) T-peel

When investigating structural adhesives these two test techniques give a reasonably fast indication of the parameters of the adhesive. Adhesive performance in shear tends to be very good, yet their peel performance does not match this.

The effect of environmental exposure of an adhesive bond is one of the most important areas of investigation when looking at the subject of bond durability.

Humidity can be especially detrimental with all types of substrates. The damaging effect of humidity depends on the diffusion of water through the adhesive layer, along the interface or through the substrate itself in the case of plastics. The reduction in joint strength may be due to the plasticisation of the adhesive or substrate by water, displacement of the adhesive from the substrate, or attack of the substrate surface. The last of these mechanisms can occur with metals with the resultant formation of weak oxides. The rate of diffusion and hence the damaging effect of water increases with increasing temperature.

Adhesive joints in service may be subjected to moist conditions and stresses for long periods. To predict the performance of an adhesive joint under such conditions, it is usual to carry out accelerated tests i.e. conditions of temperature, humidity and stress are made more severe than in practise.

For example, having subjected a large number of alternative systems for Aluminium to all possible automotive environments, it was found that the most aggressive is humidity. In a typical test, strings of lap joints may be subjected to a stress equivalent to 20% of initial strength (applied by coil springs) in a humidity cabinet maintained at 100% RH cycling between 42°C and 48°C each hour. The time the joints take to fail is measured until 50% of the specimens have broken at which time the residual strength of the remaining specimens is measured.

Other tests such as fatigue and impact tests are also necessary to find the all round performance of an adhesive system.

General Concepts

Having identified the structural integrity of the bond (i.e. its durability) as being critical it is crucially important that the best performance possible in this area be achieved. It is now accepted that the way to do this is to use some kind of pretreatment. The pretreatment needed depends on the substrate and adhesives involved.

It has already been mentioned that to make a successful joint a number of requirements have to be met. These include (i) good wettability of the substrate and (ii) the removal of weak boundary layers. Pretreatment of the substrate not only achieves these two objectives but also provides a stable substrate surface condition. This is the vital feature for enhanced bond durability.

The precise nature of the pretreatments varies depending on the substrate and service requirements. In the case of metals, chemical pretreatments tend to provide much better durability in humid environments. The pretreatment of plastics can take many forms, acid etch's and flame ionisation being two examples.

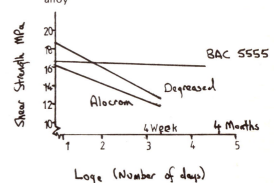

Fig 2   The effect of pretreatment on an aluminium alloy

Exposure 5% Salt Spray @ 43°C
Adhesive EC 2214 (single part epoxy)

The graph illustrates how the durability of an aluminium bond can be increased using an acid anodised pretreatment prior to bonding. BAC 5555 (1) is a phosphoric acid anodised pretreatment developed by the Boeing Company for aerospace applications. Alocrom 1200 is a paint pretreatment for aluminium which, although not designed specifically as a bonding pretreatment, doesn't improve durability to a certain extent.

It can be seen from the graph that the initial shear strength of the BAC 5555 pretreated system is lower than the degreased system but the percentage strength retention is much better.

It has often been a mistake in the past that, when an adhesive fails, the initial instinct is to look for a "stronger" adhesive, i.e. an adhesive which exhibits a higher strength at zero time. If the system in question does not exhibit any form of durability, the selection of a "stronger" adhesive alone will not bring with it a substantial increase in bond durability, it may in fact increase the rate of decay of strength retention.

A drop in initial "strength" of the adhesive joint combined with a greatly improved durability performance, achieved through the correct choice of pretreatment is a more attractive situation.

The durability of adhesive bonds to plastics presents slightly different problems. In some cases it is relatively simple to make durable bonds because of the similarity of the adhesives and substrate, yet with materials like PTFE the achievement of even moderate joint strengths presents problems.

To achieve satisfactory levels of durability in these difficult materials, it is necessary to carry out a pretreatment. A number of possible pretreatments in this area are acid etches, flame treatment, primers and Corona discharge techniques.

Acid etches and flame treatments are not really relevant for the majority of applications. The use of specially prepared primers is an increasing area of influence especially with the development of such things as plastic skin panels for automotive use. The use of the Corona Discharge technique is increasing especially with applications where the treatment of large surface areas is required. The Corona Discharge technique can briefly be described as follows.

The component to be pretreated moves over an earthed surface covered with an insulating material. About 2 mm above the component is a high voltage (typically 15 kV) electrode bar connected to a high frequency (typically 10 - 20 kHz) generator. The air gap is ionised and the stable corona discharge treats the component surface.

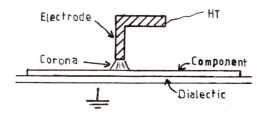

A number of years ago there was a certain amount of debate about whether the improved performance brought about by the discharge treatment was due to physical or chemical effects. There is now a reasonable consensus of opinion resulting from x-ray spectroscopy data that suggests the discharge has a chemical effect.

Whilst on the subject of pretreatments for plastics, the problems caused by additives and processing aids such as mould release agents should be mentioned. These are a frequent cause of loss of adhesion and should be removed because they create a weak boundary layer as they migrate to the surface.

Practical Example

A recent opportunity for the application of adhesives was the bonding of steel to GRP to form a stable, durable bond. Because of the requirement for a certain amount of flexibility in the joint, as well as a room temperature cure, a moisture curing polyurethane adhesive was chosen as the adhesive. This adhesive also alleviated the need for any sealing application at a later stage in the vehicle build.

This particular application called for the adhesive bonding operation to take place after the steel had passed through the electrocoat paint process. There therefore existed two potential areas of concern. The first is the adhesion-paint interface where a durable bond must be maintained but the most important is the paint-steel interface where, if the bond is not durable, no matter how good the adhesive-paint interface is, the bond will fail. This is even more important if the system is expected to carry some form of loading.

It is therefore necessary to examine not just one but two different areas for the correct pretreatment. This was done by comparing the following two systems:-

System 1

The first was steel and the full electrocoat process and primer bonded to polyester resin GRP and primer using a two part moisture curing polyurethane.

System 2

The second system was Steel and Accomet C and Electrocoat only and primer bonded to polyester resin GRP and primer using the same adhesive.

Accomet C is a metal pretreatment designed to enhance the adhesion performance prior to painting. This was applied to the steel surface in System 1 in place of the normal alkaline and phosphating process that occurs prior to paint.

The primer used was as supplied by the adhesive manufacturer and compatable with the adhesive. The primer has been the subject of tests done by B L Technology previously and found to promote bond durability when used with this adhesive.

The durability of the two systems was evaluated by exposing the simple lap shear specimens to an accelerated humidity environment, as described earlier. The shear strength of the joints after various exposure times was then measured and compared with the shear strength of the joint systems measured on a set of zero control specimens.

The results are shown graphically in Figures 3 to 4 below

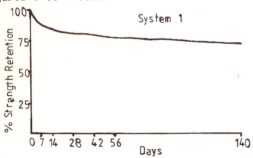

Fig 3   Percentage strength retention versus exposure

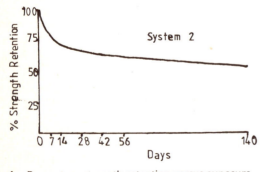

Fig 4   Percentage strength retention versus exposure

It can be seen from the graphs that System 1 performs better than System 2 in the accelerated humidity conditions. The percentage strength retention of the system with the steel pretreatment is greater than the strength retention of the system without the steel pretreatment.

If the same results are plotted against a log time scale on the same graph it is possible to see a direct comparison between the rates of fall off in strength.

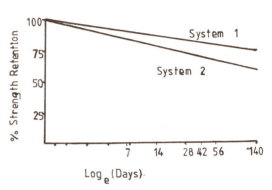

This graph clearly demonstrates how the use of a pretreatment in System 1 improves the durability of the bonding system quite significantly.

One other point to make is that of the type of joint failure observed in such experiments. When bonds are made to untreated substrates, it is common for the failure mode of the bond to be adhesive i.e. failure between the adhesive/substrate interface. With bonds made to pretreated surfaces, even after lengthy exposure to humidity exposure, the failure mode of the joints tends to be cohesive i.e. failure in the adhesive layer itself. This is a clear indication that bonding pretreatments are doing their job. The joint is then working up to the maximum performance of the adhesive which is something that can be quantified in order to assist the design engineer.

Conclusion

If all the advantages of adhesive bonding are to be fully utilised a complete understanding of the selected system which includes substrate, pretreatment and adhesive is necessary. The most crucial area in this understanding is that of bond durability. In order to achieve the maximum level of bond durability, it is essential to pretreat the substrates to be bonded. Only then can all the advantages that adhesive bonding possess be fully exploited.

References

1. Boeing Aircraft Corporation Process Specification BAC 5555. Phosphoric Acid Anodising of Aluminium for Adhesive Bonding.

C48/86

Low cost carbon fibres for high performance applications

P G ROSE, PhD, MSc, MInstP
R K Technologies Limited, Stockport, Cheshire

SUMMARY

Low count carbon fibre tow is prohibitively expensive for many applications in general engineering. The anticipated advent of low price, high performance pitch based carbon fibre is now regarded as unlikely in the forseeable future.

The substantial cost advantages in employing low priced large tow commercial PAN as a carbon fibre precursor are presented and a number of the methods of converting this to useful composities described.

1. INTRODUCTION

Since their introduction during the 1960's carbon fibres produced from polyacrylontrile (PAN) precursors have established themselves as the preferred materials in the aerospace industry, for high modulus composite structure. The term "high modulus" is used here to differentiate carbon fibre composites (CFC) from glass fibre or aramide fibre reinforced matrices (GFC and AFC resp.).

Originally a PAN tow (i.e. continuous yarn) containing 10,000 single filaments was chosen as the standard precursor, as this was the only form available,exhibiting the necessary prefered orientation vital for high carbon fibre elastic modulus (1, 2). Warp tape prepreg manufactured from this, proved however to be too thick for many aerospace laminates, in which optimum design required a large number of thin plies. Therefore, a tow with 3,000 single filaments came to be favoured and for some applications as few as 1,000. For those parts, for which 10,000 filament tow is appropriate, the practice is now to use 12,000 filaments, which are an aggregate of four 3,000 filament tows.

This preference for low tow counts by the aerospace industry is a major factor in determining the price of carbon fibre, which is widely regarded as the most important factor inhibiting its employment in general industrial applications. This, in spite of the fact that the aerospace industry has been able to realise significant cost reductions (10 - 25%) fabricating components in CFC, formerly machined from aluminium (3, 4).

Figure 1 illustrates this point with the example of a wing spar traditionally fabricated from machined aluminium ingot. Production costs are related to 1 kilogram finished carbon fibre composite, the equivalent of which weighs 1.4 kg in Dural aluminium alloy, i.e. a weight saving of 40% is achieved by using CFC. Although the composite costs four times that of the alloy, the fact that it is accurately moulded and simultaneously bonded to form the end product with only 10% material wastage, enables a 16% saving of the total production costs to be realised.

Obviously the exacting quality requirements of the aerospace industry, which demand such expensive production routes, would not necessarily pertain to the rest of the engineering industry, which might manufacture a similar component from an extruded aluminium profile or die casting, at considerably lower expense.

Nevertheless, this example like many others in the aerospace industry, illustrates that inspite of its high cost, CFC is often a more cost effective choice for highly stressed, light weight structures (5).

2. ECONOMICS OF CARBON FIBRE PRODUCTION

Figure 2 illustrates the constituent costs in the production and supply of 3,000 filament tow PAN based, high strength carbon fibre. The figures are a fair estimate for a typical process manufacturing from 100 to 200 tonnes of this grade of carbon fibre p.a.

It is apparant from this, that the precursor is a major factor in determining the cost of carbon fibre. A carbon fibre yield of 52% of the precursor input is assumed and a further 5% loss allowed for operating waste. Energy costs amount to a relatively small proportion of the total. Operating a plant with 12,000 filament tow reduces the precursor cost to 26% of the total. 1,000 filament tow increases this to over 40%.

It is on account of this high precursor cost element in the production of PAN based carbon fibre, that one has predicted such a huge potential for carbon fibre derived from pitch (6).

Petroleum residue or coal tar pitches are priced in pence rather than pounds, even if one allows for the fractionating costs in preparing a material of specific softening point.

However, such a feed stock is of very doubtful utility, as owing to its instability - it is not a chemically stable substance - it is very difficult to spin satisfactorily and continuously. After oxidation and carbonisation, it yields a highly intractible product of low strength and modulus. In order to ensure production of continuous fibres of consistent quality exhibiting practical values of strength and elastic modulus, the pitch feed-stock has to be subjected to an elaborate series of pretreatments prior to spinning.

Inimical components have to be leeched out of, or extracted from the pitch, particles filtered out, then heattreatment conducted during stirring or gas circulation and final degassing, at 380 + deg. C for 10 hours or longer. All this is necessary in order to cultivate a liquid crystal phase, or "mesophase" in the pitch, to an extent that allows a high degree of preferred orientation to be generated during the spinning process. This is a precondition for the development of practical values of strength and modulus during further processing.

Such pretreatments as reported in the patent literature (7) demand high capital investments and substantial operating costs. Indeed, the preparation costs of a spinnable mesophase pitch, even at a conservative estimate, far exceed those for the production of PAN polymer. The spinning costs for both precursors are comparable.

In view of this, it is not surprising that pitch based carbon fibres are marketed at significantly higher prices than their equivalent PAN products and have not as yet succeeded in establishing a role in the main stream applications for resin matrix composites.

Notwithstanding this, it is possible to achieve extremely high elastic moduli with graphitised mesophase pitch, values virtually matching the basal plane value for the graphite lattice and significantly higher than those practical with PAN:- 700 GPa as compared to 500 GPa. This qualifies pitch fibre for the reinforcement of brittle or very high modulus matrices e.g. graphite, ceramics and high modulus metals. We can therefore expect to see pitch preferred in these latter sectors, in which they can command a high cost premium (8).

Of the other quoted advantages of pitch, we should refer to the higher carbon yield (70% as compared to PAN at 50%) and reduced oxidation treatment. In reference to figure 2, we might therefore anticipate a cost reduction of at least 40%! This is however not the case, on account of the highly sophisticated handling necessary to cope with the extreemly fragile pitch fibre, prior to reaching 600 deg C in carbonisation.

This demands significantly higher plant and labour costs per unit of production. With pitch we are also manipulating a low count filament tow, but one very difficult (if not impossible) to handle as a broad web assemblage. Therefore, even if the cost of precursor preparation can be contained within the scope of that for PAN, realising the other advantages of pitch has not as yet proved feasible.

On the basis of a chemical engineering cost analysis, a viable process should ideally handle the pitch continuously, feeding it through the various process steps to the spinning head on line. A vertical arrangement of spinning shaft, oxidation canal and low temperature carbonisation furnace would eliminate extravagant handling aggregates. However, this would have to be a very large plant, producing approx. 5,000 tonnes p.a. of carbon fibre to be economical. This exceeds the present world demand for all grades.

A very significant reduction in PAN precursor costs can be realised by converting conventional commercial heavy textile tow to carbon fibre. This is available at at least 20% of the cost of 3,000 filament tow, enabling the price of the carbon fibre to be reduced accordingly, by at least 27%. Further economies accrue using commercial heavy tow, on account of its size, containing as it does 200, to 500,000 filaments. There can be major reductions in handling and handling equipment investments; and employing a suitable processing plant design, production costs can be easily reduced to less than half that of filament tow carbon fibre. In going over to production plant capable of coping with a minimum 500 tones p.a. output, considerable economies of scale are expected and with these we should achieve a total cost reduction of at least 60%. One can then quite credibly expect to realise carbon fibre at £10 per kg, a price at which a very wide range of general engineering applications become feasible for CFC.

Obviously the problems of dealing with carbon fibre in the form of a heavy tow have to be considered and some possibilities are dealt with below. First the actual conversion process will be described.

3. CARBON FIBRE PRODUCTION PROCESS

In figure 3 we see an outline drawing of the R.K. Carbon Fibres Limited production plant and a flow diagram showing the essential operating principles.

If filament tow precursor is to be processed, it has to be doffed upon creels and formed to a parallel web over a system of textile guides and tensioning rolls. This web is then drawn through the:

OXIDATION STAGE - comprising two ovens in which, under precisely controlled conditions of stretch, shrinkage, air flow and temperature, the tows are reacted with atmospheric oxygen. This converts the polyacrylonitrile to a thermally stable, non-flammable, carbonisable fibre of complex chemical composition (9). Oxidised PAN (RK product name PANOX) is in itself a useful product, which can be used as a substitute fibre for asbestos and in flame proof fabrics. Also as we shall see later, it can be processed to fabric, felts etc. that in turn can be carbonised for specific applications.

CARBONISATION is conducted in two phases. The web is first fed through an LT (low temperature) furnace to remove vapourisable by-products including tar, to render a fibre with approx. 90% carbon content. Further heat treatment in the HT (high temperature) furnace anneals the fibres and drives off most of the remaining non-carbon elements (nitrogen, sodium etc.). After this stage and dependent upon precursor quality, heat treatment conditions, etc. the process yields high strength grades of carbon fibre, with elastic moduli in the range 200 to 300 GPa.

In order to obtain higher values of elastic moduli than these, the fibres have to be exposed to very high temperatures (in excess of 2,000 deg C) and to this end, they are passed through an HM furnace. This treatment is in effect a crystallisation process. The fibre loses hydrogen and the small scale atomic order of the microcrystallites develops into an ordered texture, with radial and circumferential layering in the plane of the fibre and much higher basal plane preferred orientation parallel to the fibre axis (10). The elastic modulus can be increased to 400 GPa by thermal treatment alone; by hot stretching the fibres, this can be increased to 500 GPa.

For most end uses it is necessary to treat the carbon fibre so that its surface will wet with the reactive monomer or polymer melt to be used for the composite matrix and subsequently react or bond via active surface groups with it (11, 12). An oxidative medium is normally employed for this purpose. In the case of the R.K. process, anodic oxidation of the fibre in an electrolyte is used. Under optimum conditions, this method activates the surface of the carbon fibre to allow a density of surface functional groups sufficient for maximum adhesion to the matrix, while polishing it to the extent that a significant increase in tensile strength is realised. This latter effect is apparently due to the reduction of the severity of superficial stress raising flaws.

Before spooling the carbon fibre tows, they are usually sized with a small percentage of neat epoxy resin or thermoplastic, in order to protect the fibre from abrasion during winding and down-stream handling.

4. HEAVY TOW TEXTILE GRADE PAN

The processing cycle is in principle the same as above except that the precursor is drawn from standard sized cartons containing 150 to 200 kg instead of from a spool. Figure 4 illustrates this mode of operation as it is conducted for the oxidation process. A number of heavy tows are drawn to form a broad web, each being spread and tensioned to a uniform tape, before introduction to the input drive of the first oxidation oven. Commerical PAN fibre has a higher tex than standard filament tow precursor and it is supplied crimped. It must therefore be stretched during the initial phase of heat treatment.

In PANOX production, the oxidised heavy tow is collected in similar cartons to those of the delivered raw material. It does not have to be crimped for further handling, but is pretreated with an antistatic lubricant. This product can still be converted to textile yarns and fabrics using the conventional worsted or cotton routes. Usually it is stretch-broken, drafted, spun and doubled, if it is to be used as yarn. If felts or papers are to be manufactured, a cutter converter is used prior to employing the appropriate technology.

A new process has been developed to convert Panox to a form in which, once carbonised it offers a very suitable low cost reinforcement. This is an aligned, lightly needled carbon fibre mat. Panox tows are aligned unidirectionally and a light Panox scrim then applied. A further Panox tow is then lapped across this structure and lightly needled to bond the layers to a tight, strong structure. Obviously, a large number of variations are possible in lay-up angle, number of layers, structure, etc., dependent upon the intended end use. It can be carbonised, surface treated, sized and impregnated very economically on-line, in a plant especially designed for this purpose.

The advantages of the above product are manifold. It is economical to produce and to fabricate to the composite. It can be manufactured to have a given degree of through-plane (3 dimensional) reinforcement and yet adapts readily to complex profiled tooling.

Equally, PANOX yarns, papers, fabrics, knits and felts can be fabricated using high speed, economical textile production methods and the carbonisation of these conducted subsequently. Figure 5 illustrates the most common process routes for converting PANOX to a variety of products.

Heavy tow textile grade PAN, can of course be converted directly to continuous carbon fibre tow in the R.K. process. A common version consists of 320,000 filaments in the tow. This can be processed to values of elastic moduli within the same specifications as filament tow and can be employed for filament winding thick wall sections, pultruding large profile cross-sections and it can also be sized and chopped for use as a dough or injection moulding compound, with thermosets or thermoplastics.

A version of this precursor is supplied by Courtaulds PLC as a splittable tow; i.e. after oxidation, each heavy tow is divided down into 40,000 filament tows. At a small cost premium on the price of normal commercial tow, we then have a product which for many applications is as satisfactory as filament tows with 12,000 count, at 60% of their cost. It can be used in prepreg tape manufacture, filament winding and pultrusion, in a broad range of applications.

A survey of the various grades of carbon fibre and their fields of application is provided by figure 6 (see also 13 and 14).

5. MECHANICAL PROPERTIES OF CARBON FIBRE; COMPARING FILAMENT TO HEAVY TOW.

In terms of elastic modulus, equivalent grades can be produced from all values of tow count. A disadvantage however, is that with heavy tow it has not yet proven possible to achieve mean tensile strengths at the level of high strain 3, 6, and 12,000 filament tow.

Figure 7 reveals a source of this discrepency, in making a comparison between the single fibre tensile strength distributions of an R.K. 30, 3K tow count and an R.K. 320K, carbonised to the same modulus (240 GPa). The distribution of the heavy tow strengths is much broader to similar peak values but form a lower minimum level. Similarly, there is a broader distribution of cross-sectional areas in a typical batch of heavy tow PAN based carbon fibre, as compared to low count filaments. This is at least a partial explanation of lower mean strengths, for as we see in figure 8, carbon fibre tensile strength is a function of the areal cross-section of the single filament as well as being a function of the initial stretch ratio per se.

There are therefore means of ameliorating this source of discrepancy i.e. by designing the heavy tow precursor plant to ensure a constant path length for stretch and shrinkage throughout the line, from the spinnerette onward. Also, the smaller the spin-orifice, the higher the ultimate mean tensile strength (figure 8). We can increase mean tensile strength in the carbon fibre line, by a process of stretch/shrink - cycling during oxidation to reduce the ultimate tex, while maintaining constant fibre preferred orientation. We cannot however reduce the statistical dispersion of results with such a technique and have to rely upon the precursor supplier to do so.

A further source of reduced and dispersed strengths with heavy tow, might be attributed to the crimping process, damaging the precursor while in a semi-

plastic condition. Experimental trials to confirm this, have not demonstrated improved results for uncrimped, spool wound heavy tow however. Damage from any source should of course be avoided, as should included particles, bubbles or solvent residues.

A question which must be posed, is whether the commercial tow precursor is fundamentally inferior to filament tow. It is now an established fact, that all carbon fibre exhibits a guage-length effect (15); i.e. the tensile strength is observed to increase with reducing length of the fibre tested. If we test a vanishingly small volume of the fibre for tensile strength, we approach what we may name "the intrinsic strength" of the fibre.

In figure 9, the principles of the elastica test are illustrated. In this test method, a single fibre is contra-rotated at each end, to cause a loop to form. The ends are then separated, to cause the loop to decrease in size and step by step the major and minor axes measured. (The test is normally conducted in an oil film, between two thin glass slides under a microscope. The loop is photographed at regular intervals, for subsequent measurement of the above).

Knowing the modulus of the fibre, it is a simple problem to calculate the strain on the tension side of the fibre and then the strength at the point of failure.

Using the elastica test, we find that the mean breaking strains for standard types and heavy tow carbon fibre, lie in the same broad range of 2.5 to 4%. This is compared to 1 to 2% for the same fibres tested with a guage length of 10 to 20 mm. Therefore it is quite apparent that we are still significantly below the potential maximum strength possible with the material, once this is free of stress-raising flaws. For us, the vital point is that heavy tow would yield equivalent breaking strains and strength to high strain filament tow, once the problem of the disparate quality is solved.

In the table below, the mechanical properties of the various grades of carbon fibre are given.

6. CONCLUSIONS

Commercial grade heavy tow PAN is in many respects a useful precursor for carbon fibre. Carbon fibre for stapel reinforcement, filament winding, pultrusion and prepreg can be produced from it, exhibiting the same elastic properties as those of the much more expensive filament tows. At present, a lower strength level has to be accepted, but this is quite adequate for the majority of structural applications. Means for achieving comparable strength values to high strain PAN based carbon fibre are apparent and would be adopted, once the volumes being converted to composites justify the necessary plant investment.

Processing the intermediate oxidised PAN heavy tow via textile routes to mats, yarns, felts etc., followed by carbonisation, are methods of achieving

economic reinforcements for composite
products, which should find increasing
favour.

The early promise of pitch based carbon
fibre has not come to fruition and
depends upon the development of large
scale plant, which could only be justified,
once fibre applications achieve the scope
of a commodity engineering material, such
as general plastics, steel etc.

Heavy tow carbon fibre offers the next
advance in the cost effective manufacture
of high performance composite material.

REFERENCES

1. Thorne, D.J. in "Handbook of
    Composites", 474, XII (1985), Elsevier
    Science Publishers B.V.

2. Johnson, J.W. J. App. Polym. Symp. 229,
    9 (1969).

3. Conen, H. "Gestaltern und
    Dimensionieren von Leichtbau-
    strukturen" in "Kohlenstoff und
    aramidfaserverstarkte Kunststoffe"
    169, (1977). VDI Verlag GmbH,
    Dusseldorf.

4. Conen, H. and M. Kaitatzidis "Elevator
    unit for the Alpha-Jet, made from
    carbon fibre reinforced plastic", in
    "Processing and uses of carbon fibre
    reinforced plastics" 151, (1981)
    VDI Verlag GmbH, Dusseldorf.

5. Rose, P.G. "Carbon fibre composites as
    engineering materials" 77, (1985), in
    the conference volume "High
    performance fibres, textiles and
    composites" UMIST, Manchester.

6. FORSYTH, R.B. "Low cost continuous fibres
    from a pitch precursor". 20th National
    SAMPE, San Diego, (April 1975).

7. U.S. Pats, no.s 3,552,922 (1971);
    3,928,170 (1975); 4,080,283 (1978);
    4,055,583 (1977); 3,919,376 (1975).

    Further refs on processing pitch to carbon
    fibre see: Sittig, M in "Carbon and
    Graphite Fibres", Noyes Data Corporation,
    New Jersey (1980).

8. Volk, H.F. "High modulus carbon fibres
    made from pitch", 41, (1981) in
    "Processing and uses of carbon fibre
    reinforced plastics". VDI Verlag GmbH,
    Dusseldorf.

9. Rose, P.G. "Preparation of carbon fibres",
    Ph.D Thesis, Univ. of Aston, Birmingham,
    (1972).

10. Johnson, D.J. Paper 8 Int. Carbon fibre
    conference (1971), Plastics Institute,
    London.

11. Belinski, C. and G. Grenier 3, 165 (1972),
    La Reseche Aerospatiale.

12. Barton, S.S. and B.H. Harrison 13, 283
    (1975) Carbon.

13. Rose, P.G. "High strength carbon fibres
    based on polyacrylonitrile: available forms
    and properties in CRP" 5, (1981) in
    "Processing and uses of carbon fibre
    reinforced plastics". VDL Verlag GmbH,
    Dusseldorf.

14. "Fibre composites, design, manufacture and
    performance" in Composites 14, 2, April
    (1983).

15. Moreton, R. 273, 1, (1969), Fibre Sci Tech.

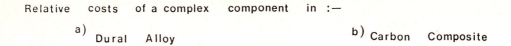

Relative costs of a complex component in :—

a) Dural Alloy   b) Carbon Composite

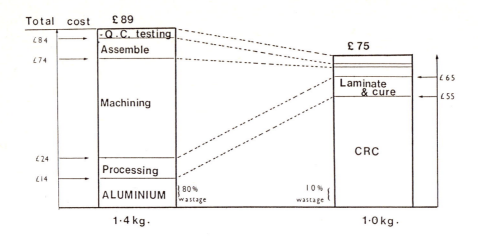

Fig 1   Relative costs of a complex component in
         (a) Dural alloy (b) Carbon composites

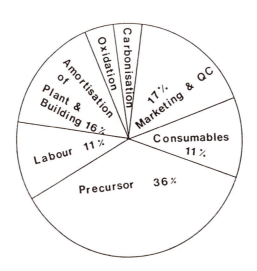

3'OOO fil. RK3O

Fig 2   Relative costs in the production of Pan based carbon fibre

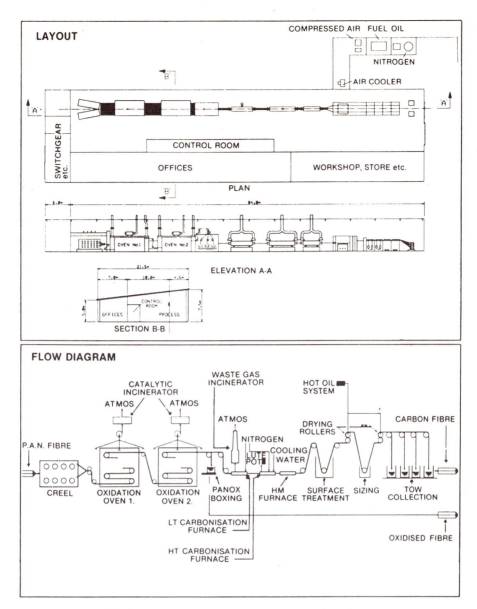

Fig 3   RK Carbon Fibres Limited production plant

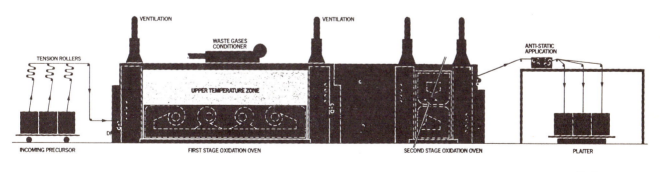

The PANOX process sequence

□ Incoming precursor is fed from handling units to tensioning controllers which form the tow into uniform bands. Facilities for hot-stretching incoming tows can be incorporated.

□ Roller drive towers at the entries and exits of the ovens control the passage of the bands through the process.

□ The tow bands pass through first and second stage multi-zone oxidation ovens of the electrically heated, forced convection type.

□ The oven environment is precisely controlled to maintain the optimum time/temperature profile. The fibre passing through the ovens has no contact with the oven interior.

□ After the process of oxidation the tows are fed to an automatic plaiter. The oxidation plant can also be operated in conjunction with an RK Textiles carbon fibre processing plant. In such an installation, the tow band would be fed from the second stage oxidation oven into successive low and high temperature carbonisation furnaces.)

Fig 4   Panox production with heavy tow commercial Pan

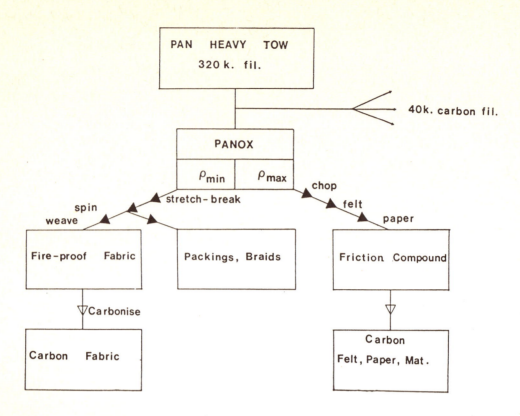

Fig 5   Product routes via oxidized Pan heavy tow — Panox

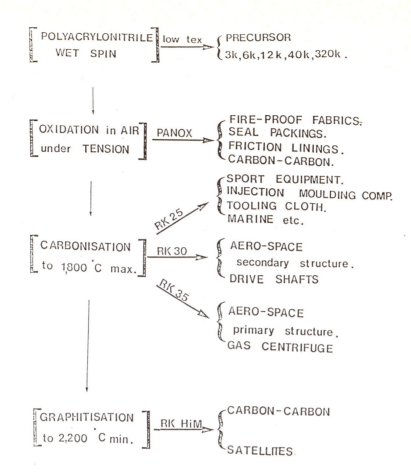

Fig 6 Application of the various grades and tow counts

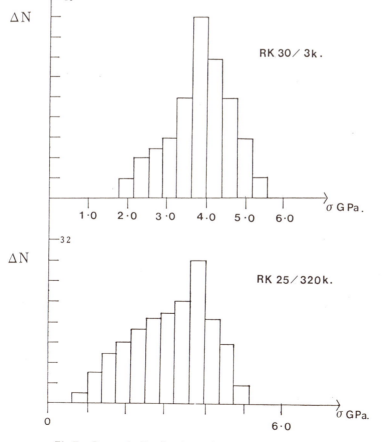

Fig 7 Strength distributions of single carbon fibres

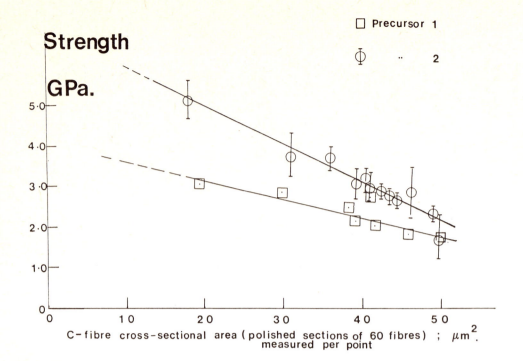

Fig 8   Effect of precursor stretch on strength

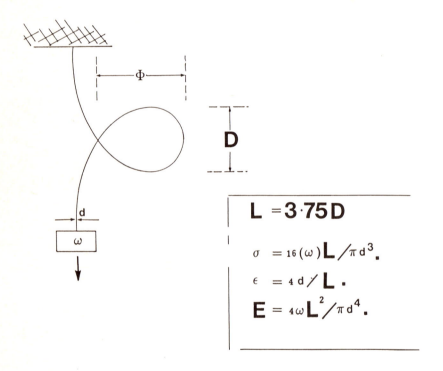

$$L = 3 \cdot 75 D$$

$$\sigma = 16 (\omega) L / \pi d^3.$$

$$\epsilon = 4 d / L.$$

$$E = 4 \omega L^2 / \pi d^4.$$

Fig 9   Elastica test

C49/86

Transport applications for fibre reinforced composites

G D SCOWEN, DipEng
GKN Technology Limited, Wolverhampton

SYNOPSIS   A new appreciation of engineering materials and their contribution to product development for demanding transport applications is currently in progress, mainly driven by technical, environmental and economic pressures. Fibre reinforced polymeric composites are exciting much attention as materials that offer solutions to many of these pressures. Products which operate in technically demanding and hostile environments, require materials with higher sophistication or high information content. Composite materials fulfil this role, leading to products with high added value. Other issues that effect the choice of materials for product development are thermo-elastic properties, processibility and cost. Finally, several product developments are reviewed with reference to their operational and materials requirements.

1   INTRODUCTION

Fibre reinforced plastic composites are a relatively new generation of engineering materials that are exciting a great deal of interest as materials that can be used for demanding product applications in the engineering sector. Transport is one of the sectors where land, sea and aerospace applications have yielded many of these exciting composite product developments.

Over the last few decades many composite product developments have been for low and medium stress applications, for example, composite non-structural body panels, cabs and internal non-functional equipment for automotive and aerospace applications. Highly stressed composite products applications have been mainly restricted to the aerospace industry. More recently, however, the automotive sector has been developing suspension and driveline composite based products, operating in highly stressed and demanding environments.

The ability to exploit fully the many attractive physical properties of composite materials for transport development depends upon the availability of design and analytical techniques to model and quantify the fibre and matrix behaviour for extreme environments, and also to quantify adequately the structural response by using finite elements and advanced C.A.E. methods.

Finally, an adequate financial and product manufacturing base is a pre-requisite to successful composite product developments.

2   MATERIAL REQUIREMENTS FOR PRODUCT DEVELOPMENT

The present trend is towards developing materials with increasing sophistication and information content. Composite materials fulfil this role extremely well. By the composites

definition, a material can be built-up from a multiplicity of fibre, matrix and core materials to fulfil adequately many materials behavioural roles in one single material. For example, low weight, low thermal expansion, high corrosion resistance and high fatigue resistance, can be obtained using fibre reinforced plastic composites.

Figure 1 shows the heuristic trends in future material developments for engineering products. The illustration is based upon Altenpohls work, Reference 1. Each point in the figure illustrates the current degree of materials sophistication and weight of material per unit of product. The arrows further illustrate the future trends for various engineering materials. The expectation in the automobile and aerospace sectors are shown for products with high value. These sectors are alive with many exciting high stress and temperature applications for polymeric composites. Some of these developments will be described later.

Materials costs play an important part in the selection of composite materials for engineering products . Materials such as steel, glass fibre and epoxy resin have generally increased in price with inflation. High performance fibres such as carbon have gradually fallen in price, mainly reflecting an increasing demand from the aerospace sector for materials with enhanced physical properties. Carbon and glass fibre are presently less susceptible than other materials to energy cost fluctuations.

The volatility in materials price movement is another important issue. For example, aluminium being one of the traditional light weight materials, has seen large fluctuations in price over the last few years. In a six month period in 1985, the price of aluminium peaked at over £1000/tonne, but only four months later a 25 per cent fall occurred. Price fluctuations such as those for aluminium have occurred since the beginning of the eighties, further illustrating the problems encountered in costing components manufactured in traditional light weight materials. During a similar two year period, there was no significant price fluctuation for glass, carbon fibre or resins such as polyester and epoxide, the basic constituents of composite materials. However, no material, whether new or traditional is immune to price fluctuations, but composite materials have undoubtedly shown good price stability over recent years.

3 COMPOSITE CONSTITUENTS AND PROPERTIES

Table 1 shows some of the composite materials that are available to the engineer for product developments. Some high performance systems are too costly for some transport product applications such as those found in automobiles, however, ´hybridisation´ consists of mixing relatively low cost glass fibre with more costly high performance fibre materials such as carbon fibre, resulting in more cost effective composite components.

The diversity of material properties and behavioural benefits achieved by using fibre composites in engineering products cannot be surpassed by traditional materials.

Figure 2 shows the specific tensile strength and the specific tensile modulus for various composite fibre constituents and traditional metallics. It can be seen that high modulus (HM) graphite or carbon fibre has a specific stiffness approximately seven times greater than structural steel. Aromatic polyamide (KEVLAR - tradename) has a specific strength five times greater than structural steel. Fibrous materials can therefore, be compounded with polymeric resins into composite materials with very attractive physical properties and behavioural characteristics, thus increasing the portfolio of engineering materials available to industry for product development.

4 COMPOSITE FABRICATION

Table 2 shows some of the main fabrication techniques currently used for the manufacture of composite products, the choice of the process being very dependent upon the performance and market requirements for a particular product application. The process and gelation times are generally determined by the geometrical complexity of the component, product size, the resin type and properties. Typically, gel times range from a few seconds to over 90 minutes.

Reinforced reaction injection moulding (RRIM) was developed from standard injection moulding techniques, the RRIM process allows for the introduction of milled fibres with an accompanying moderate improvement in mechanical properties. Resin injection or resin transfer moulding (RTM) is a promising production process, which is currently used for moderately stressed components such as car body

panels. However, the process has the capability of being extended to produce high fibre volume fraction composite products that are capable of performing in a highly stressed mode.

Resin is injected at typical pressures of between .28 and .55 Mpa. Cycle times of a few seconds to several hours for very large components are quite common.

Figure 3 shows a large capacity, high temperature autoclave for the fabrication of large aerospace components, such as aircraft wing control flaps or helicopter rotor blades. A vacuum stage assists resin flow and removal of air or gaseous volatiles given off during the curing cycle. The pressure chamber allows a build-up of pressure on the components outer surface for compaction of the composite fabrication at chamber temperatures up to 200°C. Pressures up to 2.0 Mpa are quite common, and composite components fabricated by this method, have excellent surface finish and structural integrity.

Universal helical filament winding machines, Figure 4, are particularly useful for manufacturing tubular shell type components, such as automotive driveshafts, airframe struts etc. These last two figures illustrate the large factory resource and complexity of fabrication technology required for many composite component developments.

Compression moulding and pultrusion produce structural composites with excellent structural integrity, consistent fibre alignment and high fibre volume fraction for high stress applications. Numerically controlled gantry machines for ´pre-preg´ tape laydown are also used for fabricating large aircraft wing sections. Robots are being used for fabric cutting and ´pick and place´ operations for composite materials and pre-form handling.

Thermoforming is another process that enables fibre reinforced thermoplastic composite flat sheet to be quickly formed into complex geometry with the aid of heat and pressure. Fibre filled nylon and polyetheretherketone (PEEK) are some recent material developments of interest.

This brief mention of fabrication techniques illustrate the wide range of composite product geometries possible, and the link between design and the fabrication process technology.

5 TRANSPORT APPLICATIONS - PRODUCT REVIEW

A brief review of past composite component developments in the transport sector cannot do justice to the many exciting developments that represent significant historical and technological achievement in the use of plastic materials. Some of these composite developments have had a radical effect on the attitudes of practising engineers, changing their traditional view of materials, design and manufacture.

Over the last 20 years there have been many composite component developments for lightly or moderately stressed components in the transport sector. However, composite components for highly stressed applications have been limited to the low volume specialised markets, the large capital investment required for a high volume fabrication plant is prohibitive for many companies and can only be embarked upon when the component supplier works closely with manufacturers.

Some of the most demanding and exciting composite product developments have occurred in the marine, aerospace and automotive sectors. The developments range from highly stressed components weighing a few kilograms to components weighing several hundred tonnes. Many developments have been driven forward by the following issues:

. environmental considerations
. fuel economy
. legislation

some product developments are now reviewed.

5.1 Marine

Figure 5 shows a Vosper-Hovermarine HM218 manufactured in GRP and based upon the ˆhoverˆ principle. These craft are usually classified as surface effect ships (SES). Both power and fuel consumption are approximately half that of comparable displacement craft. Internal and external noise levels are low. Normal cruising speeds of 30-40 knots and above are attainable with an accompanying high speed handling improvements compared with similar craft. Figure 6 shows the HM218 during construction of the hand layed-up GRP ribbed and clad main hull. A GRP superstructure and cabin are assembled, see Figure 7, onto the main hull, making a very low maintenance final assembly.

Figure 8 illustrates probably one of the largest marine applications of GRP composite. The HMS Brecon was the first of the hunt-class mine countermeasures vessels, her plastic hull was reinforced with 160 000 square metres of woven fabric.

5.2 Aerospace

The Boeing 757 initial development programme identified many potential composite applications. Figure 9 shows some of these possible application areas for new materials and identifies a composites solution for edge members, spoilers, flap assemblies, engine support and cowlings etc.

Although polymeric composites have been used for some years for civil aircraft cabin interior fittings, the application of composites to highly stressed components, such as fuselage and the wing has not occurred. However, the Rockwell B1 horizontal stabiliser, shown in Figure 10, represents probably the largest fibre reinforced primary structure ever built, was less costly and weighed 227 Kg less than a comparable metal stabiliser.

The Grumman composite nacelle or engine cowling development, Reference 2, for the DC6 represent a major application of composite materials in a civil aircraft. However, many of the major composite developments have been for combat aircraft, such as the composite bomb bay doors on the BA Jaguar aircraft, Figure 11.

Recent helicopter applications have yielded some interesting honeycomb sandwich composite constructions, shown in Figures 12 and 13. The Westland helicopter composite rotor, shown in Figure 14, illustrate a complex rotor tip profile geometry that gives increased aerodynamic lift. This particular design detail in conventional metal, would be more costly to manufacture.

Some of the advantages in using composite rotor blades are:

. Designed to a specific weight, the distribution of weight can therefore be controlled more accurately.

. Improved control of blade profile, hence increased aerodynamic efficiency.

. Improved damage tolerance over metals.

. Benign failure mode.

. Infinite fatigue life. Three or four blade changes are often required for metallic rotors.

The rotor composite construction clearly illustrates the importance of a material with high information content.

Although the constituent materials for the composite are more costly than metals, the compression moulding manufacturing method is less costly, and therefore, the composite rotor blades are generally cost competitive with metal blades.

The space technology sector will undoubtedly yield many new exciting composite material developments. The design requirements for lightweight components with dimensional stability over temperature ranges of between -250° and $+250^{o}F$, a truly hostile environment, can be adequately met with composite materials. Figure 15 illustrates the space shuttle orbiterˆs many advanced composite component applications: Bay doors, maneuvering engine pods and arm booms are all fabricated in carbon fibre reinforced polymer composite.

5.3 Automotive

The existence of the automotive industry depends on manufacturing cost/competitive components in high volumes. The new generation of composite components has to fulfil also these major requirements and operate in a highly stressed and hostile environment. The following requirements for successful composite product development are therefore mandatory:

. Improved composite design analysis techniques

to deal with the interaction of the environment on composite behaviour.

. Increased sophistication and control of the manufacturing process - dimensional accuracy and structural consistency for high product volumes.

. Improved laboratory and in-service testing.

. Improved quality assurance for high product volumes.

The automotive sector has for many decades, been one of the main employers of labour and producers of high volume manufactured goods in this country. The world recession has hit the motor manufacturers hard and to compete there is currently a drive towards improved productivity and improved fuel or energy economy in road vehicles. Fibre reinforced plastic composites will, therefore, contribute to improving the performance of road vehicles, now and in the future.

Over the last 20 years, there have been many composite product developments for lightly to moderately stressed automotive components. However, composite components for highly stressed applications have been limited to the low volume specialised car market. Often, the large capital investment required for fabrication plant is prohibitive for many companies and can only be embarked upon when component supplier works closely with the automotive manufacturer.

Fibre reinforced composite were first used by Ford some fifty years ago. The car consisted of 14 wood fibre reinforced phenolic vacuum moulded composite panels, bolted onto a steel structural frame, resulting in a vehicle 30 per cent lighter than a similar sized car of that era.

In 1940, Owens Corning built an experimental car in body panels fabricated in glass fibre reinforced polyester resin, probably the first such application of this composite material in an engineering structure.

The General Motors Corvette illustrated in Figure 16 was first introduced in 1953 as a low volume, two seater sports car, with glass fibre reinforced polyester body panels. This vehicle formed the basis of many exciting new composite developments for automotives, especially in highly stressed applications. For example, the introduction in 1981, of a glass reinforced composite leaf spring with 80 per cent weight saving compared with a steel spring and a high energy absorbing face bar bumper.

Figure 17 illustrates the Lotus Eclat Excel low volume sport car. The body shells are manufactured by moulding a top and bottom pair of shells. The manufacturing process consists of a vacuum assisted resin injection technique using glass reinforced polyester. There is provision at the moulding stage to incorporate the paint primer prior to the

application of the gel coat. However, using a similar paintfinishing technique, it should be possible to eventually mould-in the final paint finish.

The box beam and sandwich construction is formed by wrapping uni-directional glass fibre around low density polyurethane foam cores. During the resin injection cycle, the resin flows around the cores to produce high performance fibre reinforced beams for withstanding side collision etc. Figure 18 shows Esprit body shells assembled and under test. Although the resin injection cycle and demould times can take up to 2 hours, the potential to mould-in many components such as windows, general accessories, as well as the final paint finish, brings into focus the flexibility in design and manufacture that is possible with composite materials components.

Fibre reinforced composite wheels have been under active development over the last decade. Figure 19 shows some typical compression moulded wheels. Companies such as Pontiac and Goodyear are planning the low volume production of 100 000 units annually for a sports car application in 1987. The wheels are composed of mixtures of glass reinforced resin, along with quantities of carbon fibre in some applications which are claimed to be cost/competitive with steel wheels. The wheels offer the benefits of improved strength, stiffness and running characteristics, accompanied by greater styling freedom. Weight saving is between 30 and 50 per cent compared with steel wheels. Moulded wheels are 'true round' and therefore, wheel balancing is eliminated. The Goodyear wheel has logged more than 4 million miles on vehicle road tests through arduous terrain.

Figure 20 shows a Mini Metro composite wheel being put through its paces during road test trials at N.E.L., on behalf of an industrial consortium, Reference 3.

The GKN composite leaf spring, introduced in 1985 on the Freight Rover Sherpa, and shown in Figure 21, represents probably the first high volume application of a polymeric composite in a highly stressed and hostile environment. The spring units are manufactured in glass reinforced plastic with a possible weight saving of at least 50 per cent over the equivalent steel leaf springs. This particular development emphasises the large research, development and manufacturing resources required to put a cost/effective product using new materials into the high volume automotive market place. The extensive testing programme utilised special purpose multi-axis test rigs which applied simultaneously a combination of vertical loads, lateral loads, roll and wind-up, based data acquired from vehicles with instrumented suspension systems. The composite springs were then subjected to slalom reversals, single wheel and two kerb strikes, corrugations, pave, snatch starts and pot hole braking. Validations were also carried out on normal roads and cross country circuits. Tests were successfully carried out in typical automotive environments, such as extremes of temperature, moisture and corrosive chemicals.

The sulcated spring concept is shown in Figure 22 and is representative of a second generation suspension concept which is not purely limited by simply replacing the existing metal spring. The design concept described in Reference 4, is based on innovative design and the full exploitation of the unique material properties of fibre composite.

The geometry has several advantages over more traditional spring solutions:-

. Spring stiffness can be tailored in an x-y-z plane and allows for transverse locations of the suspension member with an accompanying vertical spring compliance, effectively a 'compliant strut'.

. Fatigue strength of the component can be maximised since the composite material is being used in bending.

. The load/deflection can be linear or variable rate, by simply varying material thickness, width or material fibre orientations.

The new range of Volvo FL7/FL10 trucks have a highly stressed composite front shock absorber bracket, shown in Figure 23. The bracket is manufactured from an X.M.C. preimpregnated E-glass woven cloth and polyester composite. The original steel bracket weighed 2.7Kg, compared with the composite version weight of only 1.8Kg.

Another commercial motor vehicle application is the Mansell-pultrex-windfoil pultruded composite side guide shown in Figure 24. This particular product development typifies the pressure of a change in the motor vehicles regulation by a 1983 amendment, that all motor vehicles in excess of 3500Kg maximum gross weight, must have side guide protection. The main design requirements and restraints for the component were;

. Minimum weight.

. Minimum cost.

. Resistance to environmental and impact damage.

. Must meet new U.K. vehicle regulations.

The E-glass reinforced polymer composite weighed 1.9Kg/metre and was about 50 per cent lighter than steel components. The composite had a minimum tensile strength of 200mpa and flexural modulus of 26 Gpa, the material gave four times more resistance to impact energy than a comparable steel side guide.

Figure 25 shows a composite automotive suspension bracket or 'wishbone'. The component development consisted of the design and manufacture of a composite suspension bracket, using glass fibre reinforced vinylester composite with a quantity of directional fibres. The composite component was fabricated using a compression moulding technique in one single stage.

The figure also shows the moulded-in web configuration which gave additional stiffness to the suspension arm structure. The design aims were:

. To achieve the correct compliance for specified ride characteristics.

. To carry the required suspension loads.

. To tailor and optimise the composite materials lay-up configuration; this was generally dictated by the manufacturing process.

Although the composite suspension bracket had a similar weight to the original metal 'wishbone', it was thought that additional weight saving would result in future development, since the mounting bush assembly could be virtually eliminated. The benefits derived from using composite materials were a corrosion resistant suspension member, and reductions in noise levels and vibration.

6 CONCLUSIONS

The choice and selection of a particular new material for component development depends on the interaction of a wide range of variables, such as:

. Market conditions and material price.

. Material properties and behaviour.

. Environment.

. Energy costs.

. Legislation.

Design, manufacture, testing and quality assurance are critical aspects for high volume manufacture of composite products.

The engineer and management must not be prejudiced by their traditional or historical view of materials.

Composites have outstanding potential for building in a high information content, thus enabling products to be developed with a high added value.

The potential of composite material was reviewed with reference to some exciting past, present and future composite component developments for transport applications.

REFERENCES

(1) ALTENPOHL, D.G. Materials and Society, 1979, 3, 315.

(2) ANDERSON, R and POVEROMO, L.M. Composite Nacelle Development. Proc. 29th Ann. Conf., The Society Of The Plastics Industry, 1984, pp 11-C-L - 11-C-5. New York: The Society Of The Plastics Industry.

(3) HENDRY, J.C. and WOOTTON, A.J. Materials
    - Design Relationships. International
    Seminar, Utilisation of New or
    Alternative Materials, Autotech ´85
    Congress. The Institution of Mechanical
    Engineers, Automobile Division.
    November, 1985.

(4) SCOWEN, G.D. and HUGHES, D. The Sulcated
    Spring, International Seminar, Autotech
    ´85 Congress. The Institution of
    Mechanical Engineers, Automobile
    Division. November, 1985.

Table 1 Some typical fibre/matrix constituents available for composite product developments.

COMPOSITE/ COMPOUND	FIBRE	RESIN	FILLER	COMMENTS
DMC	´E´Glass	Polyester	Ca Co3	Fibre length approx 15mm (low performance)
SMC-CLASS	´E´ Glass (approx 30% by wt)	Polyester (approx 30%)	Ca Co3 Mg O etc	Sheet moulded compound. CLASS C – Unidirectional continuous fibres CLASS D – Unidirectional discontinuous fibre CLASS R – Random fibre. (fibre length approx 50mm)
UMC ´E´ GLASS	Polyester	–		Combinations of random (20%) and aligned (30%) fibres (ARMCO CORP)
XMC	´E´ Glass (70% by Vol)	Polyester (approx 30%)	–	Continuous fibre orientations generally between 5 and 10° some classes have a % of random fibres

SOME HIGH PERFORMANCE COMPOSITE MATERIALS

GLASS "PRE-PREG"	Glass	Epoxide	–	CLASS E CLASS S) Higher per- CLASS R) formance than ´E´ glass
CARBON "PRE-PREG"	Carbon	Epoxide	–	CLASS: . High Modulus . High Strain . High Strength
AROMATIC POLYAMIDE PRE-PREG	Aramid	Epoxide	–	Denoted aramid or ´Kevlar´

EPOXIDE

PHENOLIC

POLYESTER

POLYIMIDE

BISMALEIMIDE   SOME HIGH PERFORMANCE RESINS

POLYETHER- ETHERKETONE

POLYSULPHONE

Table 2 Main fabrication techniques.

- R.R.I.M.

- RESIN INJECTION (OR RTM)

- AUTOCLAVING/VACUUM MOULDING

- HOT PRESS MOULDING (SMC)

- PULTRUSION (PULFORMING)

- FILAMENT WINDING

- THERMOFORMING

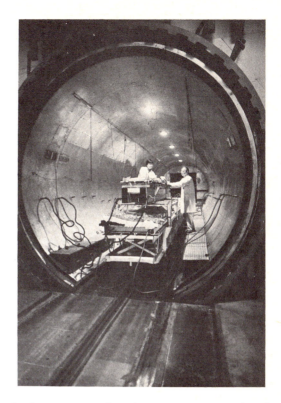

Fig 3 Large capacity, high temperature autoclave for fabricating composite aerospace components

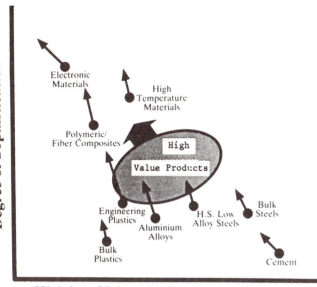

Fig 1 Degree of materials sophistication

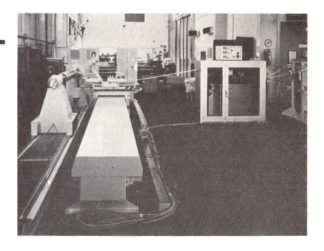

Fig 4 Filament winding machine

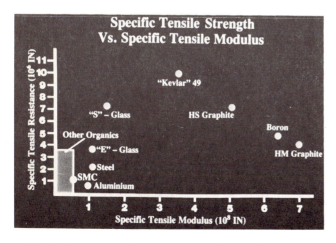

Fig 2 The relationship between specific tensile strength and specific tensile modulus for various composite constituents and metals

Fig 5   Vosper Hovermarine HM218

Fig 8   The HMS Brecon-Hunt class mine countermeasures vessel

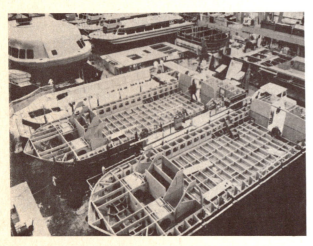

Fig 6   HM218 during fabrication of composite hull

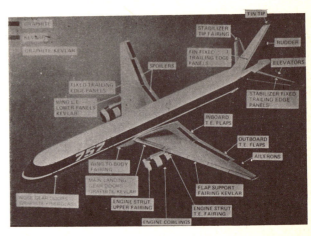

Fig 9   Advanced composites applications for Boeing model 757

Fig 7   Assembly of superstructure

Fig 10 B1 horizontal stabilizer — Rockwell/Grumman

Fig 11 BA Jaguar — composite bomb doors, support struts etc

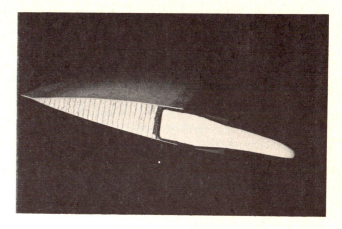

Fig 13 Fibre reinforced plastic and honeycomb sandwich composite main rotor

Fig 12 Westland combat helicopter

Fig 14 Complex composite rotor for increased aerodynamic efficiency

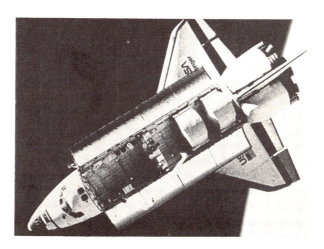

Fig 15 The Shuttle Orbiter — composite doors, engine pods, arm booms etc

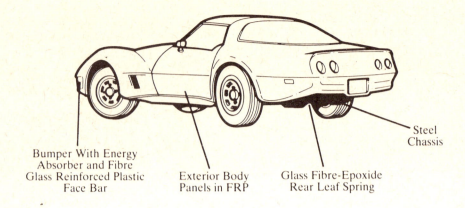

Bumper With Energy
Absorber and Fibre
Glass Reinforced Plastic
Face Bar

Exterior Body
Panels in FRP

Glass Fibre-Epoxide
Rear Leaf Spring

Steel
Chassis

Fig 16 General Motors 1980s Corvette

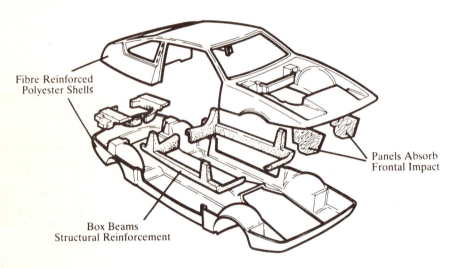

Fibre Reinforced
Polyester Shells

Panels Absorb
Frontal Impact

Box Beams
Structural Reinforcement

Fig 17 Lotus Eclat Excel

Fig 18 Lotus Esprit on test

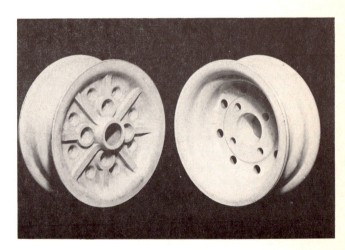

Fig 19 Composite automotive wheels

Fig 20 Composite wheel during road test

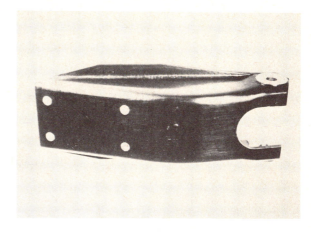

Fig 23 Composite shock absorber bracket

Fig 21 GKN composite leaf spring

Fig 24 Composite side guide for commercial vehicles

Fig 22 The sulcated spring

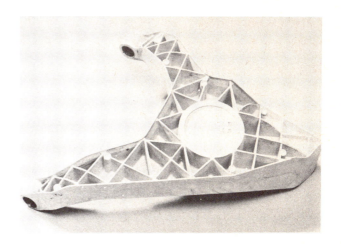

Fig 25 Composite suspension arm with moulded webs

C51/86

Development of a cost effective composite automotive exterior body panel

B R K PAIN
Austin Rover Group Limited, Cowley, Oxford
D CHARLESWORTH, BSc and **A BENNETT**, BSc
BL Technology Limited, Gaydon Proving Ground, Lighthorne

DEVELOPMENT OF A COST EFFECTIVE COMPOSITE
AUTOMOTIVE EXTERIOR BODY PANEL

Introduction

The increasingly competitive car market is
creating a demand for models which incorporate
advanced technology to give positive
improvements in fuel economy, durability and
visual appearance without reducing interior
space, passenger comfort and refinement.

In a significant number of concept and
production vehicles, the automotive industry is
adopting polymeric materials to assist in the
advancement of these aims. The advantages of
plastic skin panels are well known viz:-

-   light weight

-   corrosion resistance offering greater
    durability and enhanced resale value

-   minor impact/damage resistance

-   greater styling freedom which can
    contribute to improved performance and
    economy

-   potential reduction in warranty and
    insurance costs

-   minimisation of pedestrian injury

These advantages however have not led to a
significant breakthrough because of cost
parameters and/or failure of substrates to
satisfy in house paint process requirements in
particular the on line finishing temperatures.

The basic parameters for a plastic body
component as laid down by Austin Rover Group
are:-

-   acceptable in service performance

-   paintable on line

-   15% (minimum) weight saving over the steel
    equivalent

-   in final (assembled) component form equal
    or less than the cost of its steel
    counterpart
    During 1982 through new process
developments, a new range of polypropylene

polymers which had uprated physical properties
compared with
conventional polypropylenes in terms of modulus
and temperature resistance became available and
most significantly retained the competitive cost
associated wth polypropylene.

The potential to meet the basic parameters for a
plastic body panel thus appeared to be feasible
and the decision was made to develop a polymeric
front wing, the Maestro van variant being
selected, for a service evaluation and test bed
prior to material decisions being taken for
future models.

Material and Process Selection

The material specification derived by BL
Technology for skin panels is detailed in Table
1 with the properties of the new grades of
polypropylene listed against other potential
thermoplastic materials available at the time.

Injection moulding was selected as the most
viable plastic processing route on the basis of
the volumes/cost involved. Materials
processable via the reaction injection moulded
route had been considered but rejected prior to
this stage as were glass reinforced thermosets
produced by both injection and compression
moulding being below requirements due to
strength and impact weaknesses.

With confirmation that production quantities of
the two materials, PXC 51370 and TE 354 would be
available our initial work commenced with
internal verification of suppliers data
(detailed manufacturers data in Appendix 1).
Favourable results were obtained and work was
extended to include both standard and non
standard tests. A number of types of test
mouldings were supplied by ICI including:-

a. Box injection moulded, 400 x 250 x 60 mm,
    centre gated 3 mm thick

b. Extruded sheet, overall width 1000mm,
    thickness 2.5 x 2.9 mm

c. Injection moulded plaques, edge gated with
    "coat hanger" feed, dimensions
    150 mm x 150 mm x 3 mm

d. Maestro bumper centre section - injection
    moulded.

Material Evaluation

Bumper mouldings were used primarily to assess thermal stability in terms of 135°C paint oven temperatures where PXC 51370 proved to be superior to TE 354 (Figure 1).

The injection moulded box samples proved to be particularly useful since they showed effects from in mould stresses and, in the case of PXC 51370, orientation in the fibre reinforcement. Also by removing two adjacent sides of the box it effectively became a highly stylised model of a wing and provided a more representative form for thermal stability and environmental tests.

The complete test programme included:-

- Thermal stability at paint oven temperature (135°C).

- Stability and property measuremeents under environmental conditions to BLS 30 TP 903.

- Impact assessment at -20°C, after simulated paint line (20 min 130°C) conditions and after environmental testing.

- Comparison of general physical properties before and after the various thermal and environmental tests.

In parallel with the above work ICI Paints Division, Slough, provided input on suitable finishing systems and the test programme above was repeated on painted test mouldings.

Whilst some potential problems, notably on distortion and directionality of reinforcement effects were noted nevertheless the test results gave a good degree of confidence to proceed. Paint tests had indicated that good adhesion could be achieved and that little effect occurred with respect to other properties. However, painting of PXC 51370 plaques had indicated a potential aesthetic problem in that painted surfaces developed a very fine textured effect during processing with the result that gloss level and distinction of image test results did not meet specified requirements.

Tooling and Component Design

As a result of the preceding test work component design and tool manufacture were authorised based on polypropylene PXC 51370 at a uniform wall thickness of 3 mm.

The tool design incorporated a vertical flash feature to accommodate the injection/compression moulding route which was recommended as a means of reducing panel warpage and component distortion.

The component design required that it replace the existing steel panel and utilise the same fixing points. Some modifications were necessary in particular along the top fixing edge of the panel where the geometry did not allow for a single moulding. This required the separate manufacture of a "nose piece" to complete the upper fixing edge of the component which could be ultrasonically welded in position.

The rear fixing of the panel is by an integrally moulded bracket at right angles to the panel/door edge and is a single fixing utilising only one of three fixing holes already present.

Component Evaluation

Moulded components in PXC 51370 and a 30% glass reinforced material were evaluated for tensile, impact and thermal performance for two primary reasons.

a. To validate previous plaque evaluations.

b. To pre identify problem areas in service.

Some problems were immediately encountered in the fit of the panel compared to its steel counterpart due to shrinkage effects.

Some variation in panel thickness was noted there being a tendency to thinning towards the front or nose area, however this area is to a large measure protected by the bumper.

The thermal behaviour of the panel was satisfactory following attachment to vehicles and conventional treatment on the paint line.

Major conclusions drawn from the test results were:-

1. Injection - compression mouldings gave superior results to "straight" injection panels and showed a weight saving of 50% compared to the steel panel.

2. Impact performance was affected by the position of tooling split lines.

Wings were submitted to Engineering Department for head and shin impact tests and current metal wings were also submitted for comparison.

The depth and pattern of dents obtained on the plastic wings were similar to those on current production wings. However, it was noted that time dependence recovery of dents occurred with the plastic wing, some 50% recovery being recorded over an 18 hour period.

Stiffness measurements showed that the plastic wing was approximatey 75% less stiff compared to the steel wing.

Wings were also fitted to vehicles for track testing at speeds up to 90 mph with no evidence of excessive vibration or flutter.

Paintability

Plaque tests had indicated a potential problem with regard to texture effects in the paint finish. These indications were unfortunately confirmed with paint trials on components.

The required specification for a finished panel was:-

-Distinction of Image   Minimum 9

-Ability to withstand impact at -20°C

-Polishable surface

- Appearance equivalent to that of steel

The texture effect was particularly significant in relation to distinction of image measurements where the value fell to below 5. The low value appeared to be due to differential expansion effects between the glass fibre reinforcement, polymer and paint film and the texture effect became discernible at 60°C becoming more pronounced as the temperature was increased. Experimentation with primer coats, solvent take up and release failed to overcome the problem.

At this stage we had a functionally and economically acceptable component but with intrinsic features which resulted in an unacceptable finish.

The problem was therefore how to obtain a finished surface which would meet vehicle standards. Commitment had already been made to injection moulding and choice was therefore limited to (a) material changes with potential problems in shrinkage compared to polypropylene and (b) moulding techniques. Investigations into the coaxial dual injection sandwich moulding technique indicated that this route was feasible and that polypropylene could be retained as the component material. Modifications to the tool to comply with the sandwich moulding technique were instigated and concurrently plaque testing on sandwich moulded panels was carried out. The glass reinforced polypropylene PXC 51370 was retained for the core material with a non reinforced polypropylene PXC 41494 selected for the skin. Physically such a combination did not equate to an ideal sandwich structure since the stiffer material formed the core. However the achievement of an acceptable surface or finish was the prime objective.

Plaque mouldings quickly indicated that to overcome the texture problem the unreinforced skin material was required to be < 0.7 mm thus reducing the core on a 3 mm thick moulding to 1.6 mm which would be difficult to achieve on the component moulding. A reduction of the skin thickness to less than 0.7 mm could result in core breakthrough.

The panel thickness requirement was therefore increased to 3.5mm allowing for a 2.0 mm core. At this order of panel thickness the properties of the original PXC 51370 mouldings could be retained, (Figure 2).

A large number of wing components were moulded using the PXC 51370 - PXC 41494 core/skin combination. Extensive testing of these components has shown that their performance is satisfactory but some interesting features have emerged as follows.

a. Variations in skin: core thickness ratio occurs with the core being predominant at the injection area reversing to skin predominance at points furthest away from the injection point.

b. Extensive property tests with samples taken from mapped points covering the whole component area have indicated property variation in line with varying direction of fibre reinforcement in the core but these variations have not had a significant effect on component behaviour (Figure 3).

c. The component is capable of meeting on line paint temperatures without significant distortion and there are no apparent effects due to differential expansion between skin and core.

d. Head form impact tests on the upper area of the wing have shown an acceptable resistance to denting, recovery under low energy impacts and acceptable absorption energy relative to pedestrian impact requirements.

Shin form impact tests at the centre of the wheel arch have also been acceptable in terms of deflections sustainable and damage occuring. In general the composite wing outperforms its steel counterpart in the above tests and barrier crash tests at Thatcham have similarly given encouraging results when comparing damage results to steel.

e. Injection moulded wings require reasonably careful but not delicate handling since if simply stacked or badly supported for relatively short periods distortion occurs.

Currently 30 Maestro vans are fitted with injection moulded sandwich structure composite wings and many thousands of miles of road mileage have been accumulated without serious problems. Structural and environmental tests carried out have included:-

1. 10,000 miles at temperatures between -38°C and 0°C + max under conditions varying between heavy packed snow and freezing slush.

2. 20,000 km durability test track trial.

3. Approximately 1000 mile pave and corrugated road test.

4. Head and shin form impact testing at ambient and low temperatures.

5. Dynamic performance test.

6. 40% fronted impact at 25 mph in which damage was confined to hair line paint cracking at a severe fold in the top of the wing and some pull through at fixings in the drain channel.

In general the results have been satisfactory with no major problems and in some instances a performance superior to that of steel wings.

Main indication of the results is a need to ensure adequate bearing area for the drain channel fixing washers to ensure maintenance of fixing torque.

Finally, in a programme of this nature a combination of expertise is essential and in this respect acknowledgement is made of the input provided by those involved who included:-

DTI (Vehicles Division)

ICI Petrochemical and Plastics Division Ltd

ICI Paints Division Ltd

Tooling Products (Langrish) Ltd

Rolinx Ltd

Battenfeld (England) Ltd

Motor Ins. Repair Research Centre Thatcham

Table 1   Properties of contending materials for wing panel application

Property	Criteria	Steel	PXC 51370	TE354	"Thermo-plastic Alloys"
Flexural Modulus	1.5 GPa Min	207	3.3	1.7	1.5
Sp Gravity	-	7.8	1.05	1.00	1.2
Cost £/tonne	(at 1982 costs)	300	1200	950	2000
Impact	4.6 kgm 200 mm	OK	Borderline	OK	OK
Heat Stability	Stable at 135°C	OK	OK	Border-line	OK
Weight	< 85%	100	52	50	60

N.B.   PXC 51370   20% glass reinforced polypropylene ex ICI

TE354   Mineral filled polypropylene ex ICI

Thermoplastic   typical value for a range of technically
Alloys   competing materials

Table 2

PROPERTY	UNITS	GRADE +			
		HW60 GR20 Natural 001	HW60 GR30 Natural 001	PXC 51370 Natural 001	PXC 51438 Natural 001
Melt flow index (230°C/2.16 kg) ASTM D1238 ISO 1133		4.0	3.0	1.0	2.0
Density	g/cm^3	1.04	1.12	1.05	1.12
Tensile yield stress ISO 527) 23°C ASTM D638) 80°C (50mm/min)	MPa	72.4	86.2 56.0	39 21	60
Flexural modulus Flexural strength ISO 178 ASTM D790	GPa MPa	4.96 108	6.5 137	3.3 63	5.0 92
Notched Izod impact strength 23°C ISO 180) -40°C ASTM D256) (0.25 mm notched radius)	J/m	70 40	100 70	300 150	210 100
Drop weight impact 23°C strength* -40°C	J	6.0I/P 4.0I/P	6.9I/P 5.5I/P	13.1I/P 10.9I/P	14.8I/P 8.7I/P
Heat distortion temp ISO 75) 0.45 MPa ASTM D648) 1.82 MPa	°C	156 140	158 148	144 100	152 124

+ These materials are anisotropic and therefore the values given depend on the orientation of the glass fibres in the moulding

* ICI's instrumented version of BS 2782:306B (3mm thick injection moulded test specimens).

I/P initiation propagation

C51/86 © IMechE 1986   261

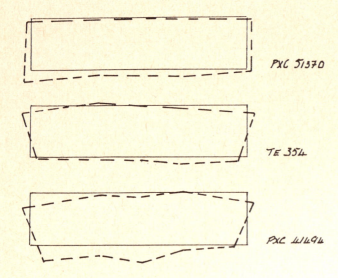

Fig 1   Injection moulded bumpers. Relative distortions after paint oven (135°) simulations

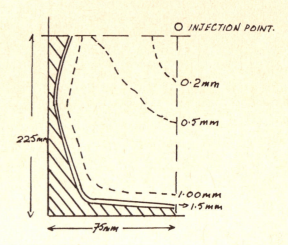

Fig 2a   Sandwich moulded plaque. Skin thickness contours
Plaque dimensions: 450 mm x 150 mm x 4.3 mm
Centre gated
Skin: PXC 41494   Core: PXC 51370

▨ _ SKIN MATERIAL ONLY.

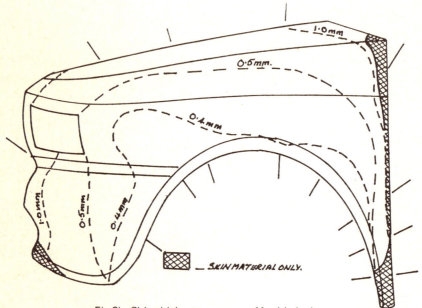

Fig 2b Skin thickness contours. Moulded wing

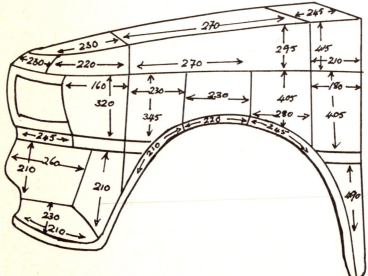

Fig 3   Typical test mapping procedures tensile loads (N) at break